Build Your Own Own Electronics Workshop

Thomas Petruzzellis

McGraw-Hill
New York   Chicago   San Francisco   Lisbon
London   Madrid   Mexico City   Milan   New Delhi
San Juan   Seoul   Singapore   Sydney   Toronto

The McGraw·Hill Companies

Library of Congress Cataloging-in-Publication Data:

Petruzzellis, Thomas.
    Build your own electronics workshop : everything you need to design a work space, use test equipment, build, and troubleshoot circuits / Tom Petruzzellis.
      p. cm.
    ISBN 0-07-144724-5
     1. Workshops.   2. Electronic apparatus and appliances—Design and construction—Amateurs' manuals.   3. Electronic apparatus and appliances—Testing—Amateurs' manuals.   I. Title.

TK9965.P4797 2004
621.3815'4—dc22
                                          2004059248

4 5 6 7 8 9 0 DOC/DOC 0 1 0 9 8

ISBN 0-07-144724-5

The sponsoring editor for this book was Judy Bass and the production supervisor was Pamela Pelton. It was set in Times Roman by Ampersand Graphics, Ltd. The art director for the cover was Margaret Webster-Shapiro.

Printed and bound by RR Donnelley.

McGraw-Hill books are available at special quantity discounts to use as premiums and sales promotions, or for use in corporate training programs. For more information, please write to the Director of Special Sales, McGraw-Hill Professional, Two Penn Plaza, New York, NY 10121-2298. Or contact your local bookstore.

*This book is dedicated to Betsy, my wife and confidant,
who has endured my wayward musings, explorations,
and stubbornness for over 25 years. Reading, writing, music
and amateur radio separate us at times, but despite
the separation, my heart grows ever fonder.*

CONTENTS

Introduction   xv

Chapter 1   The Electronics Workshop   1

            Work Mats   9
            Test Equipment Considerations   10
            Tools and Stuff   12
            Reading Schematics   13

Chapter 2   The Multimeter   29

            Basic Analog Meter Characteristics   32
            The Basic Voltmeter   34
            The Basic Ammeter   35
            Multirange Ammeters   38
            The Basic Ohmmeter   39
            ac Meters   43
            The Digital Multimeter   45
            Basic DVM Characteristics   45
            Additional DVM Characteristics   55
            Special DVM Features   56

Chapter 3   The Oscilloscope   59

            The Analog Oscilloscope   61
            Dual-Trace Oscilloscopes   62
            Oscilloscope Specifications   63
            Oscilloscope Controls   64
            Scope Input Circuits   66
            Oscilloscope Start-up   71
            Oscilloscope Signal Measurements   73
            Digital Oscilloscope   80

Chapter 4   The Function Generator   93
            All About Signal Generators and Waveform Generators   94
            Build Your Own Function Generator   103

Chapter 5   Frequency Counter   111

Frequency Counter Architecture   113
Choosing the Appropriate Time Base   115
Differences Between Resolution and Accuracy   117
How to Adjust Sensitivity to Avoid Noise Triggering 117
The Problem of Low-Frequency Measurements   118
Frequency Counter Measurement Techniques   119
Frequency Counter Terms   120
Frequency Counter Accessories   122
1 GHz Frequency Counter Project   124

Chapter 6   Test Bench Equipment   133

Multimeter   134
Vacuum Tube Voltmeter (VTVM)   135
Capacitance Meter   136
ESR Capacitance Meter   137
Inductance/Capacitance (L/C) Meter   137
Resistance/Capacitance Decade Boxes   138
Wheatstone Bridge   139
Frequency Counter   140
Audio-Signal Generator   141
Function Generator   142
Arbitrary-Waveform Generator   142
Pulse Generators   143
RF Signal Generator   143
Tone Test Set   143
Signal Injector   144
Signal Tracer   144
Clamp-on Ammeter   145
Insulation Tester   146
Cable Testers   146
TV Pattern Generator   146
Transistor Tester   146
Transistor Curve Tester   147
Tube Tester   148
RF Impedance Bridge   149
Field-Strength Meters   150
SWR or VSWR Bridge   150
Oscilloscope   150
Spectrum Analyzer   151
Logic Pulser   152

Contents

Logic Probe 153
Logic Analyzers 153
Microprocessor Programmers 154
Variac 154
Isolation Transformer 154
Battery Tester 155
Batteries and Chargers 156
Power Supplies 157
Temperature-Controlled Soldering Station 158
Desoldering Stations 158

Chapter 7   Power Supplies 159

How to Choose a Power Supply 160
Power Supply Components 162
Build Your Own Bench/Lab Power Supply 174

Chapter 8   Battery Power 185

Batteries and Charging 186
Chemical and Other Hazards 188
Internal Resistance 189
Battery Capacity 189
Discharge Planning 190
Charging/Discharging Requirements 190
Battery Charging 191
Solar Electric Battery Charger 196
An Emergency Power System 200
Battery Safety 204

Chapter 9   Electronic Components 205

Resistors 206
Capacitors 214
Inductors 220
Transformers 226
How Semiconductors Work 228
Protecting Components 230

Chapter 10   Testing Electronic Components 233

Why Testing is Important 234
Multimeter 234
Oscilloscope 234

Logical Thinking is Your Best Test Instrument   235
Making Resistor Measurements   235
Making Capacitor Measurements   239
Making Inductance Measurements   244
Continuity   245
Other Failures   246
Transformer Basics   246
Measuring Semiconductor Devices   248
Testing with an Oscilloscope   259
Build Your Own Diode and Transistor Tester   263
Simple Extended-Range Measurements   264

Chapter 11   Electronic Troubleshooting Techniques   267

Test Equipment   269
Safety   269
Documentation   270
Using All Your Senses   270
Defining the Problem   271
Simplify the Problem   272
Look for Obvious Problems   272
Open the Equipment   272
Looking Inside the Equipment   273
Troubleshooting Approaches   273
Signal Tracing   274
Signal Injection   276
Dividing and Conquering   277
The Intuitive Approach   277
Testing within Intermediate Stages   278
Testing for Voltage Levels   278
Oscillation Problems   278
Amplitude Distortion Problems   279
Frequency and Distortion Problems   280
Noise Problems   280
Distortion Measurement   281
Alignment   282
Arcing Problems   282
Contamination Problems   282
Solder "Bridges"   283
Replacing Defective Components   283
Typical Symptoms and Faults   283
Troubleshooting Hints   289

Chapter 12   Workshop Tools   293

Tools and Their Uses   294
Care of Tools   294
Tools Organization   294
Parts Organization   296
Sharpening   296
Proper Tool Use   297
Specialized Tools and Materials   299
Sources of Tools   307
Useful Parts and Shop Materials   307
Shop Safety   308

Chapter 13   Soldering   311

The Soldering Iron   312
Soldering Gun   314
Preparing and Using the Soldering Iron   315
Preparing Work for Soldering   315
How to Solder   316
Build Your Own Solder Station   321

Chapter 14   Circuit Fabrication   325

Electronic Circuits   326
Electrostatic Discharge Protection   326
Electronics Construction Techniques   327
Point-to-Point Techniques   328
Solderless Protoboard Construction   328
Perf-Board Construction   329
Ground-Plane Construction   329
The Lazy Builder's PC Board   331
"Ready-Made" or "Utility" PC Boards   332
Wire-Wrap Construction   332
Surface-Mount Construction   333
When are PC Boards the Best Choice?   335
Schematic to PC Board   335
Rough Layout—Manual Method   336
PC Board Design Software   337
Printed-Circuit Boards   338
Double-Sided PC Boards   341
Plating   341
Complex Printed Circuit Boards   342

Design Considerations	342
Board Production	343
Commercial Production Costs	344
Drilling Your PC Board	344
PC-Board Assembly Techniques	345
A Final Check	348
Chassis Fabrication	348

Chapter 15   Buying Equipment, Components, and Tools   353

Buying New Equipment	354
New or Used Equipment?	354
Guidelines for Buying Used Equipment	356
Additional Thoughts on Buying Used Test Equipment	357
Buying Versus Building Test Equipment	360
Buying Electronic Components	360
Buying Tools—New and Used	363

Chapter 16   Building Your Own Test Equipment   365

Continuity Tester	366
Logic Probe	367
Wire-Tracer Circuit	368
Zener Diode Tester	369
Visual Diode/Transistor Curve Tracer	371
Transistor Tester	372
Capacitor Leakage Tester	376
Electronic Fuse	382
Pulse Generator	386
Inductance/Capacitance (L/C) Meter	394

Appendix   403

Electronics Workbench Resources	404
Used Test Equipment Sources	407
Tools for Electronics	407
Electronics Workbench Sources On-line	407
Books—Schematic Reading	408
Equipment Manuals	408
Resistor Color Codes	409
Ohm's Law Nomograph	410
Parallel Resistance Nomograph	411
Resistor Formulas	412

Capacitor Color Codes   413
Electronic Schematic Symbols   415
Computer Connector Pinouts   420
Copper Wire Specifications   421

Index   423

INTRODUCTION

Build Your Own Electronics Workshop was created to assist newcomers in electronics who wish to learn how to build their first electronics workbench. This book is designed for electronics hobbyists, enthusiasts, technicians, and recent graduate engineers, alike, who are new to the field of practical electronics. The electronics workshop is a specially designed place where one can go to work on an electronics project such as testing electronic components, troubleshooting a piece of electronics or radio gear, or building a new electronics project or kit. The electronics workshop is a place where you can work undisturbed and leave an ongoing project to find it untouched and undisturbed when you return. Building your new electronics workshop will free you from building projects on the kitchen table and the terror of returning to the table to find your project thrown in a box, with small components tossed everywhere. This book will help you to construct an efficient new electronics workshop that is well lit and well ventilated, a work area that houses your special test equipment and tools where you can work undisturbed.

The book covers many topics such as how to locate a suitable area of your home, basement, or garage to set up your new electronics work area; whether to build or buy your new electronics workbench; lighting and ventilation; and power requirements. You will learn what the essential pieces of test equipment, components, and tools are and how to use and purchase them. You will learn how to purchase test equipment, components and tools. The book will also help you to read schematics, design and build circuit boards, as well as build your own pieces of electronic test equipment. You will learn how to solder and how to construct circuits using different practical circuit-building techniques. The book will also present methods on how to test electronic components as well as many formulas, charts, and resources for the electronic enthusiast.

Chapter 1 discusses what is needed to create your electronic workshop in your home. Many ideas for constructing your electronics workshop are presented, including layout, type of workbenches, building or buying your electronics workbench, power requirements, lighting, ventilation, and backup power concepts.

Chapter 2 discusses the multimeter, perhaps the single most important item that you will have on your electronics workbench. Topics covered include the differences between analog and digital multimeters, how to use a multimeter, what kinds of tests you can perform, and the importance of good test leads.

Chapter 3 presents the oscilloscope, the engineer and technician's best friend. With the oscilloscope you can troubleshoot just about any problem in electronics. Differences between analog and the digital scopes are discussed, then analog to digital PC cards that function as oscilloscopes are examined, followed by test leads and probes, the most important controls and how to operate them, and, finally, how to care for your oscilloscope.

Chapter 4 highlights the function generator, a very useful addition to the electronics test bench, for testing logic circuits, filters, and amplifiers. Different types of function generators are discussed, from simple signal generators to the more complex function and arbitrary pulse generators. Practical uses of function generators as well as how to build your own low-cost function generator are also covered.

Chapter 5 discusses the last of the "big four" pieces of test equipment that you should have on your electronics workbench—the frequency counter. The frequency counter is often used in conjunction with the signal generator and oscilloscope when troubleshooting and constructing new electronics circuits. In this chapter you will also learn how to construct your own 1 gHz frequency counter.

Chapter 6 takes a look at other important test equipment that you might want to obtain for your electronics workshop. These secondary pieces of test gear make solving electronics problems easier. The function or signal generator, frequency counter, battery testers, capacitor testers, LC testers, inductance meter, logic injectors, logic probes, IC testers, clamp-on ammeters, and tube and transistors testers are discussed.

Chapter 7 covers power supplies, perhaps the single most important piece of equipment you can have on your new electronics workbench. Different types of power supplies are discussed, including lab, dual-power, and dc-to-dc power supplies, as well as whether to buy or build them, and, finally, how to construct your own lab power supply.

Chapter 8 revolves around a discussion of batteries and battery power supplies. This chapter delves into the differences and similarities between various types of batteries and their characteristics, and how and when to use them, as well as how to charge them most efficiently. This chapter also shows how to build a battery charger for a 12 volt battery, a solar battery charging system, and a backup power source that uses a deep-cycle 12 volt battery and an inverter.

Chapter 9 discusses electronic components such as resistors, capacitors, inductors, transformers, and semiconductors, their basics, how and when they are used, and their characteristics. Also discussed is how to connect them in a circuit, series and parallel connections of components, and so on.

Chapter 10 explains how to test electronic components such as resistors, capacitors, transistors, batteries, SCRs, and MOSFETs. In this chapter, you will learn how to test electronic components using both the multimeter and the oscilloscope. Simple tests and tricks are presented.

Chapter 11 presents a number of electronics troubleshooting techniques and approaches to help you solve your future electronics repair problems. It begins by discussing signal tracing and signal injecting techniques for testing circuit stages in analog equipment such as oscillators, amplifiers, and power supplies, and explains how to eliminate circuit problems such as unwanted oscillations and distortion. This chapter also discusses some troubleshooting techniques for servicing digital circuits and microprocessors.

Chapter 12 is all about various hand tools that would be useful around your electronics workbench, the "must have" or primary tools, such as drills, screwdrivers, needle-nose pliers, knives, wire cutters, and hemostats, and various other types of important tools you will find invaluable in your electronics workshop. It also talks about the care and cleaning of tools.

Chapter 13 presents the topic of soldering, what soldering irons to purchase, as well as what types of solder to buy, and the soldering gun versus the soldering iron. Soldering techniques for both point-to-point wiring as well as circuit-boards are discussed. This chapter also discusses the care of soldering tools and how to make good solder joints, as well as how to construct your own solder-station controller.

Chapter 14 covers the extensive topics of circuit fabrication, from point-to-point wiring techniques to wire-wrap techniques, perf-board construction, and protoboard circuit building. This chapter also discusses circuit building, from designing circuits using CAD/CAM and layout software to different aspects of making your own circuit boards.

Chapter 15 presents the subject of purchasing test equipment, where and how to buy equipment, and what to look for. It also discusses how and where to purchase electronics components and workshop tools, and how to save money as a smart consumer. The topic of buying versus building your own test equipment is also covered.

Chapter 16 features several pieces of test equipment you can build. You can save money by building your own equipment and learn a lot in the process, not to mention the great feeling that you get when you create your test gear. This chapter will show you how to build a continuity tester, ESR capacitor tester, logic injector/probe, capacitance/inductance tester, transistor tester, Zener diode tester, electronic fuse, and a pulse generator.

The Appendix presents many useful electronics charts and formulas, and lists of companies offering all types of electronics products, including test equipment, components, circuit boards, tools, work benches, and books, which could be useful in your electronics hobby or career.

Build Your Own Electronics Workshop will give you a good introduction to the many aspects involved in setting up and building your own electronics workshop. It will be useful and serve you well, both in the present and in the future, whether electronics is your occupation or avocation.

I would like to thank the following people and companies for their help in creating this book: Judy Bass, Senior Electronics Editor for technical publications at McGraw-Hill, for her vision and helpful suggestions throughout; Neil Heckt and Almost All Digital Electronics for details of the L/C meter project; and Tony VanRoon for the information and diagrams in Chapter 7. A special thank you is due the American Radio Relay League for the use of diagrams, schematics, and information used within. Many thanks are also due the following companies for use of photos and data sheet information: Exar Corp., Tektronix Corp., Protek Test Equipment Corp., Electronic Design Specialists, Microchip Corp, and Phillips Corp.

The Electronics Workshop

The electronics workshop is an essential place for the electronics enthusiast, the electronics technician, and the electronics engineer. It is akin to the woodworker's workbench, the jeweler's workbench, and the machinist's workbench—a place set aside for a particular interest or work activity. Each particular area of interest, hobby, or skill requires a workbench—a place where a particular type of work activity is done. It is usually a place set aside in a basement or spare room, or even in a garage or attic, where you can work undisturbed; a separate area in which work can be left on the table, untouched until you can return to your building, testing, or troubleshooting a project. It is essential that a reserved space be set aside so work can be left undisturbed until you can return to it. This is the main reason why the kitchen table is not suitable as an electronics workbench, since every day you would have to set up your workbench and begin anew.

The electronics workshop is no different than any other specific work area or place where a particular type of work can be performed, except that you probably will need specific types of tools, instruments, or test gear to complete the electronics job at hand, whether it be building a particular piece of electronics circuitry or troubleshooting or repairing a piece of equipment. Working with electronics generally requires special equipment and tools for building, testing, and troubleshooting activities.

A distinction is often made between the electronics workshop and a general or conventional workshop. A general workshop might best be described as a work area where you might cut wood, assemble a wagon, or construct a wooden toy. An electronics workshop or workbench might be described as a specific area in which you work on electronic projects. Electronics projects are generally smaller projects than, say, building something with wood or a plumbing project. Working on an electronics usually requires specific tools and test equipment for electronics project work. A distinction is generally made between a general workshop, a woodworking shop, or a metalworking shop versus an electronics workbench or workshop. Although it would be possible to work on a woodworking or metalworking project at your electronics workbench, an electronics project is usually much cleaner—there are no wood chips, metal filings, or oil puddles. Electronic circuits should be free of metal chips, which are usually fatal to them, and oil can be conductive and should be kept away. So our focus will be on an electronics workbench with a clean work top with a nonconducting, antistatic mat and a notebook or two. A typical electronics workbench might contain nonelectronic tools such as soldering irons, wire cutters, and small screwdrivers, as well as test equipment such as multimeters, oscilloscopes, and frequency counters.

In order to set up your new electronics workshop, you will first need to select a free room or section of a room where you can carve out an area that you can call your own. This could be in your basement, attic, closet, garage, or even in a section of a barn. You will need to take some time and figure out where you will want to spend time working on your hobby or vocation. The electronics workbench area should be comfortable and have an adequate heating/cooling system so you can be comfortable over a period of a few hours or so at least. If you are working on a kit or troubleshooting project, you may need to spend a number of hours in an undisturbed area that is comfortable to work in. There is nothing worse than working in an area that is either too hot or too cold for any length of time. Being uncomfortable will seriously affect your overall concentration.

Next, you will need to determine how much space you will actually need to set up your electronics workshop area. You might desire a large area or an entire room but, in reality, what will your particular situation permit, that is, what will your spouse or roommate allow? Ask yourself this question: Do you want to have your electronics workbench adjacent to or near your wood or metal workshop or in a different area altogether? If the electronics workbench area is going to be separate from other work areas, then you will need less room. Since a separate electronics workbench might likely be a cleaner area than, say, a woodworking shop or a general workshop, it may be possible for it to share a room with another workbench— half the room for electronics project and the other half for crafts, sewing, or some other activity.

There is an important psychological aspect to determining space considerations for your electronics work area. Many people, especially artists and writers, believe that having a dedicated work space is essential in doing productive work, a space that is respected by others and where concentration can be maintained. This concept can be extended to work areas for metal- or woodworking and other creative hobbies. So you many want to further consider "space" as more than just a place to fix, build, or test an electronic circuit. Writers, for example, need a quiet space in which to work, often for many hours at a time, without being interrupted or disturbed by children or others. When you are in your private work space, immediate demands, such as taking the garbage out or watering the lawn, are temporarily suspended. Many view their work area as a refuge where things or projects are left undisturbed, but this also extends to privacy and a respect for your need to get something accomplished without being disturbed. When in your private space or work area, your space and projects are to be respected by others. It can be a place to get away from the noise of the family—the din of the television, the barking dog, and the kids crying—in order to concentrate on the project or tasks at hand, whether it is reading a manual, soldering a circuit, or writing about a project. The concept of "dedicated space" is extremely important, especially if your electronics workshop is the place where you earn your livelihood. Or it can be considered as a space for getting away from the family or others for a period of time. However, the "dedicated space" issue is also a matter of balance. You cannot just lock yourself up in your room forever and totally escape your family and responsibilities. Your "dedicated space" should be respected by others but not abused by you in the avoidance of others at all times.

Next, you will need to determine if you want to buy or build your workbench and whether you want to have a metal or wooden workbench. Metal workbenches are often available through Sears stores or building supply stores such as Home Depot. Metal workbenches are generally more expensive and limited to certain sizes. They have the ability to be grounded well but, on the other hand, if you are working on a "hot" chassis, that is, a circuit without isolation, then it is possible to get a nasty shock or to damage the equipment you are working on. If you plan on buying or building a metal workbench, then you should consider placing a thin foam, wood, plastic, or Formica sheet on the top to isolate the bench from the circuits that you will be working on. The steel bench shown in Figure 1-1 shows a commercial metal workbench fitted with a wood top. The workbench shown in Figure 1-2 shows a two-tier metal workbench with an insulated work top. The second tier can be used for electronic test equipment. The alternative workbench featured in Figure 1-3 shows a two-tier metal workbench with an inlaid insulated work top. The

Figure 1-1.

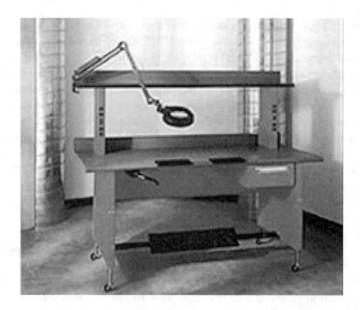

Figure 1-2.

workbench shown in Figure 1-4 shows yet another type of two-tier steel bench with insulated work surface and a full set of drawers that can be used to house special tools, schematics, fuses, jumper cables, and so on.

You might be able to locate a surplus steel workbench from office liquidator in your area or through a mail order business. Another option is to obtain and older, large metal desk with drawers, or even a large metal work table that could be used as an electronics workbench.

Figure 1-3.

Figure 1-4.

Many people prefer wooden workbenches since they are easy to build and are often available through building supply stores such as Lowes or Home Depot in the form of kits made from two-by-fours, for about $50 to $60. Wooden workbenches can be easily built from scratch if desired, using two-by-fours and wooden doors placed over a wooden frame. Another option is to use two or three sawhorses with a wooden door placed over them; this is bit less stable but can be made more secure. Another, more stable option for creating your own electronics workbench is to locate two low-cost, two-vertical-drawer file cabinets and place a six- or eight-foot-long wooden door over them. An alternative to using a door for a workbench top is to locate a surplus or low-cost overstock Formica countertop. Simply place the countertop or wooden door over the two file cabinets and secure the file cabinets to the top using recessed machine screws or large velcro pads.

Building or buying a wooden workbench will allow you to easily construct shelves above the table top in order to create separate cubbyholes for your test equipment. The shelf unit can then be placed above the workbench top and secured in place with screws if desired.

The electronics test bench shown in Figure 1-5 shows a well-organized old wood desk setup . The electronics workbench in Figure 1-6 shows a wooden door placed over wooden file cabinets with some vintage gear. The electronics test bench shown in Figure 1-7 is a wood-topped metal bench with test gear.

Another option for building your electronics workbench would be to look for a computer table that has lots of cubbyholes and/or shelves to hold test equipment. These computer tables are often available as low-cost kits that can be constructed in an evening or two. This is more of an instant approach for those who are not comfortable with building their own wooden or metal electronics workbenches from scratch or plans. Computer tables are available in many different styles and types and are available from both computer stores and discount retailers.

You will also need to consider the lighting of the area that you are thinking of adopting for your new electronics workshop or workbench area. Good lighting is important in order to see and identify small parts. Poor lighting will cause eyestrain when working over a long period of time, which will cause headaches. If possible, choose an area that has good exposure to natural lighting in order to supplement your workbench lighting. Setting up your work area near a window is preferable to a dark, unlit corner.

Figure 1-5.

Figure 1-6.

Figure 1-7.

You will also need to obtain extra lighting for your electronics workshop area. Incandescent lamps do not really light work areas very well. You should consider a bank of at least two dual-fluorescent-lamp fixtures to light up the work area in order to see well enough to work with delicate parts and to solder in confined spaces. You should also consider "spot" lighting, which will light up a specific area on your workbench, that is, the area in which you will be building, soldering, or testing. High-intensity "spot" lamps or jewelers lamps would be a good solution. Another consideration for lighting and seeing small work is to use one of those lighted magnifier lamps that mount to the side of a table or bench. These magnifier lamps are readily available through parts and tool suppliers. Also consider purchasing a magnifying glass or the Radio Shack lighted handheld microscope/magnifier (catalog no. 63-851). This device is an invaluable aid in working on electronics projects, from identifying components to looking for bad solder joints on a circuit board.

Once the appropriate electronics workshop area has been selected, and once you have built or secured your workbench, you will need to make sure that power is readily available. You may need to determine if you will be working solely with 110 volts or if you have the need to work with 220 volts or more as you are setting up your electronics workshop. You will need to ensure that there is power available alongside or above your workbench area. You should install at least two quad outlets very close to eye level above your electronics workbench. Purchase at least one and possibly two 110 volt power strips with good power "spike" protection. Power line spikes are often the cause of damage to sensitive electronics equipment. The last thing you want is to have your expensive electronic test equipment damaged by power line spikes or glitches. Power line spikes usually result from inductive loads switching on and off. Nearby loads such as freezers, vacuum cleaners, saws, and motors can cause power line spikes. Operating nearby saws and shop equipment can readily cause these spikes to be sent down the power line, directly to your sensitive test gear. Purchase the more expensive spike protection power strips such as the Tripplite Isobar series, that cost about $40 to $70. These power strips will provide an excellent level of protection for your test equipment, which through time will represent a sizeable investment.

Power protection is one of the most overlooked consideration for both computer and test equipment. Its not a fun or "flashy" purchase and is thought of as dull and uninteresting, but is not to be overlooked. After accumulating hundreds and maybe thousands of dollars worth of test equipment, why not spend a few extra dollars to protect your investment. These power protection strips can be mounted on the side or at the top rear of the workbench, so that test equipment and equipment under test can be powered through it. Mounting the power strips should be planned so that gear can be quickly plugged in and unplugged without breaking your back. One idea is to mount the power strips right near the quad outlet boxes or at the back or to one side of your electronics workbench.

After your lighting and power needs are met, you should next consider a ground rod connection at your workbench if possible, in order to provide a good ground to circuits under test or for radio equipment grounds. Be sure to connect your ground rod to the electrical power ground system to ensure a good stable ground system, free of floating ground potentials.

Next, you will need to begin thinking about where to place the pieces of test gear that you have to work with. This may initially only consist of a few hand tools and/or a multimeter and soldering iron, but as money permits, you will want to add more and more test equipment. Birthdays, holidays, and Christmas are great times to ask for those needed pieces of test equipment. Eventually you will acquire all the necessary pieces of test equipment and tools to make your electronics life easier. So make sure that your electronics workbench has the ability to expand, to house your future needs for important test equipment.

When designing your electronics workbench area, you should also consider extra space to store electronic components, batteries, chargers, fuses, notebooks, and so on. We will discuss these items in more detail later on.

Another often overlooked consideration when setting up and using your new electronics workbench is ventilation. The room chosen for your electronics workbench area should have adequate air circulation if possible. A window in the room will provide not only light for your work space but the ability to let in fresh air if needed, and this becomes a serious issue if you do a lot of soldering, cutting, sanding, Dremel work, or deburring of metal or plastic parts. You may want to consider a method for keeping a nearby window open when you are working in your electronics workshop. You may also want to consider placing a fan near the window, either to let cool air inside the room or to remove tainted air from the room, as the conditions dictate.

Many people have serious allergies to smoke, dust, and strong smells, and attention should be paid to this issue, especially if you have allergies. People with allergies may have to take additional precautions against these problems, especially if working in small or confined spaces for long periods of time. If your electronics workbench area is separate from your metal- or woodworking area, you may not have the problem of wood or metal dust but there is still the problem of solder smoke removal. Many people who have allergies are seriously affected by solder smoke. If you do a lot of soldering in a confined space for a period of hours, you may become light-headed or sick from solder fumes. To avoid getting sick or having sinus infections, it is important to consider a "spot" ventilation system for removing solder fumes. Many people who are allergic to strong odors or smells become more sensitive over time and their allergy problems get worse, so planning ahead will save you from more complex allergy problems later in life.

There are two approaches to solder-smoke or fume removal. One method utilizes a fan that sucks the fumes into a carbon filter that absorbs them. The filter material has to be replaced after a period of time.

The second method of solder-fume removal is to vent the fumes away from your work area. You could use a plastic vacuum cleaner hose held in place by a wood or metal stand near where the soldering takes place. The opposite end of the plastic hose is either vented to another room or the outside, using a small "muffin" or computer fan. Commercial fume removal systems are available from various mail order tool companies. With some ingenuity, you can design and build your own solder-smoke removal system.

Purchase a new kitchen trash can for your spouse or roommate and use the old trash can next to your electronics workbench to discard your electronic trash items.

Work Mats

Now that your electronics workbench has been constructed and everything is oriented just the way you want it, you will need to provide a clean, clear area in the center of the bench or table where you can work on building new circuits or troubleshooting circuits, running test setups, or drawing circuit schematics. In this clear center area of your electronics workbench, you should place a 12 to 18 inch piece of heavy paper or card stock or an antistatic work mat, described below. This work mat area will give you a highlighted special work area, separated from the rest of the workbench. It will make it easier to see and identify small components, parts, and hardware and will also provide a good work surface.

Once your electronics workshop or workbench has been constructed and you settle in to troubleshooting, designing, and building electronics circuits, you will likely soon encounter sensitive electronics components such as FETS, JFETS, and MOSFETS. The sensitive electronic components can be damaged very quickly if proper precautions are not taken when handling them. When working with sensitive electronic components such as MOSFETS, JFETS, CMOS transistors, and some sensitive op-amps, you need to pay particular attention to the buildup of static electricity. Make sure you are grounded when installing sensitive electronics components. Purchase an antistatic wristband, which goes around your wrist at one end while the other end connects to the ground lug of a 110 volt outlet. These are available from electronic supply houses and mail order suppliers and perhaps at your local electronics parts store. Also consider purchasing an antistatic work mat on which sensitive electronics components can be sorted and handled. The diagram shown in Figure 1-8 illustrates an electronics workbench connected to a ground system.

The antistatic work mat is the ideal place for assembling a circuit board. It can be grounded like your antistatic wristband. Static-sensitive components are usually always packaged in antistatic bags for shipment. Place the antistatic bag on the work surface area of your electronics workbench. Next, put on the antistatic wristband and ground yourself by touching a duplex outlet box. Make sure that there are no carpets under your electronic workbench to generate static electricity. Once you are positioned in your work area, you can proceed to open the antistatic package, remove the components from the bag, install the component in the circuit, and proceed to solder the components. The main caveat to remember when handling static-sensitive components is to seat yourself, with your antistatic wristband on and grounded, and to not move around from that point until the components are soldered in place. Moving around, especially on a carpet, while handling or removing components is an invitation to disaster. These simple steps will help to ensure that you will not destroy expensive sensitive electronic components before they are utilized.

The problem of static buildup, of course, is a problem associated with the cold, dry winter months but these precautions should be practiced at all times. There is nothing worse that ordering new MOSFET components for an ailing circuit for yourself, friend, or customer only to damage it during poor handling

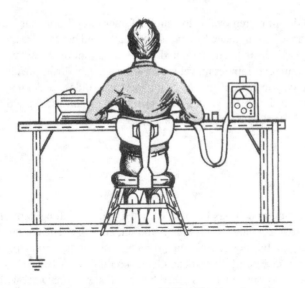

Figure 1-8. (Courtesy 2003 ARRL Handbook.)

and having to order the part over again. Once you have handled or repackaged sensitive components or finished building a new circuit, you will need to either remove the anti-static mat or place a clean, non-conducting rubber mat or heavy paper or card stock under directly under your current project before applying power to the new circuit. If your workbench is large enough, you could have an antistatic building–handling area and a separate nonconducting work area for "hot" or powered projects.

One common mistake when working on open-frame electronics chassis and new electronic circuits is having conductive paths under your project. This is not a problem when you are constructing a circuit or project, but once power is supplied to a circuit, you want to make sure that there are no cut wire leads, foil, screws, antistatic mats, or anything else under your project. You need to be constantly aware of the surface under your current "hot" or powered project.

Test Equipment Considerations

If you are just starting out in electronics and want to set up your own electronics workbench and you have a limited amount of money with which to buy or build some test equipment, where would you start? What is essential when you have a limited budget and/or equipment? The most essential pieces of electronic equipment would likely be a single- or dual-voltage power supply, a multimeter, a simple continuity tester, and a few hand tools, such as a set of screwdrivers, a pliers, wire cutter, hammer, and soldering iron.

Multimeters are very important pieces of test gear. Multimeters are relatively inexpensive now and can perform many different functions around your electronics workbench, including measuring voltage, current, resistance, and temperature and transistor gain. When purchasing a multimeter, consider buying a medium- to higher-priced meter. Many cheap multimeters can only measure dc voltages and resistance. A slightly higher-priced model might measure dc and ac voltages but not current. Look for a meter that will measure both dc and ac voltages as well as ac and dc current in a number of ranges. Current ranges up to 10 amperes are desired if possible, if you will be testing appliances. Look for a multimeter that measures dc resistances in at least three or four different ranges, from 200 ohms to 10 megohms or higher.

Higher-priced models might also feature a continuity tester, which is very useful, or perhaps an hFe or transistor gain measurement feature or a capacitance meter feature. More expensive models might also have a temperature probe option, which could be useful.

When buying a multimeter, you might also come across a meter that features an RS-232 data-logger. These meters have a computer interface and software that let the multimeter act as a computer data-logger. This feature is very useful if you want to measure voltages over time and graph them later. If you are testing circuits with intermittent problems or trying to establish if the power circuits are stable, then this type of meter might be very useful to you. For example, you could leave your recording digital multimeter on the bench to monitor the stability of voltage of an intermittent circuit overnight and have the meter log results over time.

As money permits, you could jump up to the next level and purchase a used or new oscilloscope. The oscilloscope is one of the most important and useful pieces of test gear that every electronics enthusiast should eventually purchase.

When setting up your new electronics workbench, you will need to decide if you can afford new or used test equipment. Used test gear can have many more years of useful life left on it if you purchase quality equipment from a used test equipment vendor. If you want to buy used gear, stick with name brands like Fluke, Tektronics, and HP. Of course, your other option is purchase new equipment if your budget allows. If you can afford name brand test instruments, by all means buy them; if not, select moderate priced equipment like Protek from a reputable dealer.

When buying a used oscilloscope, make sure it has been recently calibrated before you buy it; if you purchased it from a reputable test gear recycler, they will likely have specified that calibration was recently done and the equipment should have a tag with a calibration date shown. Since the oscilloscope is perhaps the most important piece of test gear on your electronics workbench, don't skimp—buy the best you can afford. A quality "scope" with relatively low "duty cycle" of home use should last many years. If you are looking to purchase a used scope, be sure to look it over carefully. Make sure that all the controls move, and only make the purchase if the oscilloscope comes with a manual. We will discuss the oscilloscope operation and whether to buy a new or used in Chapter fifteen.

If you can afford it, your next purchase after the oscilloscope might be a function generator. A function generator generates waveforms that can be used to test and analyze electronic circuits. Signal generators can be as simple a sine-wave generator, and function generators can be as simple as a sine-, square-, and triangle-wave generator. Newer, more complex function generators might also produce complex waveforms generated by an internal computer. Depending upon your budget, you could purchase either a new or used function generator. You can locate good used function generators with many more years of useful service left in them from many test equipment resellers or calibrators (see the Appendix).

The third most important piece of test equipment for your electronics workshop might be the digital frequency counter. Frequency counters can be used for many applications around your new electronics test bench. Counters can be used to test transmitters, oscillators, clocks, pulser circuits, and so on. Frequency counters have dramatically dropped in price in the last few years. It is now possible to purchase a decent, new frequency counter for the price of a major-brand used counter. For a hundred dollars or so, you can buy a good frequency counter for your electronics workbench that will serve for most all the needed applications. See chapters 2 to 6 for discussions on the specifics of the different types of test equipment that you will need for your new electronics bench setup.

Tools and Stuff

In working with electronics, as with any other specific area of interest, you will need to acquire special set of tools, equipment, and test equipment in order to perform your work. These tools or equipment are broken down into essential or first or most important items, and then into a second level of equipment that you would want to acquire as time and money permit. These second-level tools and equipment would make your life easier but you may not be able to afford them initially. The primary tools or first-level tools might include a hammer, a set of screwdrivers, an end (wire) cutter, a pair of pliers (maybe needle-nose pliers), a soldering iron, and some fuses. Secondary, or level-two, tools might include right-angle pliers, hemostat, dental picks, magnetic screw starters, solder gun, and so on. In Chapter 12 we further discuss the most essential tools and supplies that you might want or need for your electronics workbench.

Another item that will help you round out your electronics workshop is a stock of common-value fuses. Fuses come in a variety of types, sizes, and shapes. For the most part, the "AG fast blow" fuses are the most common types of fuses used for protecting circuitry. Historically, fuses came in little metal boxes with five fuses in them; today, you might find a package of two to five fuses in a blister pack. Try to select the most common values of "fast blow" fuses, such as 1, 1½, 2, 3, 5, and 10 ampere fuses for electronic circuitry, and buy a box of each value. You may also want to duplicate these values in the "slow blow" or delayed action fuses. Having these fuses available when you are servicing electronic equipment will save you much time and perhaps one less trip to the electronics supply store. Also, you could consider building the electronic fuse project described in Chapter 16. The electronic fuse project will allow you to temporarily substitute the electronic fuse for an actual fuse while you are troubleshooting a circuit that continuously blows fuses. You simply set the electronic fuse to the amperage or fuse value that was in the circuit, substitute the electronic fuse until you solve the problem, and then put the original fuse back into the circuit after you solve the problem.

You will want to acquire some electronic hardware items such as small screws, nuts, spade lugs, rubber foot chassis boxes, terminal strips, binding posts, and connectors. Screw sizes such as 2-56, 4-40, 8-32, and 10-32 are commonly used for electronic projects. For example, #6-32 screw size denotes a size 6 screw with 32 threads per inch. One of the best ways to store all small parts is to obtain some plastic storage boxes with "see-through" plastic drawers. The plastic storage boxes are readily available and often on sale at major discount stores such as Wal-Mart. These storage systems are great for organizing small parts such as electronic hardware, fuses, jacks and plugs, spade lugs, and so on.

As you set up your electronics workbench, you will find that there are many little extra items that will make your life a bit easier. One of these little items is the mini jumper wires with alligator clips at both ends. This item is invaluable for connecting one portion of a circuit to another or connecting circuits to power supplies or batteries. These jumper leads are readily available will save you a lot of time; rather than having to make up special cables for every occasion, these general-purpose cables can readily connect new circuits to circuit additions.

If you discover that you really like to design and build new or prototype circuits, you will want to investigate the ProtoBoard or Solderless Breadboard. The ProtoBoard is a relatively new development for designing new circuits. This type of development system can consist of just the ProtoBoard, which is typically a 2¼ by 6½ inch plastic block with mini holes with recessed metal clips below the surface of the block that can be connected together via jumper wires. Integrated circuits and electronic components can be inserted into the ProtoBoard and be temporarily connected together to form a complete circuit. ProtoBoards are used to build new circuits that can be debugged "on the fly," that is, as the design develops.

ProtoBoards make designing and building new circuits a simple and quick matter. Solderless Breadboards come in a number different types and sizes, and are often integrated with small power supplies or battery holders to permit them to be portable. Some advanced, self-contained ProtoBoard systems include a power supply and mini waveform generators and digital clock circuits. ProtoBoards can be transported from work to home to another hobbyist's home to codesign a project. Prototypes and design as well as circuit fabrication are discussed in more detail in Chapter 14.

Reading Schematics

Before we move on to getting our hands on electronic circuits, you should get to know and identify electronic symbols and how to read an electronic schematic. Learning how to read a schematic will take some time and experience, but this chapter will give you an introduction and a place to start. Before you can read a schematic, you will have to be familiar with electronic symbols. Figure 1-9 shows some of the more common symbols used in electronics circuits. Look over the diagram carefully and familiarize yourself with the symbols. You will see some of the more common ones very often in your electronic schematic and electronic project diagrams. See Appendix I for an extended set of electronic symbols.

The resistor is a device that resists the flow of charge. The unit of resistance is the ohm and is represented in schematics as k, which stands for kilohms. A listing of 10 k means 10,000 ohms. Sometimes, you will also see a capital M, which stands for Meg or megohm. A listing of 4.7 M means 4.7 million

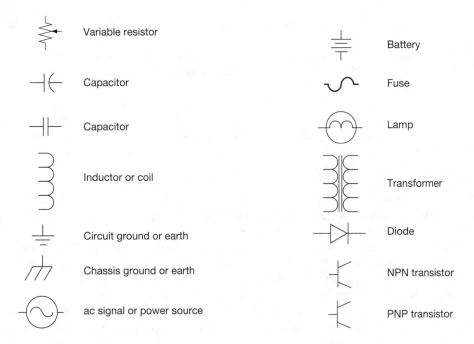

Figure 1-9. The most common electronic symbols.

ohms. Variable resistors are sometimes called potentiometers, and they come in many different forms and shapes. The symbol for a variable resistor is a resistor with an arrow near the center, to show that its value can change. Resistors can be connected in series or parallel. If resistors are connected in series, their values add up. If resistors are connected in parallel, their values are divided. Resistors come in many different shapes and sizes as well as power ratings; see the components section in Chapter 9.

The capacitor is also another ubiquitous symbol in electronic circuits. Capacitors are devices that have metal plates separated by an insulator. They are used to temporarily store an electrical charge or to couple two circuits together. The symbol for a capacitor is two parallel plates with leads connected to each plate. The unit of capacitance is the farad, but its value is so large that the microfarad is the unit used for capacitance values. Microfarads are millionths of farads, and are often abbreviated as mf, MF, or μF. Capacitors are also often listed in picofarads or pf. A picofarad is sometimes called a micromicrofarads. It is 10^{-12} farad. Capacitors are often marked with a plus (+) or minus (–) terminal, which means that the capacitor is a polarized value capacitor. Capacitors with no markings are known as nonpolarized capacitors. Capacitors can be connected in series or parallel. When connected in series, their values are divided, but when connected in parallel, their values are added together.

Inductors are passive electronic components that stores energy in the form of magnetic fields, and are also used to form resonant circuits in radio equipment. In its simplest form, an inductor consists of a wire loop or coil. The inductance is directly proportional to the number of turns in the coil. Inductance also depends on the radius of the coil and on the type of material around which the coil is wound. The unit of inductance is the henry but its value is usually too large to be used in most electronic circuits, so the microhenry is generally used to represent inductance values. It is written as μH. Sometimes, the millihenry (mH) is used. Inductors can be connected in series or parallel.

The symbol for a circuit ground or earth is shown in two different ways: a vertical lead perpendicular to a series of three increasingly smaller lines that point downward, or as three parallel lines pointing downward toward the left.

An ac signal is usually shown as a circle with a sine wave inside. A battery is shown as an alternate series of parallel lines of two lengths. Each size line represents a battery pole. Usually, the smaller line is the plus (+) and the longer line is the minus lead (–).

A fuse is depicted as a wavy line that looks a lot like a sine wave and is sometimes shown inside a rectangular box. A miniature light or lamp is shown as a coil inside of a circle. A transformer is shown as two coils parallel to each other, with two parallel lines between the coils or inductors. Two parallel lines indicate a transformer with a metal core inside. Transformers are sometimes shown without the parallel line; this denotes an air core transformer, without a metal core.

A diode is shown as a triangle pointing to a perpendicular line, with a wire or lead at each end of the device. The triangle points to the cathode lead; the lead at the flat side of the triangle is the anode. Diodes come in many different shapes, sizes, and power ratings. Diodes pass current in one direction only and are often used as rectifiers for converting ac to dc voltages.

Transistors are most commonly shown as a vertical line with two lines coming off the vertical line at 45 degree angles. One of the angled lines will have an arrow placed on the line pointing toward or away from the vertical line. A transistor with an arrow pointing toward the center vertical line is a PNP transistor, whereas an arrow pointing away from the center line represents an NPN transistor.

Figure 1-10 shows many more electronic symbols. Look these symbols over and use this diagram often to refer to symbols seen on the electronics schematics that you come across while learning about electronics (also see Appendix I). One of the most important aspects of reading schematics is understanding the depiction of terminals, junctions, and crossover points on a diagram. When two lines cross on a schemat-

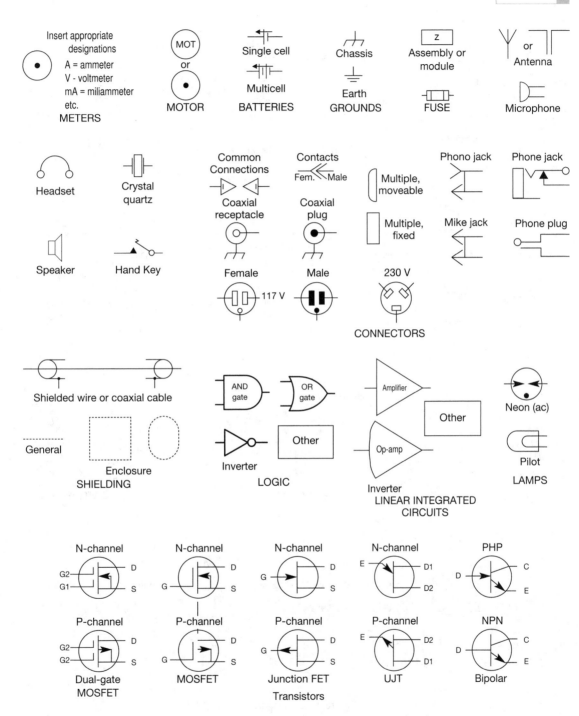

Figure 1-10. Schematic symbols used in electronic diagrams. (Continued on next page.)

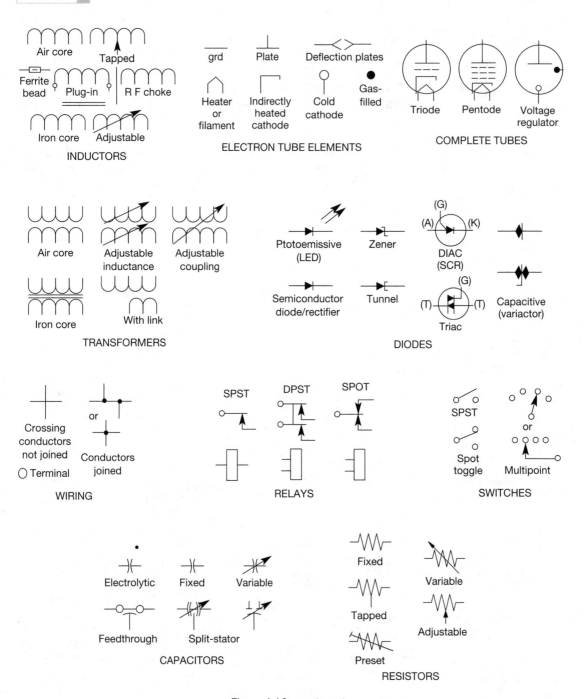

Figure 1-10 continued.

ic and they are not joined or connected together, they will simply cross over each other for diagram and line simplicity. When two circuit lines are supposed to be connected together, they can be seen as crossing each other with a small solid dot or circle. This distinction between crossing lines can mean the difference between a working circuit and a nonfunctional or damaged circuit, so pay particular attention to the joining of circuit lines on a schematic.

Before we take a look at some electronics schematics, it is important that you understand the relationship between voltage and current and resistance. This relationship is called Ohm's Law and it is the backbone of simple electronics formulas for electronics. In its simplest form, it is shown as the equation $V = I \times R$, which says voltage (V) is equal to current (I) multiplied by resistance (R). Voltage is represented as electromotive force where V is used instead of E. Current is in amperes or amps. In typical electronics circuits, milliamps may be used, equal to 1/1000 of an ampere. One milliamp = 0.001 amp and is abbreviated as MA or ma. The three most important Ohm's Law equations are $V = I \times R$, $I = V/R$, and $R = V/I$. The most used Ohm's Law power equation is $P = I \times V$, where P is power in watts or wattage, which is equal to the current of a circuit multiplied by the voltage in the circuit. Thus, a voltage of 100 volts multiplied by a current consumption of two amps would equate to a power consumption of 200 watts. Appendix I illustrates the whole set of Ohm's Law equations.

The diagram shown in Figure 1-11 depicts a very common circuit called a voltage divider. The basic voltage divider consists of two resistors, usually connected in series as shown. The total resistance is simply the sum of the two. In this example, it would be 22 k ohms plus 33 ohms = 22,033 ohms. If a volt signal is applied to the input end of R1, the 22 k ohm resistor, the current through the whole circuit would be $I = V/R = 1/22,033$ or 0.0000453864 amps, or about 0.05 milliamps. Voltage dividers are used to reduce or step-down voltages or signal levels in voltage converter circuits, amplifiers, and so on. Voltage dividers are often set up using resistor ratios of 1000:1, 100:1, or 10:1. For example, a 10 to 1 resistor divider would have a 10 ohm and a 1 ohm resistor for its R1 and R2 values.

The introduction of the operational amplifier or op-amp in the late 1960s and early 1970s transformed the electronics industry like no device before it. Op-amps began life as large integrated amplifier modules, some measuring up to three inches in length. Early op-amps were large and noisy but nonetheless were a monumental development in electronics. The op-amp permitted a simple building block approach to electronic circuit design. Early op-amps were primarily used as preamplifiers and instrumentation am-

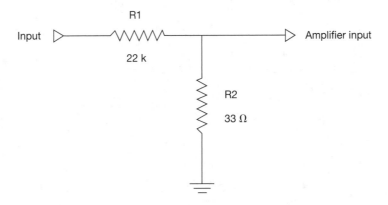

Figure 1-11. Voltage divider.

plifiers. As op-amp quality progressed and their sizes and power consumption reduced over time, op-amps were used in all types electronic circuit designs. Op-amps are now utilized in audio amplifiers, filter design, oscillators, comparators, regulators, and so on. Since op-amps are now found in almost all electronics circuits, we will spend some time identifying different op-amp configurations and applications.

Integrated circuits are used throughout much of modern electronics, so understanding how they are represented is important. Integrated circuits generally contain many individual circuits or components, and are shown schematically as functional blocks. Op-amps can be configured in many different ways, including noninverting amplifiers.

The ubiquitous operational amplifier or op-amp is shown in Figure 1-12 as a triangle with a plus (+) and minus (–) input and a single output at the point of the triangle. This type of op-amp configuration is called an inverting input amplifier circuit, since the input is fed through the minus (–) input of the op-amp. Note that the signal input is applied through the input resistor at R1. Also notice that a second resistor, R2, is shown from the minus input across to the output of the op-amp. Resistor R2 is called the feedback resistor. The plus (+) input of the op-amp is shown connected to a resistor network used to balance the input through a bias control at resistor R4. Both plus and minus voltage sources are connected to the bias network.

Simulate a drive a current through the inverting input by placing 1 volt on the input at R1 and assume that the right end has 0 volts on it. The current will be $I = V/R = 1/1$ k $= 1$ ma. The voltage output will try

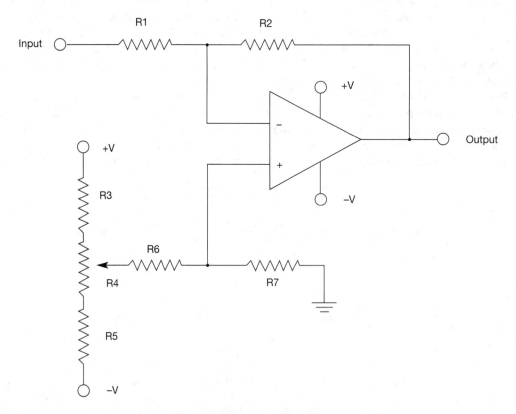

Figure 1-12. Inverting amplifier.

to counter this by driving a current of the opposite polarity through the feedback resistor into the inverting input. The required voltage to do that will be $V = -(I \times R) = -(1 \text{ ma} \times 10 \text{ k}) = -10 \text{ } V$. Thus, we get a voltage-to-current conversion, a current-to-voltage conversion, a polarity inversion, and, most importantly, amplification: $G = -(\text{Feedback Resistor/Input Resistor})$. In this example, it is shown as gain or $G = -(R2/R1)$.

This inverting input op-amp configuration is common in modern amplifiers and filters. Op-amps generally require both plus (+) and minus (–) voltages for operation; this allows a voltage swing on either side of zero.

The diagram in Figure 1-13 shows a noninverting op-amp configuration. This configuration is used when you want to maintain the same phase or polarity from the input through to the output. Notice that in this configuration the signal input is applied to the plus (+) input of the op-amp. The balance and trim capability is performed via the bias network formed by resistors R4 through R7 which connect to both the plus and minus power supply.

The circuit shown in Figure 1-14 depicts a basic differential op-amp instrumentation amplifier. In this circuit, notice that there are two separate inputs and they are referenced against the ground connection, which is separate from the inputs. This type of circuit is used in high-gain instrument amplifiers when you want to keep common-modes noise from reaching the op-amp. This configuration is commonly used for high-input impedance sensor inputs, where low leads may be used and low noise is required. Note that the

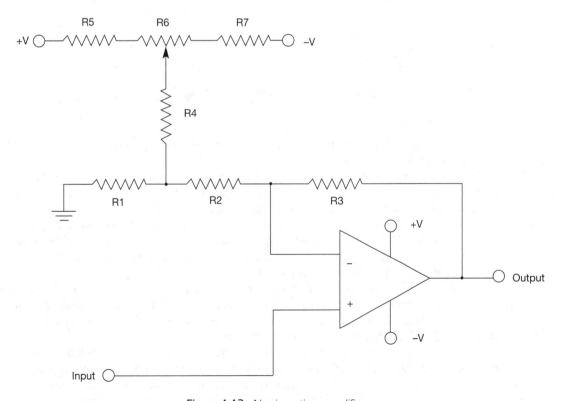

Figure 1-13. Noninverting amplifier.

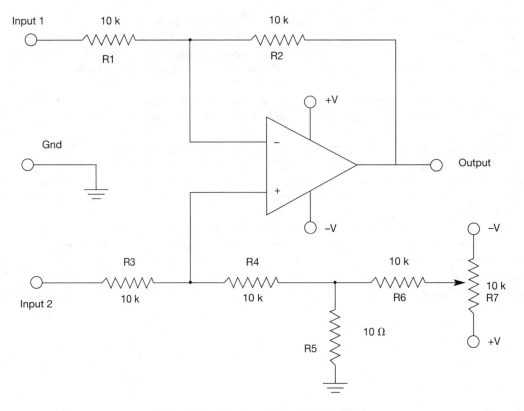

Figure 1-14. Op-amp differential amplifier.

balance or trim for the op-amp is provided through the resistor network composed of R4 through R7, which allows bias from both the plus and minus power sources.

Figure 1-15 illustrates how an op-amp can be used as a comparator to trigger an LED when a certain voltage threshold is reached. In this configuration, the plus (+) input of the op-amp is connected to a 50 k potentiometer, which is connected between the +9 volt source and ground. Note that in this configuration only a single voltage is required. The minus (−) input lead of the op-amp is connected to the voltage source you wish to monitor. In operation, you would apply a voltage to the minus input pin of the op-amp and adjust the R1 so that the LED is not lit. Raising the voltage at the input would now offset the comparator and its trip point, and the LED would become lit.

Op-amps can also be used in oscillators and function generators, as shown in Figure 1-16. In this circuit, the LM339 comparator op-amp is used to generate a square-wave signal. A resistor and the capacitor at C1 form a timing network that is used to establish the frequency of the square-wave generator. The duty cycle is set up through resistors to produce a symmetrical square-wave output.

As you can see from these examples, the op-amp is a very powerful electronic building block tool for greatly simplifying the design of electronics circuits. We have only covered a small number of op-amp applications. Op-amps are also used for filters, power regulators, integrators, and voltage-to-current and current-to-voltage convertors. For more information on linear op-amp theory and applications, look for

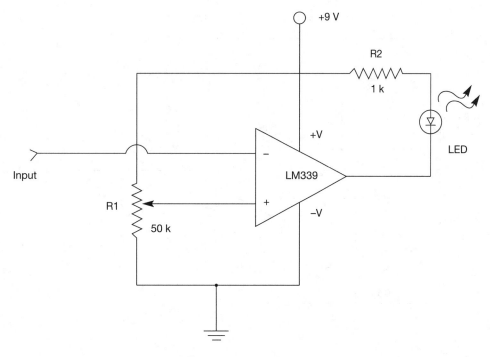

Figure 1-15. Op-amp differential comparator.

the *IC Op-Amp Cookbook* by Walter Jung or point your browser to http://w1.859.telia.com/~u85920178/begin/opamp00.htm for a discussion on operational basics by Harry Lythall.

The above examples illustrate linear op-amps, which represent nearly a half of the spectrum of integrated circuits. On the other side of the spectrum of integrated circuits are the digital integrated circuits, which generally contain multiple digital gates, switches, and memory functions all in one package. One of the more common digital integrated circuits is the quad two-input AND gate, shown in Figure 1-17. This diagram illustrates a 14 pin 74LS08 quad two-input AND gate and its truth table. The 74LS08 contains four AND gates. The inputs to the first AND gate are at 1A and 1B, represented by pins 1 and 2, whereas the output is at pin 3. The second AND gate has inputs 2A and 2B on pins 4 and 5, with its output on pin 6. The common ground connection is shown on pin 7; 5 volt power is supplied to pin 14. The third AND gate has its inputs at 3A and 3B, on pins 9 and 10, with its output on pin 8. The fourth AND gate, shown as 4A and 4B, is on pins 12 and 13, with its output on pin 11. The truth table or function table for the 74LS08 describes the input versus the output condition for each of the gates. If inputs 1A and 1B are high then the resulting output at 1Y will be high. If input 1A is low and the input at 1B is either high or low, then the output at 1Y will be low. If input 1B is low and input 1A is either high or low, the output at 1Y will be low once again.

In working with electronic circuits, you will come across many schematics as well as pictorial diagrams. Schematics are the electronic representation on paper of an electronic circuit. A pictorial diagram is a generally a physical layout diagram of the same circuit. In electronics work, you will come across both types of diagrams. They look quite different but they are really the same circuit shown differently.

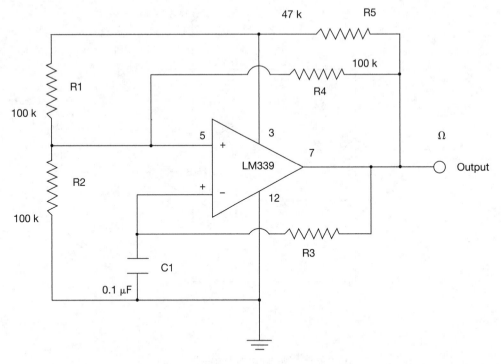

Figure 1-16. Op-amp square-wave oscillator.

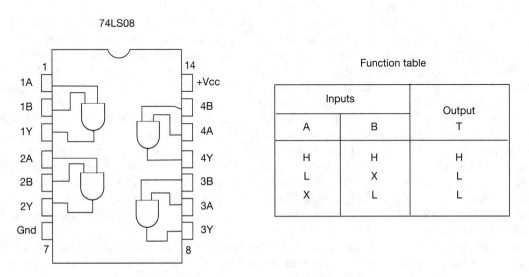

Figure 1-17. Quad two-input AND gate.

Experience in looking and comparing these two types of diagrams will help you enormously in your electronics projects.

The diagram shown in Figure 1-18 illustrates a schematic of a two-transistor audio amplifier circuit. The input jack at J1 is coupled to an input capacitor at C1. Capacitor C1 couples the microphone of the input source to the amplifier. It acts as a coupling device as well as a blocking device, which can keep any constant dc component or voltage from reaching the input of the amplifier. The resistors R1 and R2 form the input impedance and frequency response characteristics of the input to the amplifier. The NPN transistor at Q1 forms the first stage of amplification of the amplifier. Resistors R4 and R5 form the bias or voltage supply for Q1. The output of the first amplifier stage at Q1 is connected to capacitor C3, which is used to couple the first amplifier stage to the next amplifier stage. Capacitor C3 is next fed to variable resistor R6, which acts as the system volume control. Capacitor C5 couples the audio signal to the second amplifier stage at Q2. Resistor R10 is used to bias the second amplifier stage. Capacitor C7 is used to connect the output of transistor Q2 to the final amplifier stage or to the input of a transmitter, at J2. Note the ground bus or common connection for ground at the bottom of the circuit, and the power bus connection at the top of the schematic diagram.

The pictorial diagram shown in Figure 1-19 serves to illustrate the same two-transistor amplifier circuit shown in Figure 1-18. Both diagrams are electrically the same, but look quite different. The pictorial diagram shows the circuit as it would look wired and laid out in a chassis box. Compare the two diagrams so that you can see and understand how they differ from each other.

A two-op-amp guitar amplifier is shown in schematically Figure 1-20. The input to the amplifier is shown at input jack at J1. Resistor R1 serves to control the input signal or current flowing into the amplifier circuit. Resistor R1 establishes the input impedance to the amplifier circuit. R1 is next fed to the 0.1 μF capacitor at C1. Capacitor C1 is used to block low-frequency signals, couple the input to the next stage, and keep any constant dc voltage from the guitar away from the input to the op-amps. The low fre-

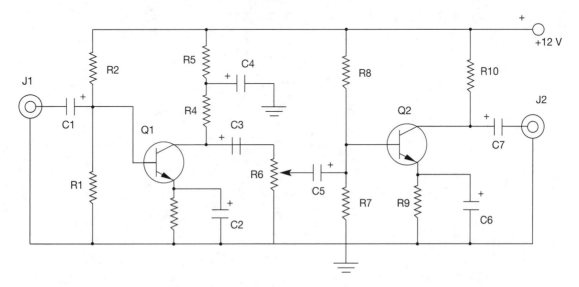

Figure 1-18. Transistor amplifier schematic diagram.

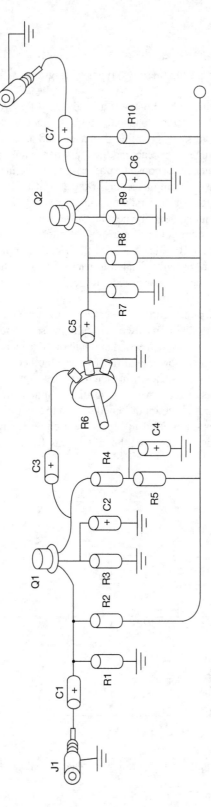

Figure 1-19. Transistor amplifier pictorial diagram.

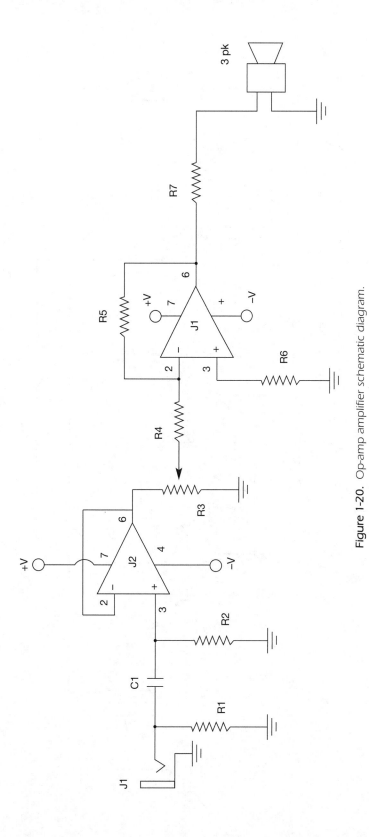

Figure 1-20. Op-amp amplifier schematic diagram.

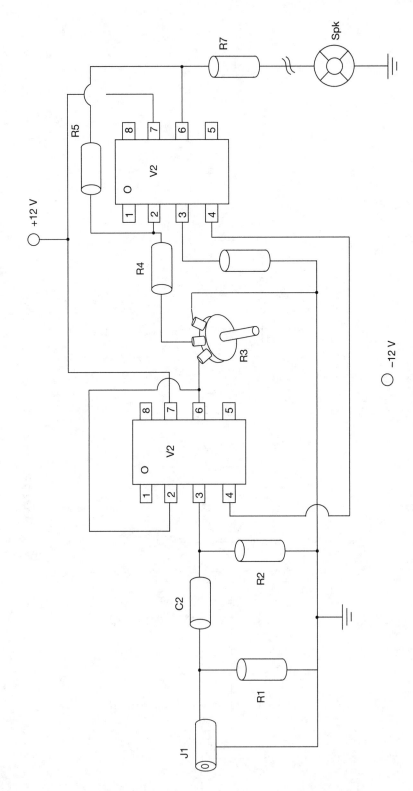

Figure 1-21. Op-amp amplifier pictorial diagram.

quencies to be blocked are dependent upon resistors R1 and R2. The triangle symbol shown at U1 represents the first op-amp amplifier in the circuit. The op-amp has two inputs, an inverting input, and a noninverting input. The input signal from resistor R2 is fed to the noninverting input of the op-amp. The connection from the input to the output pin of the op-amp forms the feedback network, which generally determines the gain of the amplifier. In this example, there is no resistor but a direct connection between the input and the output. In this example, the direct connection from the input to the output of the amplifier establishes a unity gain, or no change in the signal input. The purpose of the first stage in this example is to reduce the amount of current the guitar must supply to the amplifier. Resistor R3 represents a variable resistor, which is used as a volume control for the amplifier. The center tap on the variable resistor is fed to resistor R4, which is used to couple the first stage of the amplifier to the second stage. Resistors R4 and R5 establish the gain of the second amplification stage. This gain stage forms the real muscle of the amplifier. The output of the second op-amp at pin 6 is fed directly to a speaker through resistor R7, which is used to protect the output of the amplifier as well as to protect the speaker from too much current.

The diagram shown in Figure 1-21 depicts the same guitar amplifier circuit, but it is now shown pictorially instead of schematically. The pictorial diagram shows the circuit as it might appear physically on the circuit board. The diagram looks quite different but it is really the same circuit drawn slightly differently. You will need to recognize and become familiar with the differences between these types of drawings. The best way to familiarize yourself with the difference between schematics and pictorial diagrams is to see and compare a number of them over time. Once you get some practice, it will become second nature to you. There are a number of good books on how to read schematics that cover the topic in more depth. See Appendix II for a good book list.

The Multimeter

In this chapter, you will learn how a multimeter works and how it can be used to make basic electrical measurements at your electronics workbench, at the office, or on the job. There are two basic types of meters used to make electrical measurements. One type has a needle that deflects on a curved scale; the position of the needle on the scale indicates the magnitude of the value being measured. This type is an called an analog meter and commonly a voltmeter, ammeter, and ohmmeter, are all combined into a single inclusive multimeter or multitester called a VOM or volt–ohm meter, depending on the quantities that it measures. The analog meter mechanism is generally a D'Arsonval type movement.

Figure 2-1 shows a typical analog multimeter. The meter's main component is the meter movement, which is behind the display scale. The meter movement contains a moving needle that deflects to the right along a calibrated scale. A selector switch is used to select the mode and/or quantity the meter is chosen to measure, as well as the full-scale range of measurement. Resistors are used to scale the voltage and current readings, and several adjusting controls, are often included. Usually, one control is a potentiometer provided to zero the ohmmeter function. Test lead jacks are provided to accept test leads sets used to connect the meter to a component or circuit for measurement.

The second type of multimeter is called the digital multimeter. Any quantity that is measured appears as a number on a digital display. It is commonly called a digital voltmeter, digital multimeter, digital multitester, or DVM. A typical digital meter is shown in Figure 2-2. The digital display indicates the numerical value of the measurement taken. A selector switch selects the measurement function. Often, the main selector switch as acts as an on–off switch, applying battery power to the internal circuits of the meter. A range/function switch is usually provided for selecting the quantity to be measured. Some DVM models are autoranging and do not have a range selector. Digital multimeters often special features, such as memory hold, voice output, and semiconductor and continuity testing.

Figure 2-1   Analog multimeter.

Figure 2-2   Digital multimeter.

The Analog Multimeter

The Basic Analog Meter Movement

The basic analog VOM, with its moving needle indicator that moves across a calibrated scale, is basically a small electrical motor. This type of analog meter uses a D'Arsonval meter movement and is shown in Figure 2-3. The movable coil in the center of the movement is termed an armature, has a needle attached to it, and is located within the strong magnetic field of a permanent magnet, which is hidden under the scale plate. The D'Arsonval meter movement is essentially a micromotor in disguise; the armature shaft just drives a needle instead of a wheel. The uniform radial field about the moving coil is required to make the torque produced by the current in the coil result in a linear movement of the meter needle along the calibrated scale. As current increases, the deflection increases.

One end of the needle is attached to a spiral spring so the needle will be returned to an initial position when the current in the coil is removed. The springs are calibrated and the coil and needle are balanced so that the total assembly produces a linear deflection on the meter scale as the current in the coil is increased linearly.

The other end of the spring is attached to a zero-adjust screw located on the core of the moving coil. With the zero adjust, the initial or static position of the meter movement is adjusted mechanically to zero on the scale by varying the position of the end of the spiral spring. Extreme care must be taken while making the zero adjustment because of the delicate spiral spring construction. The complete needle assembly usually is constructed around an aluminum filament so that the assembly is very sturdy. The aluminum frame also serves as a damper.

Since the radial magnetic field is produced by a permanent magnet, the direction of the force of this field is always in one single direction. The current must pass through the coil in one direction only to

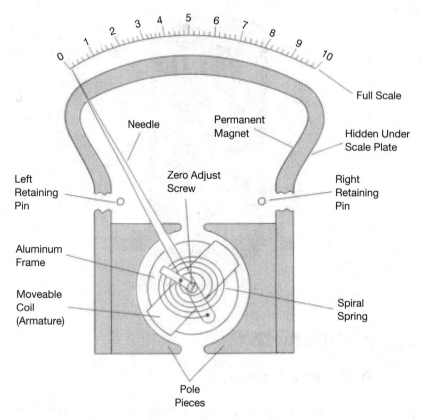

Figure 2-3 *D'Arsonval meter moving coil movement.*

cause an upward deflection of the needle. If current passes through the meter movement coil in the other direction, the needle deflects backwards against the meter's left retaining pin. One of the most common ways of damaging or burning out a meter is passing too much current through the coil in the wrong direction, but many multimeters have built-in protection against incorrect polarity connections.

Basic Analog Meter Characteristics

Meter Movement Sensitivity

The sensitivity of the meter movement depends upon the current required for full-scale deflection, which varies inversely with the current being measured. Simply stated, the most sensitive meter requires the least current for full-scale deflection. The amount of current required to deflect the needle full scale depends upon the number of turns of wire on the moving coil. When more turns are added, generally by using smaller-gauge wire, a stronger magnetic field is created to react with the permanent magnetic field. Smaller-gauge wire also keeps the mass of the meter movement low.

Measuring Meter Sensitivity

The sensitivity of a meter can be determined by measuring the current required to produce full-scale deflection of the meter movement. In order to do this, a standard current meter is placed in series with the unknown meter and a source of current is applied to it. Since the current is the same in all parts of a series circuit, the amount of current required to produce full-scale deflection on the unknown meter can be read directly from the standard meter. Meter sensitivities generally range from a few microamperes for very sensitive movements to 50 to 1000 microamperes for the average movements used for multitesters that measure voltage, current, and resistance.

Internal Meter Resistance

The internal dc resistance of a meter movement and the meter sensitivity and are usually fixed by the design. These characteristics cannot be altered unless the physical construction of the meter movement is altered. To measure a meter's internal resistance, with the full-scale current through a meter indicated by the standard current meter, a voltmeter is connected across the unknown meter to measure the voltage drop across its internal resistance. The meter's internal resistance can then be obtained by using Ohm's Law:

$$R_m = V_m/I_m$$

where V_m is the voltage across the meter and I_m is the current through it for a full-scale reading. A meter that is designed to have full-scale deflection with a very low current will have high sensitivity and high internal resistance.

Note that a sensitive meter movement can be easily destroyed by connecting an ohmmeter across it in an attempt to measure its internal resistance. The ohmmeter has an internal battery and can supply current. The current from the ohmmeter may be 100 times that required to deflect the meter full scale.

Meter Accuracy

There are many factors that affect a meter's accuracy: the design of the meter; the quality of its parts, the care in its manufacture, the accuracy of its calibration, the environment it is used in, and the care it has had since being put into service.

The accuracy of a meter used in everyday applications is usually between 0.01% and 3%. The accuracy of a typical D'Arsonval movement commonly used in all types of analog measuring instruments is about 1% of the full-scale reading. If a voltmeter has a 10 volt scale and 1% accuracy, any reading is going to be accurate to 0.1 volt. The advertised accuracy of the meter usually refers to the percent of full-scale reading on any range.

The accuracy of a voltage reading for a meter that has a stated accuracy of 1% is referenced against the voltage to be read for three full-scale ranges: 1 volt, 5 volt, and 2.5 volt. If a voltage of 2 volts is measured, it will be accurate to ± 0.1V (5%) if measured on the 10 volt range, ± 0.05V (2.5%) if measured on the 5 volt range, and ± 0.025V (1.25%) if measured on the 2.5 volt range. As a result, a meter reading taken at the low end of the scale is going to have a greater percent of absolute error than a reading near full scale. To obtain the best accuracy when making meter measurements, first choose the meter range so that the deflection is nearest to full scale. The closer the reading of the meter to full scale, the less the absolute error.

Meter Damping

To prevent meter overshoot when the meter coil and needle assembly move, the meter must be damped. The aluminum frame is used to perform this damping. It is effectively a short-circuited, single-turn loop within the meter magnetic field. A loop of wire moving in a magnetic field will have a voltage induced in the loop. Any movement of the meter armature moves the frame, and the induced voltage causes a current in the frame. As with any current in a conductor, the frame current develops a magnetic field that interacts with the strong magnetic field of the permanent magnet and offers a slight opposition to the movement of the coil. The law states that the resultant current in a wire moving in a magnetic field produces a magnetic field that opposes the original magnetic field that generates the voltage that produced the current.

The Basic Voltmeter

Adding an additional series resistance to the basic meter circuit allows you to convert a basic meter movement into a direct current (dc) voltmeter. The series resistor is called a multiplier. The basic circuit is shown in Figure 2-4. The value of the multiplier resistor R_x determines the voltage range for full-scale deflection. By changing the value of the multiplier resistor, the full-scale voltage range can be changed.

Multiplier Resistance

The value of the multiplier can be found by using Ohm's Law and the rules for a series circuit. Since the multiplier resistor R_x is in series with the basic meter movement, the current I_m passes through R_x. If V equals the full-scale voltage of the voltmeter, then, according to Ohm's Law,

$$V = I_m(R_x + R_m)$$

Multirange Voltmeters

In configuring a typical multimeter to measure voltage, a range selection switch is used to select one of the multiplier resistors that provide a voltmeter with a number of voltage ranges. Figure 2-5 shows a multirange voltmeter using a four-position switch and four multiplier resistors. The multiplier in each range was calculated using the simple equation for R_x given earlier.

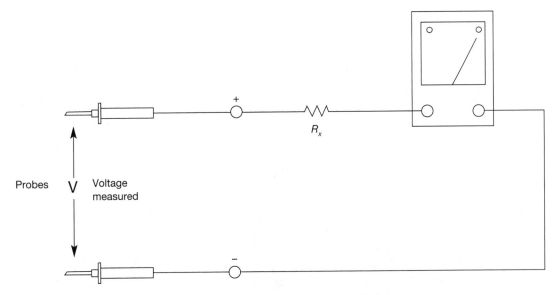

Figure 2-4 Voltage measurement circuit.

Voltmeter Sensitivity

The diagram shown in Figure 2-6 shows how a voltmeter can be is used to make a voltage measurement. In the example shown, the meter is monitoring the voltage across resistor R2. The voltmeter is connected across or in parallel with the resistor R2 to measure the voltage. The voltmeter must have a current I_m through the meter movement to produce the deflection. This current must be supplied by the circuit and reduces the current through R2 as shown in the diagram. Note that the resistance of the voltmeter ($R_x + R_m$) should be much higher than the resistance of the device to be measured. This reduces the current drawn from the circuit by the voltmeter and, therefore, increases the accuracy of the voltage measurement. The value of the voltmeter resistance (referred to as input resistance) depends on the sensitivity of the basic meter movement. The greater the sensitivity of the meter movement, the smaller the current required for full-scale deflection and the higher the input resistance. When expressing the sensitivity of the voltmeter, the term "ohm's per volt" is used exclusively. It is referred to as the voltmeter sensitivity. The sensitivity of the voltmeter, written as S, is easily calculated by taking the reciprocal of the full-scale deflection current of the basic meter movement: $S = 1/I_m$.

The Basic Ammeter

Ammeters are utilized for measuring current in a circuit, which can range from microamperes to tens of amperes. The simplest ammeter contains only the basic meter movement. It can measure values of current up to its sensitivity rating or full scale. Meter movements that require a large current for full-scale deflection are not practical because of the large amount of wire they would require on the coil and, thus, the

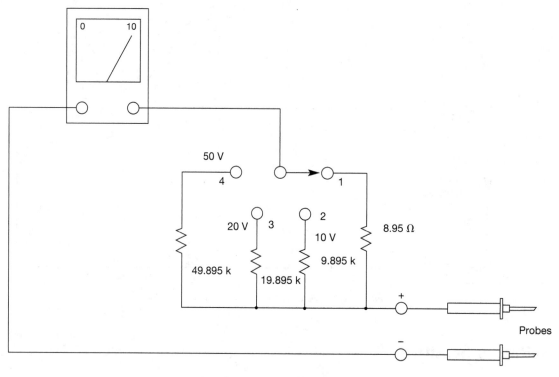

Figure 2-5   Multirange dc voltmeter.

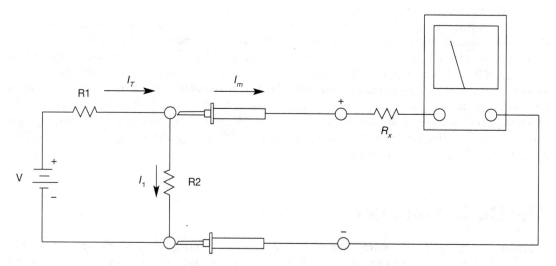

Figure 2-6   Actual voltage measurement.

large mass. Therefore, highly sensitive meter movements are desensitized by placing low-value resistors called "shunts" in parallel with the meter movement to extend their current handling range. For example, if a meter movement requires 100 microamperes for full-scale deflection and the meter is to display a 10 milliampere (10,000 microamperes) scale, then the shunt placed across the meter must have a resistance value capable of carrying 9900 microamperes.

Simple Ammeters

Figure 2-7 shows a single shunt across a basic meter movement, used to extend the range of the meter movement. Shunts are easily calculated by applying Ohm's Law ($V = IR$) and the principles of a parallel circuit, if the meter movement's internal resistance R_m is known. The equation for the ammeter shunt resistor Rs is:

$$V_s = V_m$$

The voltage across the shunt V_s is equal to the voltage across the meter V_m, therefore, V_s is equal to the current through the shunt I_s times the shunt resistance R_s. V_m is equal to the current through the meter movement I_m times the meter resistance R_m:

$$I_s \times R_s = I_m \times R_m$$

Now, solving for R_s gives,

$$R = (I_m \times R_m)/I_s$$

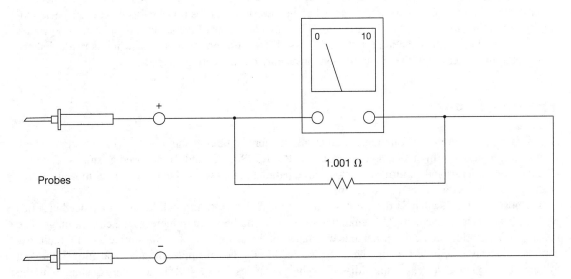

Figure 2-7　Extending meter range by using a shunt.

Then, substituting $I_t - I_m$ for I_s will give

$$R_s - (I_m \times R_m)/I_t - I_m$$

As an example, If $I_m = 1$ milliampere, $R_m = 1000$ ohm's, and I is at the full-scale range of 1 ampere, then the shunt resistance is

$$R_s = 0.001 \text{ A} \times 1000 \text{ ohm's}/(1000 - 1) \text{ mA}$$

$$R_s = 1 \text{ V}/999 \text{ mA}$$

$$R_s = 1/0.999 \text{ A}$$

$$R_s = 1.001 \text{ ohm's}$$

The use of this meter shunt equation will provide the correct value of resistance to be added in parallel with the meter movement to increase the current-handling capability of the meter.

Multirange Ammeters

Simple Meter Shunts

A basic meter movement has a known or established sensitivity and a given amount of current must pass through it to produce full-scale deflection. By using suitable shunts, the range of current that a particular movement can measure is now increased. Essentially, this is the process used in the construction of multi-range ammeters. A four-range ammeter is shown in Figure 2-8. The meter leads are used to connect the meter in the circuit to measure a current. Note that the lowest range (position 1) uses the meter's basic sensitivity of 1 milliampere (1 mA) with no shunt. The suitable values of the shunts are shown; these values may be verified using the equation for R_s as discussed. The fifth position on the switch is marked short, which takes the meter out of the circuit without physically disconnecting its leads.

The Ring Shunt

Another type of shunt that is commonly used in multimeters is the ring shunt or "universal" shunt, which is shown in Figure 2-9. In this circuit, the resistors, R2, R3, R4, R5, and R6 are used as shunts. They are all connected in series and, in turn, are connected in parallel across R1 and the basic meter movement. R1 is always in series with the meter movement.

The position of the switch S1 determines which of the resistors R2 through R5 also are placed in series with the basic meter movement. The remaining resistors form the shunt to increase the current range. The resistance needed for each current range is calculated in the same way as in the simple shunt circuit, except that the resistance in series with the meter is added to R_m. For example, Sl in is in position 3, the 150 milliampere (150 mA) scale. For this full-scale value, series resistors R4, R5, and R6 are in shunt with the total series resistance represented by R1, R2, R3, and the internal resistance of the meter R_m.

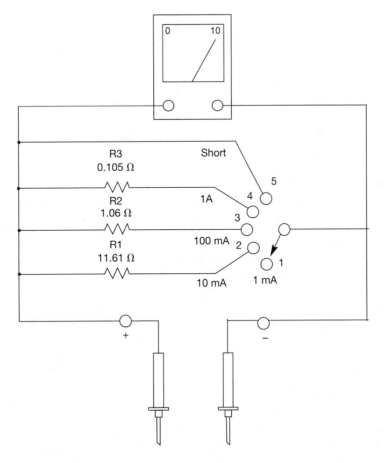

Figure 2-8 Multirange ammeter.

The ring type shunt eliminates the problem of high contact resistance which can cause errors in simple shunt meters. It also eliminates the need of a special type of make-before-break switch on the range switch of the ammeter. The make-before-break switch protects the basic movement from high currents as the selector switch is switched to different ranges. Examining the circuit will show that full measured current would pass through the basic movement if the selector switch opened the shunt circuit on switching.

The Basic Ohmmeter

Electrical resistance is measured in ohm's, and the instrument used to measure resistance is called an ohmmeter. In accordance with Ohm's Law, note that the resistance R in a dc circuit can be determined by

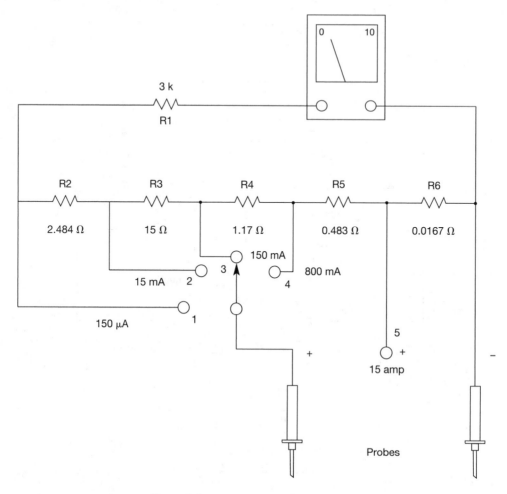

Figure 2-9 Ring shunt metering circuit.

measuring the current in a circuit as a result of applying a voltage. The voltage divided by the current is the resistance value or

$$R = V/I$$

The resistance will be one ohm if there is one ampere of current in a circuit when one volt is applied. By applying a known voltage to a circuit and measuring the current, the meter scale can be calibrated to read in ohm's of resistance. This is the basic principle of an ohmmeter.

A basic series ohmmeter circuit is shown in Figure 2-10. The ohmmeter circuit applies voltage from a self-contained battery to a series connection consisting of a known resistance (R_k) to the unknown resistance (R_x). The meter movement then determines the value of the unknown resistance by measuring the current in the circuit. Ohmmeters are very useful in servicing electronic equipment by making quick measurements of resistance values. Resistance values that can be measured with the ohmmeter vary from a fraction of an ohm to 100 megohms or more. The accuracy is seldom better than 3%; therefore, they generally are not suitable for measurements that require high accuracy.

Ohmmeters are self-contained, in that they contain their own power supply and depend entirely upon their internal calibrated circuits for accuracy, and, therefore, they should be used only on passive circuits or circuits that are not connected to powered circuits. Trying to measure a resistor in a circuit with power supplied can destroy the calibration as well as destroy the ohmmeter.

The Series Ohmmeter

Observe the diagram shown in Figure 2-10. Note that when the test leads are open (R_x equals infinity), there is no current through the meter. The meter needle at rest indicates an infinite resistance. When the test leads are shorted ($R_x = 0$), there is a full-scale deflection of the needle, indicating a zero resistance measurement. The purpose of R_a is to allow adjustment to zero on the scale to compensate for a changing battery potential due to aging, and for lead and fuse resistance. The purpose of R_k is to limit the current through the meter circuit to full scale when the test leads are shorted. It is convenient to use a value for R_k

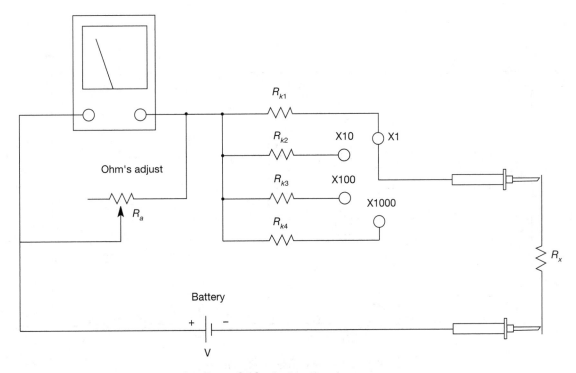

Figure 2-10 Basic series ohmmeter.

in the design of the series ohmmeter such that when an unknown resistance equal to R_k is measured, the meter will deflect to half scale. Therefore, when the leads are shorted together, the meter deflection will be full scale. Using this criterion, multiple ranges for the ohmmeter can be provided by switching to different values of R_k as shown. Each R_k sets the half-scale value for the respective resistance range, and provides the scale multiplier.

The Shunt-Type Ohmmeter

The shunt-type ohmmeter is often found in laboratories because it is particularly suited to the measurement of very-low-value resistors. The circuit for a shunt-type ohmmeter is shown in Figure 2-11. In this circuit, the current has two paths: one through the meter and one through R_x, if there is one. If the leads are open from A to B ($R_x = \infty$), then the meter movement reads full scale (adjusted by $R1$). This is marked as infinity. If the leads are shorted, the meter current drops to zero and the resistance is marked as 0 ohm's. When a value of known resistance R_x is placed across the terminals A and B, it causes the meter to deflect to some point below full scale. For example, if R_x equals R_m, then the meter movement would read half scale. Note that the scales for the two types of ohmmeters are exactly opposite. The series ohmmeter scale has 0 ohm's on the right and the shunt-type ohmmeter has the 0 ohm's on the left. This becomes a good way to identify the type of ohmmeter you might be using.

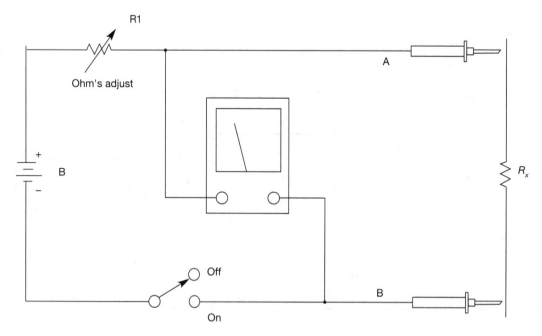

Figure 2-11　Shunt-type ohmmeter.

ac Meters

The easiest way to measure ac voltages with an analog meter is by combining a highly sensitive D'Arsonval meter movement with a rectifier. The purpose of the rectifier diodes is to change the alternating current to direct current. The D'Arsonval meter movement is a dc movement because of the permanent magnet that sets up the field for the moving coil. The method of using a rectifier to convert the dc movement for measuring ac is very attractive because the D'Arsonval meter movement has a much higher sensitivity than other types of meters that may be used to measure ac, such as an electrodynamometer or the moving-iron-vane-type meters.

Bridge-Type Rectifier

The circuit illustrated in Figure 2-12 makes use of germanium or silicon bridge rectifier circuit, which provides full-wave rectification. The D'Arsonval meter movement provides a deflection that is proportional to the "average" value of the dc current, so once the ac input is rectified by the bridge, the movement can be used to display the representative ac voltage. In practice, most alternating currents and voltages are expressed in effective values. As a result, the meter scale is calibrated in terms of the effective value of a sine wave, even though the meter is responding to the average value. Effective value is also referred to as RMS (root mean square) value.

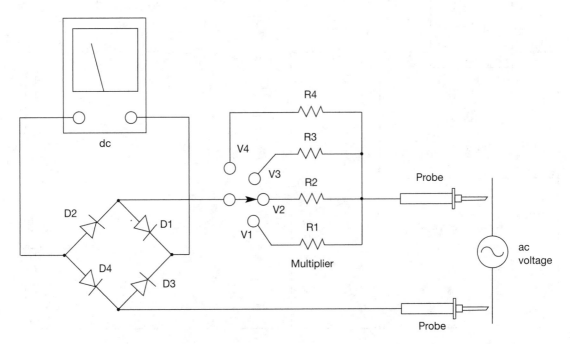

Figure 2-12 ac voltmeter with a bridge rectifier.

A conversion factor that will relate the average and the RMS values of a sine wave may be found by dividing the RMS value by the average value:

$$E_{rms}/E_{avg} = 1.11$$

Note: this conversion factor is valid only for sinusoidal ac measurements.

Multirange ac Voltmeters

Multirange ac voltmeters, like dc voltmeters, use multiplier resistors. A typical circuit used for commercial ac voltmeters is shown in Figure 2-13. This particular circuit uses rectifiers to convert the ac voltages

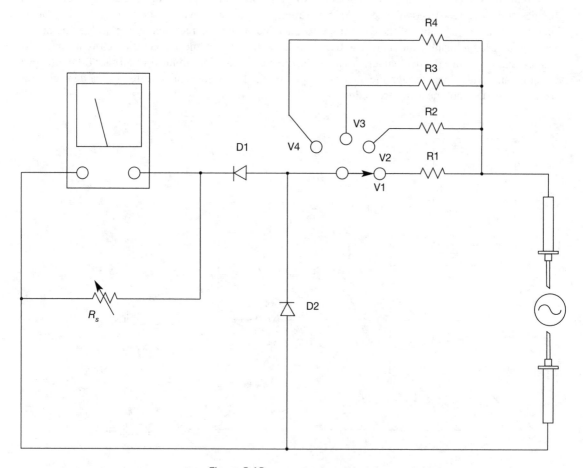

Figure 2-13   ac voltmeter circuit.

to dc in just a little different way. There is current through the meter movement: D1 conducts on the positive half-cycle, whereas D2 conducts on the negative cycle to bypass the meter. R_s is connected across meter movement in order to desensitize the meter, causing it to draw more current through the diode D1. This moves the operating point on the diode curve up into a more linear portion of the characteristic curve. Multiple ranges can be provided by having various multiplier resistors for the required ranges. ac meters normally have lower internal resistance than dc meters with the same meter movement.

The Digital Multimeter

The Basic DVM

A typical digital multimeter is shown in Figure 2-2. A digital multimeter generally has a single rotary selector switch that is used to select the functionality of the multimeter so it can act as a dc voltmeter, dc ammeter, ac voltmeter, ac ammeter, or ohmmeter. The rotary selector switch functions as a range switch to select the proper full-scale measurement with each function. Digital multimeters are also available in automatic or autorange models that can auto select the proper range for the input value to avoid destroying the meter.

Comparison of Analog to Digital Meters

Unlike the analog meter that has a needle deflecting along a scale, each of these digital multimeters has a digital LCD or LED display. The measurement value is displayed as a number with at least four digits. On the lowest full-scale range, the measurement is read on the display to an accuracy of three decimal places.

Unlike the analog meter, digital multitesters have an on–off power switch. They contain electronic circuits to produce their measurement value rather than an electromechanical meter movement. As a result, they need internal batteries to supply power for the electronics as well as an energy source to supply current for resistance measurements when the meter is used as an ohmmeter. The on–off switch connects the power source to the circuits. Like the analog meter, the digital multitesters have function selector switches and jacks to accept test leads. Digital multimeters also have range selector switches; often, range selection is selected automatically.

Basic DVM Characteristics

There are several major differences between analog and digital multimeters. Digital voltmeters or DVMs are basically just voltmeters, so when DVMs need to serve as ammeters or ohmmeters, special input conditioners have to be used so that the DVM is actually measuring voltage. A second difference between analog and digital meters is that the DVM has a high internal resistance on all functional ranges. The third and fourth differences are due to the digital display. Because the display is digital, there is no parallax reading error, the error due to interpolation between scale marks, as on an analog meter. Because the display is digital, there is no parallax reading error, the error due to interpolation between scale marks, as on an analog meter. Because the scale is digital, the conversion accuracy is within plus/minus one digit on any of the scales used. The accuracy of the display remains constant over all ranges and does not vary. As a result, overall accuracies of DVMs are typically 0.05% to 1.5% as compared to 3% to 4% for analog meters. A fifth characteristic relates to the special functions that are often available on new DVMs, such as circuit continuity testing and semiconductor junction testing.

How a DVM Works

A block diagram of a typical digital voltmeter is shown in Figure 2-14. Either dc or ac voltages can be measured. Let us look at measuring a dc voltage first. The test probes measure the dc voltage, bring it into the DVM through a signal conditioner, and couple it to a circuit called an analog-to-digital or A-to-D (A/D) converter. The A/D converter accepts a voltage and changes it to a digital code that represents the magnitude of the voltage. The digital code is used to generate the numerical digits that show the measured value in the digital display.

Analog-to-Digital (A/D) Conversion

Let us see how the digital codes are generated in the A/D converter. A block diagram is shown in Table 2-1. The table shows the BCD code that is output for each of the four digits when a dc voltage between 0 and 2 volts (1.999 volts) is applied at the input. The resolution of the DVM is one millivolt, which means

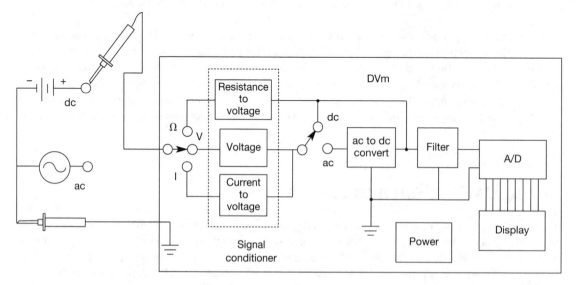

Figure 2-14   Basic digital voltmeter.

Table 2-1   A/D converter digit resolution

Voltage	Digit Code 1		2		3		4
1.999	0001	1	1001	0	1001	0	1001
1.900	0001	1	1001	0	0000	0	0000
1.800	0001	1	1000	0	0000	0	0000
1.700	0001	1	0111	0	0000	0	0000
1.667	0001	1	0110	0	0110	0	0111
1.600	0001	1	0110	0	0000	0	0000
1.500	0001	1	0101	0	0000	0	0000
1.400	0001	1	0100	0	0000	0	0000
1.300	0001	1	0011	0	0000	0	0000
1.234	0001	1	0010	0	0011	0	0100
1.200	0001	1	0010	0	0000	0	0000
1.100	0001	1	0001	0	0000	0	0000
1.000	0001	1	0000	0	0000	0	0000
0.900	0000	1	1001	0	0000	0	0000
0.800	0000	1	1000	0	0000	0	0000
0.700	0000	1	0111	0	0000	0	0000
0.600	0000	1	0110	0	0000	0	0000
0.578	0000	1	0101	0	0111	0	1000
0.500	0000	1	0101	0	0000	0	0000
0.400	0000	1	0100	0	0000	0	0000
0.300	0000	1	0011	0	0000	0	0000
0.200	0000	1	0010	0	0000	0	0000
0.100	0000	1	0001	0	0000	0	0000
0.001	0000	1	0000	0	0000	0	0001
0.000	0000	1	0000	0	0000	0	0000
decimal pt	for scale		DP1-2 V		DP2-20 V		DP3-200 V

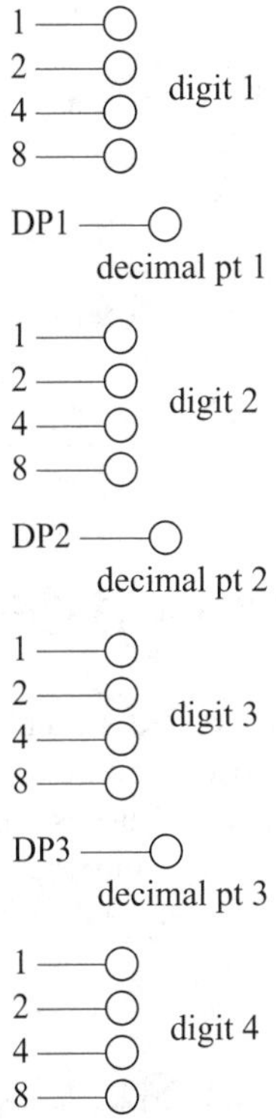

Right-side schematic labels:
- 1, 2, 4, 8 — digit 1
- DP1 — decimal pt 1
- 1, 2, 4, 8 — digit 2
- DP2 — decimal pt 2
- 1, 2, 4, 8 — digit 3
- DP3 — decimal pt 3
- 1, 2, 4, 8 — digit 4

that a new four-digit BCD code is generated for each change of one millivolt in the input voltage. Table 2-1 does not show each 1 millivolt four-digit code but lists the codes that would be generated for the four digits for each one-tenth of a volt. In addition, it shows some special example values (0.001, 0.578, 1.234, 1.667, and 1.999) to help understand what the A/D converter output is like. There is a separate four-digit BCD code for each millivolt, but, for simplicity they are not all included in Table 2-1.

If the input voltage to the A/D converter is 1.234 volts, the four-digit BCD code that would be generated at the output would be 0001 0010 0011 0100 as shown. Each of the four-bit BCD codes representing a digit is coupled to a decoder, and the proper numeral displayed for the respective digit. Individual logic gates in the A/D converter control the decimal point. If a logic 1 level is output, the decimal point will be on. As the scales change, the decimal point energized changes. The display techniques and the outputs of the A/D converter shown in Tables 2-1 and 2-2 are not the only way a DVM can produce the conversion and display it. There are also scanning techniques and multiplexing techniques, so that the digit codes are transferred in sequence along fewer bus lines, but the more direct way was chosen to make it easier to illustrate and explain the basic concepts.

Table 2-2   Seven-segment display

Digit	A	B	C	D	E	F	G	BCD
0	•	•	•	•	•	•		0000
1		•	•					0001
2	•	•		•	•		•	0010
3	•	•	•	•			•	0011
4		•	•			•	•	0100
5	•		•	•		•	•	0101
6			•	•	•	•	•	0110
7	•	•	•					0001
8	•	•	•	•	•	•	•	1000
9	•	•	•			•	•	1001

The A/D Conversion Process

There are a number of different types of techniques used to perform the analog-to-digital conversion process. Some of the more familiar A/D techniques used in multimeters conversions are the Dual-slope method, successive approximation sampling, staircase integrating, continuous balance, and voltage-to-frequency converters. Our discussions will be directed towards the staircase and the dual-slope converter types, since these methods demonstrate the basic concepts, which are relatively easy to grasp.

The Staircase Converter

Figure 2-15 shows the block diagram of a staircase converter. It consists of a comparator, a clock gate G, a clock generator, a binary counter, and a digital-to-analog converter. The output is coupled to a digital display. The digital-to-analog converter does the opposite of the analog-to-digital converter. It takes the digital code output from the stages of the binary counter and converts it to an analog voltage. Each time the binary counter increases its count by one, the output voltage V_0 increases by a millivolt. V_0 is one input to the comparator; the input voltage V_{in} is the other input to the comparator.

When the input voltage V_{in}, is first applied to be measured, V_0 is zero, and the output of the comparator is at a logic 1 level because V_0 is not equal to V_{in}. Since the comparator 1 output is an input to the AND clock gate G, the clock signal appearing on the other G input will appear on the gate output and feed into the binary counter. The binary counter counts the clock pulses, and through the D/A converter begins increasing V_0 by 1 millivolt per count. Thus, the name staircase converter.

When V_0 is equal to V_{in}, the comparator output drops to a logic 0 level, turns off the clock pulses through G, which stops the count, and holds equal to V_0 equal to V_{in}. The digital code at the counter is converted to the necessary code for display of the numerals that represents the value of the voltage. Note that it takes time to reach the point where $V_0 = V_{in}$, therefore, measurements can be done only at a maximum rate.

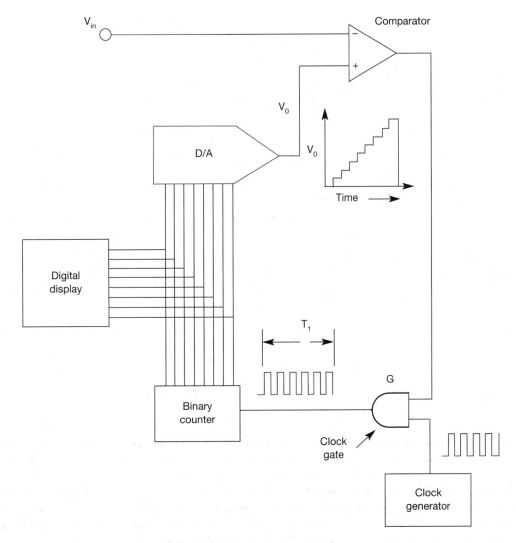

Figure 2-15 Staircase analog-to-digital converter.

The Dual-Slope Converter

A block diagram of a dual-slope A/D converter is shown in Figure 2-16. It consists of an operational amplifier A, connected as an integrator, a comparator, a logic gate (G) for gating the clock signal, a counter for counting clock pulses, a reference voltage, and control logic circuitry. The output of this circuitry the feeds a digital display. As with the staircase converter, the magnitude of the input voltage measured is determined by the number of clock pulses counted. The resultant counter code is then converted to the ap-

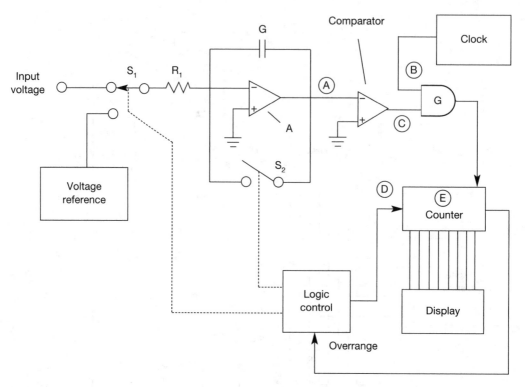

Figure 2-16 Dual-slope analog-to-digital converter.

propriate code to display the proper digits on the digital display. The number of pulses counted is determined by a start count control signal (D), which allows the counter to start counting, and a comparator output signal (C), which stops counting by gating off the clock pulses.

A dual-slope A/D operates as follows. In Figure 2-16, switch S_2 shorts out C_1 so that there initially is no charge on C_1. When a measurement is made, S_2 is open, and S_1 is connected to the input voltage for a period of time t_1 shown. In time t_1, capacitor C_1 charges at a constant rate and to a magnitude of voltage determined by the input voltage. At the end of t_1, Sl is switched to a reference voltage that is opposite in polarity to the input voltage. This reference voltage discharges capacitor C_1 at a constant rate. The time that it takes to discharge C_1 back to the initial zero level land just a few millivolts more, because the comparator output must switch states, is proportional to the input voltage magnitude.

When the discharge time starts (at the end of t_1), the start control (D) starts the counter. At the end of the discharge time, the comparator output (C) goes to a low logic level, turns off gate G, and stops the counter.

Digital DVM Displays

A DVM can display numerical readings using three or four different methods. One common method of displaying a multimeter's output is the use of a seven-segment array of elements to form the digits. The array elements may be vacuum fluorescent, electroluminescent, or plasma display elements. Historically, other displays such as ten-character neon displays (NIXIE) and the seven-segment neon displays have been used for meter displays, but the disadvantages of these displays are higher cost combined with the need for a high-voltage power supply. Neon, as well as fluorescent displays, also tend to generate radio frequency interference or broadband noise.

The light-emitting diode (LED) has been one of the most popular displays for DVMs use because of its high brightness, excellent contrast, and low cost. Unfortunately, their power consumption is high compared to the more modern LCD displays. The liquid crystal display (LCD) has become very popular in recent years for portable use, since its power drain is very low and cost of these displays has dropped considerably. Newer LCD displays do not wash out in bright sunlight as did the older models. Unfortunately, many LCD displays types will freeze at fairly moderate temperatures and become completely useless, a disadvantage that is sometimes overlooked.

In a typical LED display, the power source is connected to one lead of each LED segment. The second lead of each LED segment is excited by current passing through it, which is driven by a seven-segment decoder chip, grounding the LED segment to be lit. The decoder establishes the proper segments that must be grounded. The decoder has an input digital code that represents the numeral required.

A very common code used in digital systems to represent numerals is a four-bit binary-coded decimal (BCD) code. The BCD code for the numerals from 0 to 9 is shown in Table 2-2. This code, if fed into the seven-segment decoder, causes the decoder to ground the proper segments of the digital display to display the proper numeral.

Half-Digit Display

A unique term used to describe the capability of a DVMs display is called the "half-digit," which is used to indicate a reading beyond the meter's full-scale readings. This is called overranging. If a DVM is classified as a 3½ digit DVM, it means that the full-scale reading is displayed in three digits and that the digit to the left of the three digits for full-scale is restricted in range. For example, 0.999 would be the full-scale reading when the DVM is measuring 1 volt on the 1 volt range. The ½% digit specification means that the DVM can display a measurement up to 1.999. The digit to the left of the three full-scale digits is restricted to a one. A 4½ digit DVM would have the capability to read a value to 19999. A 3¾ digit DVM would have the capability to read a value as high as 3999. Therefore, the ½ or ¾ digit specifies the overranging that the DVM can read when set on a particular range.

ac Measurements

The A/D converter is designed to accept a dc voltage and convert it to a digital code so the corresponding digits can be displayed. Refer back to the diagram shown in Figure 2-14, note that if the input to be measured is an ac voltage, a circuit in the DVM converts the ac voltage to a dc voltage, filters it, and couples it to the A/D converter. It isn't absolutely necessary, but the dc voltage also passes through the filter in case there are some noise spikes on the input dc voltage.

Signal Conditioners

Analog-to-digital converters in digital multimeters need dc voltages at their input to be allow the multimeter to act as a voltmeter, ammeter, or ohmmeter. Each particular function requires that a different type of signal conditioner be used to convert the measured quantity into a dc voltage. The three different types of conditioners are shown in the block diagram in Figure 2-14.

Voltage Input Conditioner

Since the range of voltage that the A/D converter can handle is generally limited by digit zero to two volts. Measured voltages greater than two volts will need to be attenuated, and voltages less than one-third to one-fifth of a volt will need to be amplified. The voltage conditioner shown in Figure 2-17 is made up of a voltage divider that provides the attenuation and an operational amplifier that provides the amplification. There are five voltage ranges: 0.2, 2, 20, 200 and 2000 V. The gain of the operational amplifier is set by the equation

$$A = -R_f/R_i$$

where R_i is the input resistance. The minus sign means that the output signal is 180 degrees out of phase from the input signal. R_f and R_i (made up due to R2, R3, and R7) are chosen so that A equals one on the 2 V range. With unity gain for the amplifier ($A = 1$), the input voltage is reproduced directly at the ammeter out-

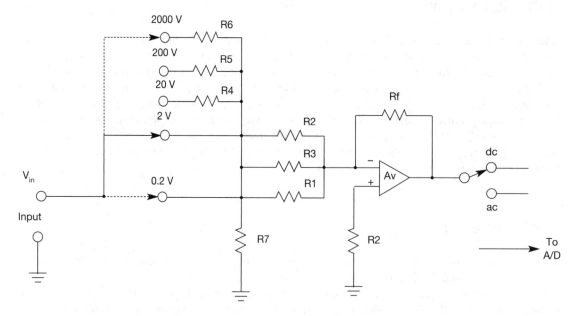

Figure 2-17   Voltage input conditioner.

put and coupled to the A/D converter. On the 20V, 200V, and 2000V ranges, R3 and R7, are chosen such that the gain of the amplifier is unity again, just as for the 2V range (RF does not change). The ratio of R4, R5, and R6 to R7 is set so that with 20 volts, 200 volts, and 2000 volts, respectively, on the input, the input voltage to the amplifier will be 2 volts. For the 0.2 V range, R1 is chosen so the gain of the amplifier is ten.

The voltage conditioner will work equally well for dc and ac voltage measurements over the frequency range specified for the DVM as long as the stray capacitance in the circuitry is kept to a minimum. Of course, the ac voltage at the output of the signal conditioner must be converted back to dc before it is coupled to the A/D converter.

ac-to-dc Conversion

Many VOMs, multitesters, and DVMs that have rectifier-type circuits have scales that are calibrated in RMS values for ac measurements, but actually are measuring the average value of the input voltage and are depending on the voltage to be a sine wave. These instruments are in error if the input voltage has some other shape than a sine wave.

There are DVMs that measure the true RMS value of input voltages regardless of the shape of the waveform. They measure the dc and ac components of the input waveforms; therefore, the measured value is the heating power of an ac voltage that is equivalent to the heating power of a dc voltage equal to the RMS value—the definition of an RMS voltage. Check the specifications of the DVM that you are using to determine if it measures true RMS voltages. If it does not, be wary of the measurement if the input voltage is not a sine wave or if it has a dc component.

Current Input Conditioner

The signal-conditioning circuit for measuring current is shown in Figure 2-18. The current conditioner changes the current to be measured into a voltage by passing the unknown current through a precision resistance (R1, R2, R3, or R4) and measuring the voltage developed across the resistor. The range switch determines the resistor used for each range. When the proper current range is selected, the proper resistance is selected so that the voltage out of the current conditioner will be within the range required by the A/D converter. The resistors used are a special type of high-power precision resistor used as current shunts.

The position of the ac/dc switch determines the route the voltage output from the current conditioner takes to reach the input of the A/D converter. A high current range, such as 10 amps, usually is measured using a special input jack to which a special resistor is connected. An operational amplifier with a fixed gain prevents loading of the current-sensing resistors by the A/D converter.

Resistance Input Conditioner

The basic circuit for the signal conditioner that the DVM uses for resistance measurements is shown in Figure 2-19. The voltage V and the resistance form a constant current source for the unknown resistor. A constant current through a given resistor will produce a set voltage drop. For example, if the constant current is 1 milliampere (0.001 A) through a 1000 ohm resistor, the voltage drop across the 1000 ohm resistor is 1 volt. That is exactly what the DVM does to measure resistance; it measures the voltage across an

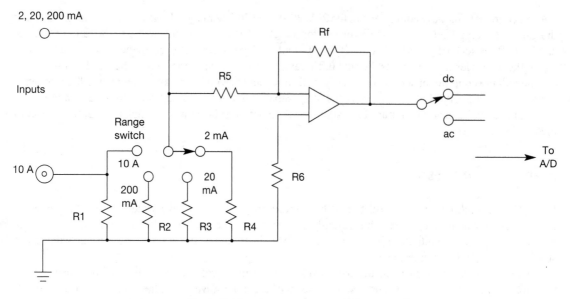

Figure 2-18   Current input conditioner

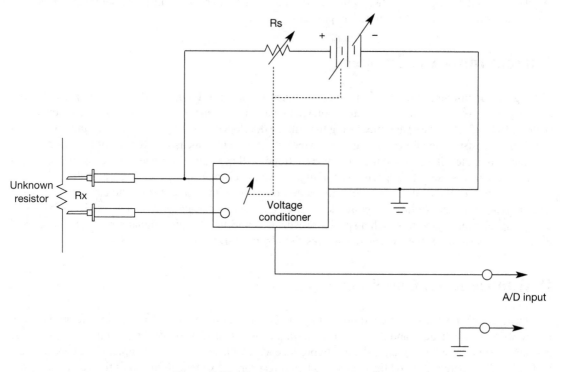

Figure 2-19   Resistance input conditioner.

unknown resistance when there is a known constant current through it. The voltage source, the series resistance, and the voltage range are changed as the full-scale range of resistance to be measured by the D/A is changed. No zero ohm's adjustment is necessary.

Additional DVM Characteristics

Accuracy and Resolution

Simple accuracy specifications for DVMs are given as "plus/minus percentage of full scale, plus/minus one digit." The "plus/minus one digit" portion of the specification is caused by an error in the digital counting circuit.

The "plus/minus percentage of full scale" includes ranging and DVM conversion errors. The resolution of an instrument is directly limited by the number of digits in the display. A 3½ digit DVM has a resolution of one part in 2000 or 0.05%, which means that it can resolve a measurement of 1999 millivolts down to 1 millivolt. A 4½ digit instrument has a resolution of one part in 20,000, or 0.005%.

A 2½ or a 3 digit DVM is considered approximately equivalent to a good VOM multitester as far as accuracy and resolution are concerned. Accuracy generally lies between 0.5% and 1.5% and resolution to 0.5%. The 3½ and 4½ digit meters generally have accuracy of one order of magnitude higher; that is, between 0.5% and 0.05% with 0.05% resolution. A resolution and accuracy of this amount will generally suffice for most service work today. A 4½ digit or 5½ digit DVM generally indicates an accuracy of 0.05% and better with resolution of 0.005%. These are indeed considered laboratory instruments. They usually are specified with a "plus/minus percent of reading, plus/minus percent of full-scale, plus/minus one digit" specification. They also may have specifications that qualify the accuracy at temperatures other that 25°C.

A DVM has essentially the same accuracy on ac that it does on dc voltage measurements, while the accuracy of an analog meter is most assuredly less accurate on the ac voltage measurements. Accuracy will also depend upon the frequency response or bandwidth of the DVM, and on the ac wave shape when the meter does not measure true RMS voltages. Input impedance DVMs have an input impedance of at least one megohm and more commonly 10 megohms. This holds true on dc measurements and on ac measurements over the frequency range specified for the DVM.

Range

Full-scale range is specified in one of two ways: (1) a full-scale range with usable overrange capabilities specified as a percentage, typically 100%; or (2) full scale specified as the maximum possible reading encompassing all usable ranges, often 1.999. For example, a DVM may be specified as having 1 volt full scale with 100% overrange, thus indicating useful operations to 2 volts; or the same DVM may be simply specified as having a 2 volt full scale. Some ranges may not be used to full scale. For example, the higher voltage ranges may be limited because of the voltage breakdown of internal components. A DVM with a 1999 volt full-scale range may not be used to read over 1000 volts because divider resistors cannot withstand over 1000 volts and break down. The capability may be even lower on ac because of peak voltages.

Overranging was instituted to take full advantage of the upper limits of a range. It also provides better accuracy at the top of a range. In other words, overranging allows a DVM to measure voltage values above the normal range switch transfer points, without the necessity of having to change ranges. It allows the meter to keep the same resolution for values near the transfer points. The extent to which the overrange is possible is expressed in terms of the percentage of the full-scale range. Overranging from 5–300% is available, depending on the make and model of the DVM.

Meter Response Time

Response time is the number of seconds required for the instrument to settle to its rated accuracy. The response time consists of two factors: (1) the basic cycle rate of the AD converter, and (2) the time required to charge capacitances in the input circuit. Instead of response time, some manufacturers simply give a number of conversions per second.

Protection

Meter protection circuits prevent accidental damage to the DVM. The protection circuit allows the instrument to absorb a reasonable amount of abuse without affecting its performance. The specification of input protection indicates the amount of voltage overload that may be applied to any function or range without damage. A separate dc limit may be indicated to cover input-coupling capacitor breakdown. Overloads from sources outside the specified frequency range of the instrument may not have as great a protection range. The current-measuring circuitry is usually protected by a fast-blowing fuse in series with the input lead.

Special DVM Features

Autoranging

Autoranging automatically adjusts the meter's measuring circuits to the correct voltage, current, or resistance ranges. One technique is for the DVM to start on the lowest range and automatically move to the next-higher range when autoranging takes place. A special feature of some autorange DVMs is a manual control function that overrides the autorange. This is useful for a series of measurements that are made within a specific range.

On some meters, if the upper limit of the range being held is exceeded, the meter produces a series of beeps if the buzzer switch is on. A portion of the display will flash, indicating an overflow error. Switching to the "auto" position releases the manual control.

Autopolarity

The automatic polarity feature further reduces measurement error and possible instrument damage because of overload due to a voltage of reverse polarity. A + or – activated on the digital display indicates the polarity and eliminates the need for a polarity switch setting or reversing leads.

Reading Storage

Many digital multitesters have a "hold" feature that is operated remotely by means of a special "hold" button on the meter or included on one of the test leads. It is particularly useful when making measurements in a difficult area because readings can be captured and held, then read when convenient.

In some DVMs, the hold signal sets an internal latch that captures the data. When the A/D gets to the point in its cycle where data is to be displayed, the state of the internal latch is sampled. New data is not transferred while the latch is in the hold state. In other DVMs, the hold feature stops the instrument clock. The last value displayed on the LCD remains displayed until the ground is removed from the input terminals.

Conductance Measurements

Some meters display the reciprocal of resistance; that is, a conductance measurement. The conditions and connections for conduction measurements are the same as for resistance. The output, however, is displayed in siemens, the reciprocal of resistance (previously called mhos).

Semiconductor Testing

A diode check or semiconductor check function appears on some DVMs. Most DVMs have an ohmmeter voltage of greater than 1.5 volts. Thus, continuity for most diodes and transistors in the forward-bias direction may be checked. Some DVMs have a special feature when in the diode check position that displays a voltage value that is essentially the forward-bias voltage of the PN junction.

Audible Continuity

Often, DVMs have an audible continuity function. With the DVM set to measure resistance and the continuity switch activated, when the probes are touched together the DVM emits an audible sound. This electronic sound is used to look for short circuits or for tracing an open circuit. Any time the continuity circuit resistance is less than a minimum amount, usually 200 or 300 ohm's, the DVM emits an audible sound when the circuit is complete.

Speech Output

A special feature recently included in some DVMs is the new speech output feature. A speech button is provided to say the actual DVM reading in clear English while displaying the measured value.

The Oscilloscope

Engineers and technicians alike will tell you that the single most useful piece of test and design equipment is the oscilloscope or "scope." The oscilloscope is a very effective measuring instrument that provides significantly more information about a signal than an ordinary analog or digital voltmeter. The oscilloscope essentially "paints" a picture of the waveform's characteristics. Oscilloscopes can measure and display voltage relative to time, as well as frequency and signal waveforms, allowing engineers and technicians to effectively interpret specific details that can help isolate faults as well as optimize circuit performance. Adjustments to circuits often require setting up a signal using an oscilloscope, so that certain characteristics of the signal can be "fine-tuned" for best performance. Figure 3-1 shows an analog oscilloscope.

Oscilloscopes are broken down into two major classifications: analog and digital. The analog scope was the first and original instrument, designed to measure real-time signals and display them on a CRT. More recently, the digital scope has made its way to the forefront. The digital scopes have many more features such as data logging, delay, and storage features. Newer digital oscilloscope are much more accurate, smaller, and lighter and will often run on batteries and utilize LCD displays. Some scopes take the form of a card plugged into a PC—the PC becomes the scope and the monitor is used for the display. Even if they don't use a PC for display, many scopes can attach to a PC and download their data for storage and analysis using advanced mathematical techniques. Many high-end scopes now incorporate nontraditional functions such as fast Fourier transforms (FFTs). This allows limited spectrum analysis or other advanced mathematical techniques to be applied to the displayed waveform.

There are a variety of oscilloscope models to choose from, depending on the type of measurement task at hand. A general-purpose scope is usually adequate for most basic measurements; however, specialized waveform analysis requires the use of advanced digital models that provide added capabilities such as delayed time base measurements, high-frequency operation, and simultaneous multiple channel displays.

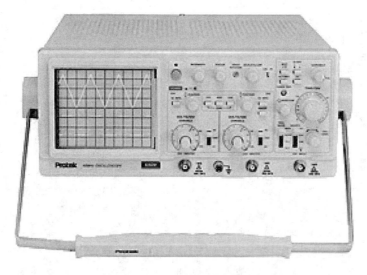

Figure 3-1   Analog oscilloscope.

The Analog Oscilloscope

Figure 3-2 is a simplified diagram of a dual-channel triggered-sweep oscilloscope. At the heart of nearly all scopes is a cathode-ray tube (CRT) display. The CRT allows the visual display of an electronic signal by taking two electric signals and using them to move (deflect) a beam of electrons that strikes the screen. Unlike a television CRT, an oscilloscope uses electrostatic deflection rather than magnetic defection. Wherever the beam strikes the phosphorescent screen of the CRT, it causes a small spot to glow. The exact location of the spot is a result of the voltage applied to the vertical and horizontal inputs.

All of the other circuits in the scope are used to take the real-world signal and convert it to a animated display on the CRT. To trace how a signal travels through the oscilloscope circuitry, start by assuming that the trigger select switch is in the "internal" position

The input signal is connected to the input coupling switch. The switch allows selection of either the ac part of an ac/dc signal or the total signal. If you wanted to measure, for example, the RF swing at the col-

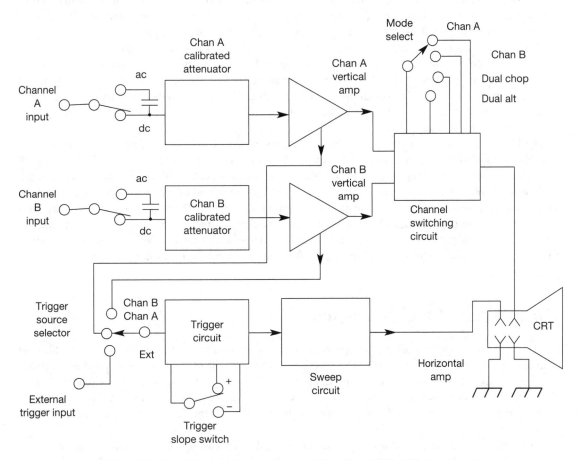

Figure 3-2 Dual-trace analog oscilloscope. Courtesy 2003 ARRL Handbook.

lector of an output stage (referenced to the dc level), you would use the dc-coupling mode. In the ac position, dc is blocked from reaching the vertical amplifier chain so that you can measure a small ac signal superimposed on a much larger dc level. For example, you might want to measure a 25 mV, 120 Hz ripple on a 13 V dc power supply. (Note that you should not use ac coupling at frequencies below 30 Hz, because the value of the blocking capacitor represents a considerable series impedance to very low frequency signals.) After the coupling switch, the signal is connected to a calibrated attenuator. This is used to reduce the signal to a level that can be tolerated by the scope's vertical amplifier. The vertical amplifier boosts the signal to a level that can drive the CRT and also adds a bias component to locate the waveform on the screen.

A small sample of the signal from the vertical amplifier is sent to the trigger circuitry. The trigger circuit feeds a start-on pulse to the sweep generator when the input signal reaches a certain level. The sweep generator produces a precisely timed signal that looks like a triangle. This triangular signal causes the scope trace to sweep from left to right with the zero-voltage point representing the left side of the screen and the maximum voltage representing the right side of the screen.

The sweep circuit feeds the horizontal amplifier that, in turn, drives the CRT. It is also possible to trigger the sweep system from an external source (such as the system clock in a digital system). This is done by using an external input jack with the trigger select switch in the "external" position.

The trigger system controls the horizontal sweep. It looks at the trigger source (internal or external) to find out if it is positive or negative and to see if the signal has passed a particular point. It is important to note that once a trigger circuit is "fired" it cannot fire again until the sweep has moved all the way across the screen from left to right. In normal operation, the "trigger level" control is manually adjusted until a stable display is seen. Some scopes have an "automatic" position that chooses a level to lock the display in place without manual adjustment.

The horizontal travel of the trace is calibrated in units of time. If the time of one cycle is known, it is possible to calculate the frequency of the waveform. If, for example, the sweep speed selector is set at 10 μs/division, count the number of divisions (vertical bars) between peaks of the waveform (or any similar well-defined points that occur once per cycle) you can find the period of one cycle, in this case, 80 μs. This means that the frequency of the waveform is 12,500 Hz (1/80 μs). The accuracy of the measured frequency depends on the accuracy of the scope's sweep oscillator (usually approximately 5%) and the linearity of the ramp generator. This accuracy cannot compete with even the least-expensive frequency counter, but the scope can still be used to determine whether a circuit is functioning correctly.

Dual-Trace Oscilloscopes

Dual-trace oscilloscopes can display two waveforms at once. This type of scope has two vertical input channels that can be displayed either alone, together, or one after the other. The only differences between this scope and the analog scope are the additional vertical amplifier and the "channel switching circuit." This block determines whether we display channel A, channel B, or both simultaneously. The dual display is a true dual display (there is only one electron gun in the CRT) but the dual traces are synthesized in the scope.

There are two methods of synthesizing a dual-trace display from a single-beam scope. These two methods are referred to as chopped mode and alternate mode. In the chopped mode, a small portion of the channel A waveform is written to the CRT, then a corresponding portion of the channel B waveform is written to the CRT. This procedure is continued until both waveforms are completely written on the CRT. The chopped mode is most useful where an actual measure of the phase difference between the two waveforms is required. The chopped mode is usually most useful at slow sweep speeds (times greater than a few microseconds per division).

In the alternate mode, the complete channel A waveform is written to the CRT followed immediately by the complete channel B waveform. This happens so quickly that it appears that both waveforms are displayed at the same time. This mode of operation is not useful at very slow sweep speeds, but is good at most other sweep speeds.

Most dual-trace oscilloscopes also have a feature called "X-Y" operation. This feature allows one channel to drive the horizontal amplifier of the scope (called the X channel) while the other channel (called Y in this mode of operation) drives the vertical amplifier. Some oscilloscopes also have an external Y input. The X-Y operation allows the scope to display Lissajous patterns for frequency and phase comparison and to use specialized test adapters such as curve tracers or spectrum analyzer front ends. Because of frequency amplifiers, the X channel is usually limited to a 5 or 10 MHz bandwidth.

Oscilloscope Specifications

Oscilloscopes have several important operating specifications that you should be familiar with. An important one is vertical deflection, specified as the minimum and maximum volts-per-division settings the scope has and the number of steps that the range is broken down into. For example, a typical scope can range from 5 millivolts per division to 5 volts per division, broken down into 10 steps. A variable control is often available to provide adjustment between voltage steps. Another specification, the time-base (or sweep) range, is the range of time-base settings that the scope is capable of along with the number of steps that are available within the range. A range of 0.1 μs/division to 0.2 s/division in 20 steps is not unusual. The horizontal sensitivity always has greater flexibility than the vertical sensitivity. A variable control usually accompanies the time base for measurements between steps.

Bandwidth is a very important specification. Bandwidth is essentially the range of frequencies that the scope is capable of displaying accurately, usually rated from dc (0 Hz) to some maximum frequency. For an inexpensive scope, the bandwidth may cover dc to 20 mHz, whereas a more expensive model may reach up to 150 mHz or more. Good bandwidth is more expensive than any other feature. For example, 100 mHz oscilloscopes can easily exceed $1200. A 20 mHz unit, on the other hand, can be bought for under $400.

Simply stated, the maximum input is the maximum voltage that can be applied to the oscilloscope input. A maximum input of 400 volts (dc or ac peak) is common for many simple scopes. More sophisticated units can sustain better than 1000 volts. The range can be artificially expanded by the use of attenuating probes.

Another important specification is accuracy, typically the accuracy of the screen display for voltage and time measurement. Although oscilloscopes are handy for making rough readings, they are not as accurate as voltage or frequency meters. As an example, a digital VOM can reach an accuracy of ±0.25% at full scale. That means that if you are measuring 1 volt (full scale), the meter may read as little as 0.9975 or as much as 1.0025 volts. An oscilloscope, on the other hand, can typically provide ±3%. That same 1 volt measurement on an oscilloscope could range from 0.97 to 1.03 volts (not counting parallax errors in reading the trace on the screen).

A scope's input impedance is the effective load that the scope will place on a circuit (rated as a value of resistance and capacitance). To guarantee proper operation over the bandwidth of the scope, it is a good idea to select a probe with characteristics similar to those of the particular scope. Most oscilloscopes have an input impedance of about 1 megohm with 10 to 50 pF of capacitance.

It is also important to consider the operating modes of a scope. They determine how selected signals will be displayed. For example, on a multichannel scope, channels could be displayed independently, together (in the "dual" mode), or even summed algebraically in "add" mode. There are more options than those listed here and the number depends upon the particular scope.

The number of trigger sources a scope can work with is another consideration. As mentioned earlier, most scopes offer triggering from one of their signal inputs, the AC line, or an external signal.

Tied to that specification are the number of trigger modes, which determine the way a trigger source is applied to the sweep generator. An "auto" mode allows the trigger source to run the time base continuously. "Normal" mode is used for unusual waveforms. There may be other modes as well, depending on the sophistication of the scope.

Oscilloscope Controls

The oscilloscope can be very intimidating to a newcomer to electronics. Most oscilloscopes have at least twenty or more front panel controls to adjust. In order to familiarize you with the oscilloscope, a description of the major controls follow. It is important to understand what these controls are used for.

Power Switch

All oscilloscopes have a "power on/off" switch of some kind. Some use a toggle switch, whereas others might use a pushbutton switch. Regardless of the type used, you must turn the instrument "on" before anything will be displayed on the screen. Most analog instruments have a pilot light to indicate that the unit is "on." Digital models may have a power indicator, but many battery-powered portable units do not, relying instead on the screen. It is always best to turn the oscilloscope off rather than simply pulling the plug, since this provides a clean shutdown of sensitive internal circuitry.

The Graticule

The oscilloscope screen is composed of an 8 × 10 matrix of squares called the graticule, which resembles graph paper. There are always eight "major" divisions vertically and 10 "major" divisions horizontally. The graticule is further broken down into 5 "minor" divisions per major division. Additionally, there are usually "rise-time" measurement marks and associated tags. Consult the instrument's operating manual for specific details on taking rise-time measurements.

It is important to treat the graticule with the same respect you would give a precise measuring instrument such as a steel rule. It is only as accurate as the care taken to properly read it. If you fail to exercise care in reading the graticule, the accuracy of your measurements will suffer accordingly.

Back-Light or Graticule Illumination

The graticule is a symmetrically lined or gridded transparent front glass or plastic plate that allows you to see and compare the oscilloscope's line trace. Depending on whether the scope is an analog or digital model, some type of graticule or display illumination may be provided. Analog scopes frequently use incandescent lamps to provide this illumination and often provide some level of control from a potentiometer or rheostat. This allows the operator to vary the brightness of the graticule depending on ambient light conditions. Normally, advancing the control clockwise increases the level of illumination, whereas turning it counterclockwise decreases it.

Many digital oscilloscopes incorporate an LCD back-light to assist the technician in making measurements more readable, especially under dimly lit situations. Usually, the back-light is not adjustable—it may only be turned "off" or "on."

Beam Intensity and Display Contrast

In the case of analog oscilloscopes, an "intensity" control is provided. It allows the operator to set the amount of trace illumination. Make sure that you set this control to the minimum necessary brightness in order to prevent the possibility of the electron beam burning the phosphor of the CRT (cathode ray tube). As the scope warms up, it is not uncommon to note an increase in brightness of the trace perhaps after 15–30 minutes of operation.

Digital scopes do not have an "intensity" control, but instead use a "contrast" control, much like that found on calculators and LCD computer screens. The "contrast" adjustment is generally accessed through one of the instrument's setup menus. Consult the operating manual for how to adjust the display contrast.

Unlike analog oscilloscopes that depend upon an electron beam to display the signal's waveform on the screen, digital scopes that use an LCD display are immune to any possibility of screen degradation from changes in contrast settings. You may set the LCD to any contrast level without fear of burning the screen. This is an important feature of digital oscilloscopes that use an LCD screen. There are trade-offs, however, that will be discussed in a later section.

Focus

The "focus" control only applies to analog oscilloscopes that use a CRT. There is no equivalent adjustment found on digital models using a liquid crystal display. Because the beam emitted from the electron gun within the CRT can be distorted, a front-panel "focus" adjustment is provided. To adjust the focus of the instrument, simply rotate the control clockwise and counterclockwise until the sharpest trace is displayed on the screen. A good starting point is with the control in the "midway" position.

Trace Rotation

Most analog oscilloscopes have a front panel adjustment called "trace rotation," usually accessed using a small screwdriver or alignment tool, to ensure that the trace is parallel to the horizontal (x-axis) graticule markings. Normally, this adjustment is performed rarely once the instrument is positioned on the test bench. In the case of scopes used in field applications, the adjustment may need to be performed more frequently. Digital oscilloscopes using an LCD screen do not normally have any user-adjustable trace rotation parameters since this type of display operates on a completely different principle. You should confirm that the displayed trace is in fact parallel to the graticule markings to be sure the instrument is properly calibrated. If the trace is tilted, the unit may require service.

The Vertical Section—Vertical Position

This control positions the oscilloscope trace to a point specified by the operator in the "Y" or vertical direction. Place the control in the "midway" position on analog models as a starting point.

Digital oscilloscopes can use either a variable control or a pushbutton adjustment, depending on the model. Portable scopes frequently use pushbuttons to save front panel room. In the case of two-channel and four-channel oscilloscopes, there is usually a separate control for each channel. Digital models may share a common control or pushbutton for multiple channels.

Scope Input Circuits

ac–dc Coupling Switch

All oscilloscope models incorporate some type of coupling switch—you can usually choose dc or ac coupling. Usually, the coupling switch is set to the "dc" position. All signals within the instrument's bandpass will be applied directly to the internal vertical amplifiers. The dc input allows the instrument to be used as a simple dc voltmeter or to display a dc level that an ac signal may be riding on.

Setting this switch to the "ac" position will block any dc component by inserting a capacitor, thus allowing the operator to only "see" time-varying signals such as an alternating current signal. This setting is useful, for example, when observing the ripple on a dc power supply. This mode is also very useful for studying small ac signals that are superimposed on larger dc levels. To closely examine the small ac variations, you would need to amplify them. Using dc coupling, you would be amplifying the dc component too, pushing the whole waveform off the face of the CRT. With ac coupling, the dc level would be blocked, so that only the ac portion of the signal would be amplified.

Once the signal is coupled, it is sent to a variable-gain amplifier circuit. Gain settings are usually scaled in volts per division (or volts/div for short). The greater the volts/div setting, the lower the sensitivity of the display, so larger signals can be displayed, but resolution is reduced. A smaller volts/div setting will increase the sensitivity of the reading and a smaller signal will fill the screen, but more detail is provided. The amplified signal is now sent along to the vertical driver to control the vertical deflection of the electron beam.

Digital scopes may use a system of "soft" keys that allows more than one function to be accessed from a single pushbutton. In these situations, it is generally easy to change the coupling mode since this is a routine task compared to less frequently used functions such as setting display contrast.

Ground Switch

The "ground" switch effectively disconnects any inputs to the vertical section of the oscilloscope. In addition to disconnecting the signal, the input to the vertical amplifier is grounded to eliminate noise that may be present from outside sources. This allows easy setting of the trace baseline using the vertical position control. On dual- and four-channel scopes, each channel is independent of the other and has its own vertical input ground switch.

Vertical Mode

In multichannel scopes, depending upon the instrument, some type of "vertical mode" will be provided. This switch selects Channel 1, Channel 2, or a combination of the two.

With many dual-channel scopes, it is also possible to mathematically add the value of the signal applied to channel 1 and channel 2. The result is the algebraic sum of the two signals, displayed as one waveform on the screen.

Many analog oscilloscopes include a means of selecting how the screen displays the two channels simultaneously. The categories are "chop" and "alternate." Depending on the frequency of the displayed waveforms, "chop" or "alternate" must be selected by the operator. In the case of slower signals, the "chop" mode is preferred, whereas higher-frequency signals are best displayed using the "alternate" mode. Additional information will be provided in a later section. It should be noted that "chop" and "alternate" displays are meaningless when it comes to digital scopes using LCD technologies. The operator will find these functions omitted from this class of oscilloscopes.

Vertical Attenuator Control—Volts per Division

Each channel will have it's own "vertical attenuator" control that sets the gain of the vertical amplifier. Because this amplifier must be accurately calibrated, it is always a rotary switch, or in digital models it can often be a pushbutton that changes the range through software. Generally, most oscilloscopes have vertical amplifier settings that range from 2 mV per major division to 10 volts per major division. This assumption is based on the assumption that a standard, 1 × oscilloscope probe is connected to the vertical amplifier input. The range sequence of vertical amplifier sensitivity is usually 2, 5, and 10. This translates to the following typical volts per division settings: 2 mV/div, 5 mV/div, 10 mV/div, 20 mV/div, 50 mV/div, 0.1 V/div, 0.2 V/div, 0.5 V/div, 1 V/div, 2 V/div, 5 V/div, and 10 V/div.

Considering the graticule has eight divisions from bottom to top, this places a limit on measurement deflection of 80 volts if the volts per division control is set to 10 V/div.

Look at the input connector on the front panel of the instrument. It is often noted that the vertical input has a maximum voltage limit (peak) It is therefore not possible to directly measure high-voltage signal levels, even though the vertical amplifier may suggest that this is possible. Damage to the instrument will result if the maximum allowable input voltage is exceeded!

The use of attenuating probes is required whenever measurements in excess of the allowable maximum to the instrument are required. Consult the manufacturer for available attenuator probe options.

Vertical Attenuator Control

Usually, a secondary control is provided to take the amplifier out of calibration over a finite range. This feature is useful when making certain measurements. The control is typically a part of the "vertical attenuator" control residing on a concentric shaft. It is customarily marked with the words "CAL" or "calibrate" and an arrow pointing in the direction of the calibrated position.

The calibrate control should always be kept in the "CAL" or "calibrate" position when making ordinary measurements. It should only be taken out of "CAL" under special circumstances.

It should be noted that this control can cause extreme measurement errors if inadvertently moved from its "CAL" position. The operator should check that any controls bearing the "CAL" or "calibrate" labels are in their proper positions before making any measurements with the scope. Failure to do so will likely result in erroneous measurements that can cause serious ramifications.

Channel 2 Invert

All two- and four-channel oscilloscopes provide a means of inverting at least one channel relative to another. In the typical two-channel model, "Channel 1" is usually fixed and not invertible. "Channel 2," however, is usually provided with an "invert" switch that flips the polarity of the signal applied to the screen. The ability to invert "Channel 2" is particularly useful for comparing the input and output of an amplifier, whereby the output is inverted 180 degrees from the input. If the amplifier input is assigned to "Channel 1" and the output assigned to "Channel 2," easy comparative measurements are possible, since "Channel 2" can be inverted by the oscilloscope. This allows the input and output signals to be displayed with the same phase, even considering the 180 degree phase shift created by the amplifier. Such measurements will be covered in a later section.

The Horizontal Section Horizontal Position

This control positions the oscilloscope's trace in the center of the screen with respect to the "X" or horizontal direction. Place the control in the "midway" position on analog models as a starting point. Digital oscilloscopes can use either a variable control or a pushbutton adjustment, depending on the model. Portable scopes frequently use pushbuttons to save front panel room.

Horizontal Sweep Rate—Seconds (time) per Division

Similar to the vertical sensitivity control, the horizontal sweep rate assigns the value for each division on the horizontal or "X" axis. Rather than thinking of this control as affecting the gain of an amplifier (as is the case of the vertical sensitivity control) consider this control to be changing the time base or sweep rate. Higher sweep rates cause the electron beam to scan the screen more quickly than slower sweep rates.

A rotary switch controls the sweep rate in analog models, whereas a pushbutton switch is frequently used with digital oscilloscopes. With general-purpose oscilloscopes, only one time base is provided. However, high-performance models are available that contain a second time base for specialized measurements. For the purposes of this book, we will only consider models having a single time-base control.

Depending on the bandwidth of the oscilloscope, the sweep rates can easily extend into the microsecond range. In considering a general-purpose model having a bandwidth of 50 MHz, the sweep rate varies from 1 sec/div to 1 μs/div using the time/division rotary switch.

Activating the 10 × multiplier increases the sweep rate accordingly. All sweep rates must now be multiplied by a factor of 10. For example, if the time/div control is set to 1 ms/division, each horizontal division is now represents 10 ms.

Since what you are actually measuring is the period (T) of the waveform, to obtain the frequency value, the value of the period must be divided into "1," or $F = 1/T$, where T is the signal's period.

Example. We measure the signal's period to be 16.67 milliseconds using an oscilloscope. We subsequently calculate the frequency of the waveform to be 60 Hz, as shown in Figure 3-3. Using the formula above, the signal's period (T) is divided into "1" since "T" is 16.67 ms; this is expressed as 16.67×10^{-3} seconds. Dividing this into "1" yields a frequency of 60 hertz (Hz). Years ago, frequency was expressed as "cycles per second" rather than "hertz." Using the term "cycles per second" may make more sense as to why the signal's period is divided into "1"—we want to find out how many cycles can "fit" into one second. A scientific calculator can speed up frequency calculations, particularly those involving milli- and microseconds.

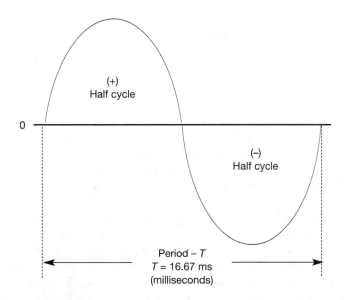

Figure 3-3 *Frequency versus time. Note: frequency is calculated by dividing the frequency into "1" or F = 1/T.*

The sweep can also be disabled for specialized measurements. This effectively stops the beam from moving under the control of the horizontal oscillator. Consult the Scope's operator's manual for details on disabling the sweep.

Horizontal Sweep Rate Control

Usually, a secondary control is provided to take the horizontal oscillator out of calibration over a finite range. This feature is particularly useful when making phase-shift measurements. The control is typically a part of the "sec/division" control residing on a concentric shaft. It is customarily marked with the phrase "CAL" and an arrow pointing in the direction of the calibrated position.

As is the case of vertical circuits, the calibration control should always be kept in the "CAL" or "calibrate" position when making ordinary measurements. It should only be taken out of "CAL" under special circumstances that will be outlined later. It should be noted that this control can cause extreme measurement errors if inadvertently disturbed from the "CAL" position. The operator should check that any controls bearing the "CAL" or "calibrate" label are in their proper position before making any measurements with the scope. Failure to do so will likely create incorrect measurements.

The Trigger Level Control

The "trigger level" control is important in that is tells the electron beam when to start the trace. The level control is used in conjunction with the "slope" selector switch. The slope can be set to either "+" or "–," depending on the signal being measured.

By rotating the "trigger level" control, it is possible to set the beginning of the trace at a convenient reference point on the graticule. Sometimes, the trigger level control is used in conjunction with the "horizontal position" control to set up the signal precisely on the display.

Trigger Source

Normally, the "internal trigger" can provide a stable triggering source based upon the signal being measured. It is derived from the signal applied to the vertical amplifiers in the scope. For internal triggering, the trigger circuit uses a sample of the signal that will be viewed on the scope from the input stage. When the input signal reaches a user-set "trigger level," the trigger circuit activates the time-base generator and a sweep begins.

However, under certain complex signals, especially when using the scope in the dual-channel mode (or multiple channels in the case of a four-channel scope), it is sometimes difficult to get the instrument to properly "sync" to the signals. In this case, the operator may benefit from using an external signal.

The second major triggering source is called "line"(or "60 Hz"). The trigger signal is derived from a 60 Hz signal provided by the scope's power supply. This is most often used for looking at signals related to the power-line frequency such as power supply ripple. Line triggering will not necessarily provide good triggering in all situations.

Most scopes can also accept an external trigger signal. External triggering is primarily useful if a sweep must be started when some external event occurs. A minimum signal of 100 mV is usually needed to activate the sweep.

To use the "external trigger," the oscilloscope must be supplied with an appropriate synchronization signal at the appropriate "BNC" connector. This signal may be a signal related to the signals being measured, but may not be a replica of the signals themselves, that is, a signal related in time to the measured signals such as a master oscillator signal.

Some oscilloscopes have a trigger option called "line." This trigger mode automatically synchronizes the signal to the ac power line. It is useful for any measurements relating to utility-supplied power. Measurements relating to ac motors, certain industrial controls, transformerless equipment, or other circuits that depend upon the power line frequency can often be easily triggered using the "line" option. Use caution when connecting the scope to any ac line-powered equipment, as lethal voltages could be present. Complex or high-frequency signals are best triggered in the normal mode. In this mode, only the user determines the best voltage level for triggering.

The manual mode only allows one sweep for each press of the "trigger button." This mode is handy when working with irregular events or very low frequency signals.

Other specialized trigger options may include "TV-vertical" and "TV-horizontal." Both trigger modes are designed for viewing television signals at either the line or field rate. Such measurements are beyond the scope of this book and will not be covered. Consult an advanced video systems textbook for details regarding the measurement of complex video signals.

Oscilloscope Start-Up

The first operation that you must perform after the scope is turned on is to locate the trace if it is not already visible. First, increase the trace intensity and set the triggering to automatic. Adjust the horizontal and vertical offset controls to the center of their ranges. Be sure that the triggering mode is set to trigger from one of the input signals, then adjust the trigger level until you see a flatline trace.

Many oscilloscopes are equipped with a "beam finder" mode that compresses the horizontal and vertical ranges. It forces the trace onto the CRT and gives you a rough idea of its approximate location. With or without this feature, once you find the trace and move it into position with the offset controls, alter the focus and intensity to obtain a crisp, sharp trace.

Probe adjustment is a quick end straightforward operation. It requires a low-amplitude, low-frequency, square-wave input (usually 1 kHz, 300 mV square wave with a 50% duty cycle) that can be provided by just about any waveform generator. Many scopes have a built-in calibration-signal generator to supply the test signal. Connect the probe to the test-signal output, then adjust the vertical and horizontal sensitivity so that one or two complete cycles of the signal are clearly shown on the CRT.

If the corners of the waveform are excessively rounded, there may not be enough capacitance in the probe. Spiked corners suggest too much capacitance. In either case, the probe is not matched properly to the scope. Slowly adjust the capacitance on the probe until the corners of the square wave are crisp and sharp. This indicates a good calibration. If you cannot achieve a clean square wave, try a different probe. Repeat the calibration each time a new probe is used on the scope, or when the probe is removed to a new scope. The scope is now ready for use.

Oscilloscope Probes

Oscilloscopes are usually connected to a circuit under test with a short length of shielded cable and a probe (see Figure 3-4). An important part of the measurement process is selecting the proper probe for the measurement task at hand. Usually, the instrument is shipped with one or more probes, depending on how many channels the scope displays. The most common probes are 1× models, although some manufacturers supply 10× or even 1×/10× switchable probes. At low frequencies, a piece of small-diameter coaxial cable and some sort of insulated test probe might do. Unfortunately, at higher frequencies the capacitance of the cable would produce a capacitive reactance much less than the one-megohm input impedance of the oscilloscope. In addition each scope has certain built-in capacitances at its input terminals (usually between 5 and 35 pF) These capacitances cause problems when probing an RF circuit with a relatively high impedance.

One of the most important reasons for using a probe rather than a simple clip-lead is to minimize the pickup of stray signals that can affect the accuracy of the measurement process. All oscilloscope probes are manufactured using shielded cables to prevent this problem form occurring.

One thing to keep in mind is the oscilloscope's input impedance. This translates to circuit loading. Loading can be resistive, capacitive, or inductive. Most scopes have an input impedance of 1 Mohm There is also a capacitive loading effect, usually printed on the front panel of the instrument. A typical capacitive loading effect is about 20 pf. This loading can affect your measurements to an extent depending on the frequency of the signal being measured, as well as other factors.

Unfortunately, there is a price to pay in terms of signal attenuation. Since the signal is attenuated 10 times, small signals might not be of adequate strength to be viewable on the scope's display. It is advisable to use the probes supplied with the instrument for the best performance. In the case of speciality applications, most oscilloscope manufacturers offer specialized probes as accessories. Probes are classified as voltage or current sensing. In sensitive measurement applications, probes can be further classified as being active or passive.

Active probes are substantially more expensive than passive probes and are generally reserved for difficult measurement tasks requiring extreme accuracy. General-purpose measurements can usually be

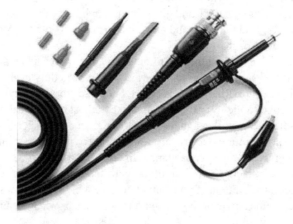

Figure 3-4   Oscilloscope probes.

made using passive probe models. This includes low-frequency amplifier testing as well as motor control, power supply, digital logic, and microprocessor applications in circuit design, test, and troubleshooting. traditionally, attenuator probes must be compensated using the adjustment provided on the attenuator unit or the probe itself.

Many probes offer the operator a variety of attachment options. Most probes have a sort of "clip hook" that permits clipping the probe to wires, terminals, and so on. For fine measurements on printed circuit boards, the nose of the probe is usually removable, revealing a sharp, piercing tip.

Oscilloscope Signal Measurements

In the following subsections, several types of signals will be presented with details related to specific oscilloscope controls. Using this information, the appropriate amplitude and period measurements will be determined. With the assistance of a scientific calculator, the corresponding frequency will be calculated, based on the measured period of the waveform.

Voltage Measurements

The first step in all voltage measurements is to set the zero-volt trace (or baseline) where you want it. To ensure that the oscilloscope is displaying zero volts, move the input coupling control to the ground position. That disconnects the scope input from the probe and connects it to ground. Adjust the vertical-offset control to place the trace where you would like the zero volts indication to be (often the centered horizontal axis on the CRT is used).

To measure dc, set the coupling to the dc position, then select the appropriate vertical sensitivity. As a general rule, set your sensitivity to a high scale to start with, then carefully increase the sensitivity (reduce the volts per division) after the signal is connected. That prevents the trace from "jumping off" the screen when the signal is first applied.

For example, with the vertical sensitivity set to 2 volts/div, each major vertical division on the screen represents 2 volts. A positive 4 volt signal will then appear two divisions above the zero axis (2 divisions × 2 volts/div = 4 volts). If the input is a negative voltage, the trace would appear below the zero axis but would be read the same way.

ac signal magnitudes (for sine. square, triangle, etc.) can also be read directly from the scope. The key factor to remember in ac voltage measurements is that the scope measures in terms of peak values. A regular ac voltmeter, on the other hand, measures in terms of rms (root mean square), so the reading on a scope will not match the reading on an ac voltmeter. To convert rms values to amplitude, simply multiply the rms value by 1.414. To convert an rms reading to a peak-to-peak quantity multiply rms by 1.414 × 2.

The peak voltage of a sine wave can be read directly from the CRT by measuring the number of divisions there are between the zero axis and the positive (or negative) peak of the signal. For example, if the peak is 2 divisions above the zero line, and the vertical sensitivity is set to 5 volts/div, then the signal would be 10 volt peak (2 divisions × 5 volts/div). The peak-to-peak voltage can be calculated by multiplying the peak voltage by 2. It can also be measured by counting the divisions from the negative peak to the positive peak. If there are 4 divisions peak-to-peak at 5 volts/div then the signal is 20 volts peak-to-peak (4 divisions × 5 volts/div). This signal would show up on an ac voltmeter as 7.07 volts rms.

Time Measurements

The oscilloscope is a handy tool for measuring such parameters as pulse width and period. Once the period is measured, a signal's frequency can be calculated. Duty cycle can also be calculated based on the high and low times for each cycle.

In order to measure the overall cycle time of the signal, adjust the horizontal sensitivity until at least one full cycle of the signals is shown. Simply multiply the number of divisions in one full cycle by the horizontal sensitivity setting. If the horizontal sensitivity is set to 1 ms/div and one complete cycle occupies 2 divisions, the period of the signal is 2 ms (2 divisions × 1 ms/div).

Since frequency is the exact inverse of the period. the frequency of the signal can be easily calculated by dividing the period into 1. For the above examples, a period of 2 ms (0.002 s) would yield a frequency of 500 Hz (1/0.002 s).

The duty cycle of a square wave is the percentage of time it spends high. To calculate the duty cycle, divide the on time by the total period × 100%. For the example used, if the on time is 1 ms, the duty cycle would be 50% (1 ms/2 ms × 100%).

Phase Measurements

Multichannel oscilloscopes can be used to measure the phase relationship between two signals of equal frequencies (typically, sine waves). One signal is sent to one scope input to control the horizontal driver and the second signal controls the vertical driver. Both channels must be set to the same sensitivity. The time-base control is then rotated to the X-Y setting. The combination of signals causes them to draw an ellipse on the screen. Phase can be determined from the shape of the figure.

Sine-Wave Measurement

Figure 3-5 illustrates a sine-wave trigonometric function. The ac power line supplies electrical energy in the form of a sine wave with a frequency of 60 Hz in the United States. It is assumed that the base controls of the oscilloscope are properly set, including trace intensity, vertical and horizontal position, and focus. Trigger level is assumed to be set so that the start of the wave begins at the first major division as indicated.

Both the vertical attenuator (volts/div) and horizontal sweep rate (sec/div) controls affect how the signal is interpreted. Pay close attention to what settings are used by each of these controls because they are the "essence" of the measurement process. It is assumed a 1× (direct) probe is used..

In Figure 3-5, the sine wave has a vertical deflection of 8 divisions. Because the vertical attenuator (volts/div) control is set to 5 volts/div, this indicates the sine wave peak-to-peak amplitude is (8 div) × (5 volts/div); or 40 volts peak-to-peak. Another way to express this is 40 Vp-p

By dividing the peak-to-peak measurement in half, the signal's "peak" amplitude is obtained. In this case, the peak voltage (amplitude) is 40 volts/2, or 20 volts peak. This can be expressed as 20 Vp.

Similarity, the waveform occupies 10 divisions horizontally. In this case, since the horizontal sweep rate (sec/div) control is set to 0.1 ms/div, it means that the time for one complete wave is (10 div) × 10.1 ms/div) or 1.0 ms. The sine wave in the example has a period (t) of 1 ms (0.001 sec). Always remember that the vertical attenuator (volts/div) control only affects amplitude measurements, whereas the horizontal sweep rate (sec/div) control relates to time (period) measurements. Technicians and engineers do not

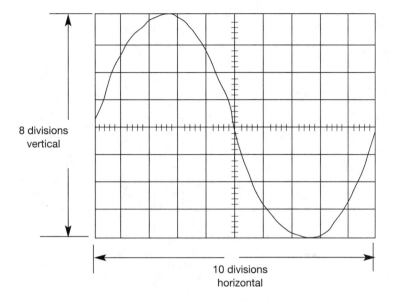

8 divisions
vertical

10 divisions
horizontal

Figure 3-5 Graticule with sine wave.

usually refer to the period of a signal but to it's "frequency" instead. Frequency is the number of cycles per second that a signal produces. A signal's period represents the time necessary to complete one cycle. In other words, in the example, the sine wave completes one cycle in 1 millisecond (1 ms).

Fortunately, the mathematics involved in converting a signal's period into frequency is quite simple. Divide the signal's period into "1." Don't forget to keep the units (milli, micro, kilo, mega, etc.) attached or it will introduce extreme errors.

It will be easier to remember how to perform the calculation by remembering the following formula:

$$f = 1/t$$

In the example, since $t = 1$ ms, or 1×10^{-3} sec, plugging this into the equation results in

$$f = 1/1 \times 10^{-3} \text{ or } 1000 \text{ Hz}$$

You can convert this using engineering notation, if desired, by attaching the appropriate unit prefix (kilo, mega, milli, micro, etc.). For the example, 1000 Hz converts to 1 KHz.

Triangular Wave Measurement

A triangular waveform is depicted in Figure 3-6. A triangular wave can be analyzed in a fashion similar to a sine wave. In following example, the oscilloscope and function generator are set up so as to display one complete cycle. Period and amplitude measurements are then made. To expand on the oscilloscope's capabilities, this example will make use of a 10× attenuator probe instead of the standard 1× (direct) probe used in the preceding examples.

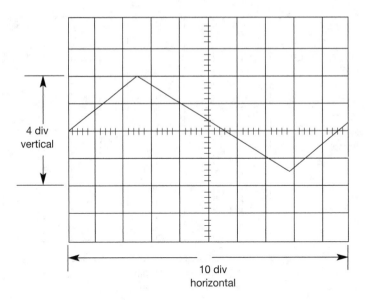

4 div
vertical

10 div
horizontal

Figure 3-6 Triangular wave.

Problem. Determine the peak-to-peak voltage of the signal. Calculate the peak voltage. Determine the period of the waveform. Calculate the frequency of the waveform.

Solution. First, determine how many divisions peak-to-peak are occupied by the signal. In this case, the signal occupies 4 divisions vertically.

The vertical attenuator (volts per division) control is set to 2 volts/div. The horizontal time base (seconds per division) should be set to 20 μs/div. In this example, the vertical attenuator (volts per division) control is set to 2 volts/div. The horizontal time base (seconds per division) should be set to 20 μs/div. In this example, a 10× attenuator probe is used; it will only affect the amplitude measurements, not those related to time (period).

Because a 10× probe is used, the actual vertical attenuator (volts per division) must be multiplied by 10. This translates 2 volts/div into 20 volts/div. Since it was determined the signal's vertical deflection was 4 divisions, the actual amplitude (voltage) is (4 div) × (20 volts/div) = 80 Vp-p. The peak value of the signal can be calculated by dividing the peak-to-peak value by 2. In this case, 80 V/p-p = 40 Vp. The peak value for a waveform can also be determined by reading it from he scope graticule. In this example, the peak vertical deflection is 2 divisions, so the peak voltage is determined as follows:

$$(2 \text{ div}) \times 20 \text{ volts/div} = 40 \text{ Vp}$$

Square-Wave Measurement

A square-wave signal is illustrated in Figure 3-7. For a square wave, follow the same process used in the previous examples. Be careful to locate the "full cycle" point of the signal because this can be slightly

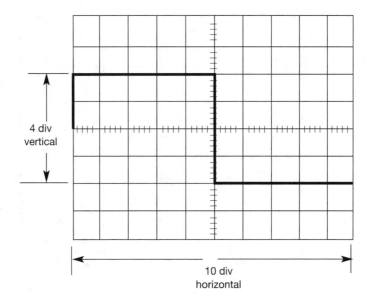

4 div
vertical

10 div
horizontal

Figure 3-7   Square wave.

more difficult in the case of the square wave a it is more difficult to see the rising and falling edges of the pulse since they are very fast. Assume that a 1× (direct) probe is used for the measurements. Using the vertical attenuator (volts per division) control setting, it is a simple task to measure the "high" and "low" points of the pulse.

In this example, the "high" voltage level is 8 volts. This arises because the reference (ground) trace is "parked" at a point 2 divisions up from the bottom of the graticule. Assuming that this is the "0" reference point, the "high" point of the square wave is deflected 4 divisions upward from this point. With a vertical attenuator (volts/div) setting of 2, this indicates each major division is worth 2 volts. Four major divisions of vertical deflection worth 2 volts per major division, which translates to 8 volts.

Note that "peak" and "peak-to-peak" really have no meaning here. The square wave generally has it's "low" point resting at "0 V," although it can be offset in some situations. It is more common to see the "on"–"off" behavior of this signal being described as "digital" in nature.

The frequency of the square wave is determined just like for the other waveforms, using the horizontal time-base (seconds per division) control setting to determine the period. The frequency is then calculated by inverting the period, paying attention to unit prefixes (milli, micro, kilo, etc.).

Pulse Width Measurement

Pulse width measurements are performed by measuring the time a digital signal is "on." A pulse-wave scope trace is shown in Figure 3-8. Depending on how the logic is set up, a logic "high" or "low" can be referenced as the "on" time.

Pulse width is easily measured using the oscilloscope. In the example below, it is assumed that the log-

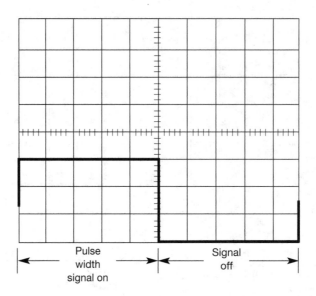

Figure 3-8   Pulse-width waveform.

ic "high" is the "on" time of the signal. In the example, the pulse occupies 5 divisions horizontally. Since the horizontal time-base (seconds per division) control is set to 20 qs/div, the pulse width is (5 div) × (20 μs/div)= 100 μs.

The pulse amplitude is determined using the vertical attenuator (volts per division) control setting. Since this control is set to 2 volts/div, the amplitude is calculated as (3 div) × (2 volts/div) = 6 volts.

Duty Cycle Measurement

Duty cycle measurements are a logical spin-off from pulse width measurements. They tell (in percent) how long a pulse is on compared to the signal's period. In other words, duty cycle = [pulse width (PW) signal period (*t*)] × 100. A typical duty-cycle wave-trace is illustrated in Figure 3-9.

Assuming that the horizontal time-base (seconds per division) control is set as indicated, the duty cycle for the pulse is easily determined using a scientific calculator. Pulse width can never exceed 100% since this is mathematically impossible

Advanced Features

Modern electronics has made many more tools available on today's oscilloscopes. Although most of these extra features are currently expensive, those costs will eventually decline. Let us look at several of these advanced features.

In an ordinary scope, voltage and time must be measured by eye, using the graticule marks on the CRT. That has been the tried and true approach since the days of the first vacuum-tube scopes. With the advent

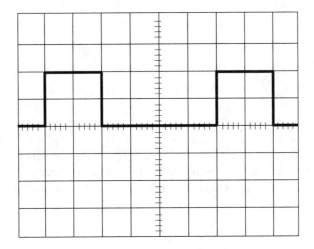

Figure 3-9 Duty cycle waveform.

of microcomputers, however, a set of on-board measuring markers (called cursors) can be included in the screen display to aid in the waveform analysis. Voltage, time, and frequency cursors are the most commonly available.

Voltage cursors are two horizontal bars that can be independently placed at any location on the screen. The on-board microcomputer automatically calculates the distance between the cursors, then multiplies that by the volts per division setting to produce the voltage reading. The resulting reading can then be displayed for quick end easy reference.

The time and frequency cursors work in much the same way as the voltage cursor. Two vertical bars can be located anywhere on the screen. The microcomputer can calculate and display time by multiplying the distance between the cursors by the time-base setting. Since frequency is just the reciprocal of time, frequency can be calculated and displayed with just one extra step. The choice between time or frequency cursors is usually switch selectable.

Oscilloscope Limitations

Oscilloscopes have fundamental limits, primarily in frequency of operation and range of input voltages. For most purposes, the voltage range of a scope can be expanded by the use of appropriate probes. The frequency response (also called the bandwidth) of a scope is usually the most important limiting factor. At the specified maximum response frequency, the response will be down 3 dB (0.707 voltage). For example, a 100 MHz, 1 V sine wave fed into a 100 MHz bandwidth scope will read approximately 0.707 V on the scope display. The same scope at frequencies below 30 MHz (down to dc) should be accurate to about 5%.

A parameter called rise time is directly related to bandwidth. This term describes a scope's ability to accurately display voltages that rise very quickly. For example, a very sharp and square waveform may appear to take some time in order to reach a specified fraction of the input voltage level. The rise time is usually defined as the time required for the display to show a change from the 10% to 90% points of the input waveform. The mathematical definition of rise time is given by

$$t_r = 0.035/BW$$

where

$$t_r = \text{rise time, } \mu\text{s}$$

$$BW = \text{bandwidth, MHz}$$

It is also important to note that all but the most modern (and expensive) scopes are not designed for precise measurement of either time or frequency. At best, they will not have better than 5% accuracy in these applications. This does not change the usefulness of even a moderately priced oscilloscope, however. The most important value of an oscilloscope is that it presents an image of what is going on in a circuit and quickly shows which component or stage is at fault. It can show modulation levels, relative gain between stages, and oscillator output.

Digital Oscilloscope

The classic analog oscilloscope just discussed has existed for over 50 years. In the last 15 years, the digital oscilloscope has gone from a specialized laboratory device to a very useful general purpose tool, with a price attractive to an active experimenter. It uses digital circuitry and microprocessors to enhance the processing and display of signals. These result in dramatically improved accuracy for both amplitude and time measurements. When configured as a digital storage oscilloscope (DSO) it can read a stored waveform for as long as you wish without the time limitations incurred by an analog type of storage scope.

The simplified block diagram shown in Figure 3-10 and the photograph shown in Figure 3-11 depict a digital oscilloscope. As the signal goes through the vertical input attenuators and amplifiers, it arrives at the analog-to-digital converter (ADC). The ADC assigns a digital value to the level of the analog input signal and puts this in a memory similar to computer RAM. This value is stored with an assigned time, determined by the trigger circuits and the crystal time base. The digital oscilloscope takes discrete amplitude samples at regular time intervals. The digital oscilloscope's microprocessor next mathematically processes the signal while reading it back from the memory and driving a digital-to-analog converter (DAC), which then drives the vertical deflection amplifier. A DAC also takes the digital stored time data and uses it to drive the horizontal deflection amplifier. The vertical signals are often referred to as "8-bit digitizing," or perhaps "10-bit resolution." This is a measure of how many digital levels are shown along the vertical (voltage) axis. More bits give you better resolution and accuracy of measurement. An 8-bit vertical resolution means that each vertical screen has 2^8 (or 256) discrete values; similarly, 10 bits resolution yields 2^{10} (or 1024) discrete values.

It is important to understand some of the limitations resulting from sampling the signal rather than taking a continuous, analog measurement. When you try to reconstruct a signal from individual discrete samples, you must take samples at least twice as fast as the highest-frequency signal being measured. If you digitize a 100 MHz sine wave, you should take samples at a rate of 200 million samples a second (referred to as 200 megasamples per second). Actually, you really would like to take samples even more often, usually at rate at least five times higher than the input signal.

Very fast signal changes between sampling points will not appear on the display if the sample rate is not high enough. Place analog and digital scopes next to each other, with the inputs tied together. Input a voltage spike into in the input amplifier of both scopes and you will see the analog scope display the

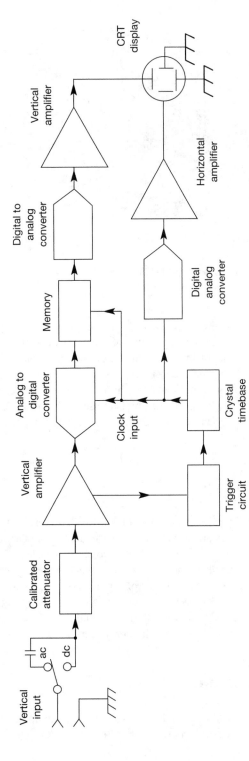

Figure 3-10 Block diagram of a digital oscilloscope.

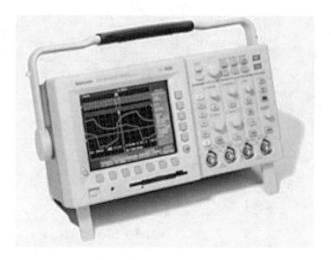

Figure 3-11   Digital oscilloscope.

"spike" but it will not be visible on the digital scope. The sampling frequency of the digital scope is not fast enough to store the higher-frequency components of the waveform. If you take samples at a rate less than twice the input frequency, the reconstructed signal has a wrong apparent frequency; this is referred to as *aliasing*. The result is that the scope reconstructs a waveform with a different apparent frequency. Many older digital scopes had potential problems with aliasing. Newer scopes use advanced techniques to check themselves. A simple manual check for aliasing is to use the highest practical sweep speed (shortest time per division) and then to change to other sweep speeds to verify that the apparent frequency does not change.

A/D Oscilloscope PC Card

The new alternative to the digital oscilloscope is the analog-to-digital oscilloscope PC card (see Figure 3-12). The price of the PC oscilloscope has become quite reasonable in the last few years and it has become a viable alternative the digital oscilloscope. The new PC scope cards have multiple input channels, so that you can look at multiple data channels at once. The major caveat in buying a PC scope card is the difference in scope speed parameters. The specifications of digital scope cards are often different from that of analog scopes, so make sure you are comparing apples to apples. The sampling rate of digital scopes is often slower or inferior to higher-priced, brand-named analog and digital oscilloscopes. If you have a computer on your "test bench" or electronic workbench, then you may want to consider this alternative.

Digital Storage Oscilloscope

Conventional oscilloscopes are versatile enough to tackle most typical applications, but low-frequency irregular or single-shot signals will be difficult to see. For low-frequency signals, the time base must be ex-

Figure 3-12   Analog-to-digital oscilloscope PC card.

tended so far that you will only see a dot moving across the CRT. Unique or single-shot signals can be almost impossible to capture properly on an ordinary scope. A storage oscilloscope overcomes all those limitations. One can retain a trace on its CRT for several seconds to several days, depending on the particular storage method used.

The oldest form of storage was simply a camera that could be fitted to the CRT but this method was bulky, awkward, and difficult to use. The CRT could not be viewed conveniently when the camera was attached, and timing the shutter to the event was partly a matter of luck. It also took time to develop the film before the signals could be reviewed.

Early storage scopes used an enhanced phosphor layer that could be energized to hold the image of the trace. Although that method added a great deal of convenience, the excitation required to sustain the image made the CRT very bright, which tended to burn the trace into the phosphor over time.

Digital circuitry has vastly improved the performance and reliability of storage functions. Since any input signal can be digitized (converted to a digital form), it can be held as binary data in the scope's memory. The stored signal can then be written and held on the CRT. The digitally stored image can be displayed continuously, just as if it was an external input signal.

The latest trend in oscilloscope is the digital storage oscilloscope or DSO, often called the digital scope. The digital oscilloscope has come a long way in the last few years, with sample speeds over 3 megasamples per second. DSOs have become the primary debugging tools for high-speed digital circuits, since they can sample so rapidly, have large acquisition memories, and can trigger on virtually all types of fault conditions involving multiple channels.

Digital oscilloscopes represent a significant departure from the older analog oscilloscope, preform many more specific operations, and have many more features and applications than the conventional oscilloscope. Digital oscilloscopes can cost significantly more than a simple analog bench scope. They are among the most expensive oscilloscopes. Therefore, you will need to take a close look at all the specifications and features before purchasing a digital oscilloscope. Table 3-1 is a comparison of DSO specifications.

Mainframe DSO power is now available in the newest portables. DSOs now can include memory depths of 1 Megasamples per channel and offer processing options such as histogramming or FFTs on up to 6 million contiguous samples. DSOs trigger capabilities are stronger than ever. Emphasis is on minimizing dead time between triggers and triggering on the most elusive events. DSOs are not just for single

Table 3-1   *Comparing DSO specifications**

Front end characteristics	**Display**
Maximum transient sample rate	Display size (diagonal) (number inches)
Maximum repetitive sample rate	Display pixel resolution (number pixels x
Analog bandwidth	number pixels)
Time-base range maximum	Display type (Color, monochrome,
Time-base range minimum	etc.)
Volts/div (range)	Multiple zooms per trace (Y/N if Y, number?)
Custom vertical rescaling (Y/N)	Multiple grids for full 8 bits (Y/N)
Vertical resolution (number of bits)	Multiple cursors per screen (Y/N if Y, number?)

Front end characteristics
Maximum transient sample rate
Maximum repetitive sample rate
Analog bandwidth
Time-base range maximum
Time-base range minimum
Volts/div (range)
Custom vertical rescaling (Y/N)
Vertical resolution (number of bits)

Display
Display size (diagonal) (number inches)
Display pixel resolution (number pixels x number pixels)
Display type (Color, monochrome, etc.)
Multiple zooms per trace (Y/N if Y, number?)
Multiple grids for full 8 bits (Y/N)
Multiple cursors per screen (Y/N if Y, number?)

Storage characteristics
Number of channels
Maximum samples on each channel
Maximum samples on 1 channel
Reference memories (number kbytes)
High-density DOS disk (Y/N)
Hard disk (Y/N)
Built-in printer (Y/N)
PC card (Y/N)

CPU performance
CPU (Model and clock speed)
Math coprocessor (Y/N)
Maximum record size on math (Y/N)

Measurements and DSP
Pulse parameters (number total/number viewable)
Statistics on parameters (Y/N)
Chained math operations (Y/N if Y, number?)
Histogramming (Y/N)
Advanced mathematics (Y/N)
FFT (Y/N)

Trigger and acquisitions modes
EDGE TRIGGER (Y/N)
HOLDOFF by TIME (Y/N)
HOLDOFF by EVENTS (Y/N)
PATTERN (Y/N)
GLITCH (Y/N)
INTERVAL (Y/N)
STATE QUALIFIED (Y/N)
DROPOUT (Y/N)
EXCLUSION (Y/N)
Video trigger (Y/N)
Maximum number of triggers/second
Trigger segments with time stamps (Y/N)
Trigger pass/fail with masks and (Y/N)
   parameters
External trigger input (Y/N)
PEAK DETECT with timing (Y/N)
Roll mode acquisitions (Y/N)
Display update rate

*DSOs vary in performance in a variety of ways. Each manufacturer provides a certain degree of standard features, but their different design schemes produce unique performance strengths and weaknesses. Compare and evaluate.

transient capture anymore; they can now provide stand-alone measurements and tests that apply waveform analysis right in the instrument. As prices are beginning to drop, many DSO users are now purchasing four-channel models as the cost difference is low and seeing what is happening on more than two channels is so beneficial.

Many digital storage scope manufacturers are now offering many specific measurement capabilities and the scopes are becoming extremely computational intensive as well (see Table 3-2, which lists primary DSO functions). Deeper memories, multiple channels, high-resolution displays, and enhanced DSP routines tax the fastest CPUs and smartest firmware. The latest designs use multiprocessor architectures. Functions are increasing, so the user interface is gaining increasing importance. There is virtually no standardization as to the method of giving a user control of a DSO. As you look at the scopes, you will see combinations of knobs, buttons, touch-sensitive screens, mice, computer interfaces, and even programmable buttons on the probes. As a general rule of thumb, the more menus, the fewer the knobs.

Digital scopes with floppy disk drives have become the rule rather than the exception. Almost all new units offer them, either as options or as standard features. Most store front panel setups, screen shot graphics you can import into word processing programs, and waveform data archives. Some manufacturers are now offering built-in hard drives that are removable and, thus, transferable (via PC card interfaces) to laptops or PCs.

Some models of digital oscilloscopes offer channels interleaving so that, for example, in a two-channel model, twice the sample rate is achieved when using just one channel compared with both channels being used. Some DSO models also now offer memory interleaving so that, for example, in a two-channel mod-

Table 3-2 *Primary DSO functions*

There are five primary DSO functions:

1) **Capture** the signal
2) **View** the signal
3) **Measure** the signal
4) **Analyze** the signal
5) **Document** the signal

Capture = consider sample rates, memory depth, bandwidth, trigger, number of channels, display update rate, and/or dead time between acquisitions.

You should have a predetermined knowledge of the highest-frequency signal you need to digitize, what its full-scale amplitudes are, and whether or not you need to capture single-shot or repetitive waveforms. If these issues are unclear to you, review them with your sales engineer.

View = consider ADC resolution, display resolution, display size, DSP results, and zoom expansions.

Measure = consider pulse parameter requirements, cursors, and statistics.

Analyze = consider DSP, pass/fail testing, mask comparisons, and eye diagrams.

Document = consider printing, saving screen shots to disk, and data transfers.

el, twice the memory depth is achieved when using just one channel (and twice the recording time) compared with both channels being used. DSO acquisition memory depths are expanding, with 50 K (typical minimums) up to 8 million samples (maximum) available.

The enthusiasm surrounding the new digital oscilloscopes is due their ability to be upgraded via firmware and hardware expansions, but not all models are expandable. Table 3-3 lists DSO form factors as well as DSO manufacturers.

Many manufacturers of analog oscilloscopes have begun to discontinue their high-bandwidth models. Some manufacturers are offering hybrid analog and digital scopes. DSOs have become the highest bandwidth oscilloscopes available and are usually less expensive than their analog model equivalents. Several

Table 3-3   DSO form factors

There are five available DSOs form factors:

1) **PC Card.** Analog-to-digital conversion on a card that uses a PCs CPU and memory. PCB style. Interesting due to low cost. Look out for noise problems from some PC backplanes.
2) **Stand-Alone Card.** DSO on a card for embedded systems. PCB style. Interesting due to low cost, small size, fewer noise issues, and greater functionality.
3) **Handheld.** Portable capabilities. Portable style, battery operated for field measurements.
4) **Portable.** With some amount of upgrade capabilities. Portable style. Performance approaching mainframes and low cost due to high competition, high volume, and large-scale integrations.
5) **Mainframe.** Typically with plug-ins that determine performance. Lab style. Highest cost but highest performance and greatest versatility.

U.S. manufacturers of DSOs are:

Fluke Corporation
800 443-5853
Everett, WA
http://www.fluke.com/

Hewlett-Packard
800 452-4844
Palo Alto, CA
http://www1.hp.com/

LeCroy Corporation
800 553-2769
Chestnut Ridge, NY
http://www.lecroy.com/

Tektronix Corporation
800 426-2200
Beaverton, OR
http://www.tek.com/

DSOs go well beyond 1 Ghz, but note that the price of these models is not presently within the realm of the home electronics enthusiast.

The Analog-to-Digital Converter—The Heart of the Digital Oscilloscope

One of the most important factors when considering purchasing a digital oscilloscope is the ADC or analog-to-digital converter. The first consideration is the speed of the ADC. Be sure to avoid short-record-length DSOs that can only sample at maximum rates for short periods of time. Ideally, the sample rate value should be displayed on screen all the time. Be certain that you can sample fast enough in real time (single-shot mode should be available on all channels so you can record without aliasing).

Another point of confusion when considering a DSO is the sample rates specified for repetitive versus single-shot acquisitions. "Real time" refers to single shot. RIS stands for random interleaved sampling. RIS is sometimes called ET or equivalent time sample mode and can only be used with repetitive waveforms. It is also called "random repetitive sampling."

The sample rate speed of the DSO's analog-to-digital converter is a very important specification. It is the minimum time between each sample. For instance, a 500 megasample/sec sample rate relates to 2 ns per point resolution. Multiply the number of sample points by the sample period to determine a DSO's maximum recording time at the maximum sample rate.

Many people confuse the sample rate speed with the bandwidth. Bandwidth is simply the analog front-end performance (preamplifier and sample and hold circuitry).

ADC resolution refers to full-scale resolution. An 8 bit digitizer will divide full-scale input voltages by 255 counts. Thus, the minimum discernible sampled value on a 1 volt, full-scale, 8 bit DSO would be 1/256 or 0.00390 volts per step. If you have repetitive waveforms, you can increase your vertical resolution with averaging. If you have transient waveforms, you can increase your vertical resolution by low-pass filtering each sweep for enhanced resolution.

ADCs vary in accuracy; not all ADCs are created equally. The most common figure of merit is effective bits, which relates the number of correct bits of a given ADC's actual measurements to the ideal.

Trigger

If you can't trigger on the waveforms you need to see, you have a problem! This should be the most important part of the demonstration if you are considering the purchase of a new DSO. There are various trigger capabilities, and a combination of those that are easy to use when capturing your waveforms is the most desirable.

Know how frequently you need to trigger. The maximum trigger rate is a key specification that is not often published in manufacturers' specifications as there can be many variables. Evaluate and compare!

Some DSOs have a memory segmentation feature that lets you trigger very rapidly and fill just a portion (one segment out of n segments) per trigger. Most DSOs will let you trigger on the width of pulses, the intervals between pulses, the logical or pattern conditions between inputs, after specific delays by events or time, dropout conditions, and so on.

Look for trigger icons that relate how the current trigger selection is working. This is very helpful if you are looking at a screen dump later and trying to reacquire the same trigger conditions. DSOs are valuable tools for looking at video signals, but not all DSOs offer a video trigger as a standard feature.

DSOs can almost always capture single-shot events but not always with the amount of pre- or posttrigger delay you might need. If your application requires capturing a lot of transient waveforms, look into the span of the trigger delay as an important specification.

Bandwidth

Bandwidth is the amount your signal will be attenuated by the DSO's front-end amplifier, and is specified at the –3 dB point as a function of input frequency. Note that bandwidth ratings are at the input to the amplifier and that your probes might also attenuate your signals. If you are looking at signals > 50 MHz, you should use an FET probe. Remember that –3 dB is down in amplitude by almost 30%. You probably do not want a 30% error in your amplitude measurements. Consider purchasing one of the-higher bandwidth DSOs so your measurements will be accurate.

A few notes of caution. Many DSOs have bandwidth ratings that reflect their best performance but only in certain voltage ranges. Also, some DSOs reduce sample rates by the number of channels activated. This could cause aliasing by changing the relationship of how fast the DSO is sampling versus the bandwidth of the signals you are digitizing.

Many DSOs have analog bandwidth specifications far greater than their single-shot sample rate's Nyquist (0.5 sample rate) frequency. This is so that when repetitive waveforms are viewed, the maximum-bandwidth signals can be seen.

Digital Signal Processing (DSP)

DSP is doing math on the waveforms so additional information can be obtained. Some instruments really slow down when DSP is being performed. Ask for benchmark specifications on the key functions you need. Don't waste your time looking at DSOs that just aren't fast enough.

Another DSP concern is that some DSOs don't process an entire waveform because of poor CPU power. Make sure the data you need processed is really being processed! Another DSP concern is that you may wish to do a series of functions. How many functions can be chained varies from model to model. Potential DSP functions include:

Arithmetic	Add, subtract, multiply, and divide any traces
Averaging	Remove random noise, improve resolution
Enhanced resolution	Provides smoothing, improves resolution at reduced bandwidth
Functions	Integrate, differentiate, envelope, and so on
FFT	Spectral analysis of any trace

Histograms	Distributions of measured values
Mask testing	Comparisons of live waveforms to masks
Measurement testing	Comparisons of live waveforms to measurements
Trending	Time series of measured values

Displays

Ideally, the DSO will compress an entire sweep of acquisition memory onto a single screen using a min/max algorithm. The benefit of this is that you don't have to page through screens to see the interesting details in your data. Min/max compaction makes faults obvious.

Ideally, the display will be large enough so that you can see the waveforms and measurements clearly. Look out for small diagonal measurement displays that put measurements on top of the waveform data.

DSO displays are typically specified in terms of resolution and diagonal size. The higher the resolution, the easier it will be to see fine details and the better your publications that have imported DSO screens will appear. The larger the diagonal size, the greater the chance of being able to see critical information on screen all at the same time versus pages of menus. Ideally, you should have all the information on the screen that you want in your report. Consider things like trigger parameters, icons, measurements, input and timebase settings for each trace, sample rate, cursors and their measurements for each trace, and clock/calendar. That is a lot of information on screen.

Ideally, you should have the ability to expand or zoom in on different parts of your waveforms to see details more clearly and to limit your measurements to within a given region of data. This means that multiple expansion windows are best.

Ideally, the graticule should be done in software and allow multiple traces to be displayed, each within their own grid. This preserves the full-scale voltage input ranges for best accuracy.

Archive and Memory

The new digital storage oscilloscopes often have a means for storing waveforms. This important feature allows you to keep an archive or database of measurements. For example, you can take measurement of "test points" in specific pieces of equipment and use them later to compare a known reading against the current reading in the event of an instrument failure. Floppy disk drives were the first method of storing waveforms and they are convenient. The scope's record format should be MS-DOS and should come with a file formatter so you can convert to ASCII and then back to binary.

Another method of storing waveforms is the hard drive. Hard drives are becoming more available in the newer DSOs and offer the same kind of convenience that is realized in personal computers. More recently, digital oscilloscopes are now available with memory storage cards. They are more expensive than floppy disk drives, but they are very fast—up to 200 times faster than a floppy. If you plan on using the memory card in your personal computer to analyze the data, you will have to obtain a low-cost PC card reader.

Many new digital oscilloscopes have internal printers and/or plotters. These hard copy devices are best for instantly showing the world your measurements and waveforms. Plotters are great for producing elegant color plots that are most impressive when displayed on overhead projectors.

Digital Storage Oscilloscope Applications

Due to expanding functions and capabilities, DSOs lend themselves to a wide variety of application areas. Like a computer, the more applications you can use an instrument, for the greater value it has and the easier it is to justify its cost. Here are a few things for you to consider. The most obvious one is to use a DSO as an oscilloscope. Typical applications are electronic circuit design and debugging and troubleshooting faulty or intermittent circuits.

Another common application is as the front end of a data acquisition system. DSO's cost per channel has become very competitive. Many people find that the triggering flexibility, deep memory, "live" view of waveforms, and fast transfer rates make the DSO a great candidate. If your experiment is short lived, it is nice to have a DSO left over when you are done rather than a black box that gets shelved and forgotten. Typical applications are research experiments, process monitoring, and flaw detections.

DSOs lend themselves to being fully integrated into automated test equipment (ATE) systems. For example, the DSO can be under remote control from a host computer and conduct its business by computer command. Typical applications are incoming quality assurance of components, manufacturing/production functional tests, final tests, and system tests.

DSOs can be used as card-level or portable rack-mounted form factors and embedded into systems that require analog-to-digital conversion and data analysis. Here the DSO is used as a system component and eliminates the need and time for engineering custom devices.

DSOs will displace and replace many dedicated instruments in the future, such as DMMs, spectrum analyzers, impedance analyzers, time-interval analyzers, frequency counters, pulse counters, and power meters.

Caring for Your Oscilloscope

Since your oscilloscope will most likely be the most expensive piece of test gear in your workshop, you should take care to see that it will provide many years or service. You should always store the instrument in a clean, dry environment whenever possible. Cover the scope when it is not in operation, or store it in its carry case if it is a portable model, to prevent the buildup of dust and dirt.

Because of the sensitive internal electronics, keep the instrument cool. Whenever possible, don't store it in a vehicle because inside temperatures can soar. If you must transport the unit, allow it to reach ambient temperature before applying power. Electronic components are rated to operate between a range of temperatures and can suffer damage if operated outside of these limits.

Cold is another concern, especially with LCD display equipment. The display can become cloudy when exposed to extremely low temperatures. Allow the instrument to reach room temperature before switching it on.

Transporting an oscilloscope requires the same care as any sensitive electronic instrument. Don't drop or otherwise abuse the instrument as that could cause internal damage. Calibration accuracy can also be affected since many oscilloscopes use variable components such as potentiometers for certain calibration adjustments. Finally, keep the instrument out of direct sunlight. Ultraviolet (UV) radiation in the sun's rays can deteriorate painted surfaces and plastics. This is a particular concern when it comes to digital scopes that use LCD display screens.

Never place the oscilloscope in an operating position that continually exposes it to sunlight, such as near a window. Although it might be tempting to locate your test bench near an outside window for the sake of a nice view, it is best to avoid such locations. Over time, even fluorescent lighting can cause deterioration of plastic surfaces, causing a shift in color. This is especially obvious on light-colored plastics.

Buying a Used Scope

Most electronics hobbyists will end up buying a used scope due to price, since used scope will likely be less expensive than a new one. It's probably best to start out with a plain analog oscilloscope (unless you have special requirements that demand a digital one). As it stands now, you can get the best bandwidth for the buck with analog devices. However, this may not be true in the future. You can get good used analog scopes at very good prices. If field service is a prime concern, consider exploring portable scopes before deciding on your purchase. Most portable models will perform adequately when used on a test bench and will provide the added flexibility of offering "on-the-go" capability. It may make more sense to buy this type rather than a standard bench model that is larger, heavier, and overall less portable as a result of its intended use.

Even though bench models claim to be portable, they usually afford no protection for the instrument's front panel display and knobs, making them less than desirable in field applications. When choosing a portable model, consider one that comes with a carry case option.

The bandwidth is the oscilloscope's main specification. It dictates how high a frequency it can measure. For example, if a bandwidth is specified at 100 MHz, this means that the response at this frequency is down by 3 dB or 30% (typically). So, if you are expecting to see a 1 volt signal a 100 MHz, you will see 0.7 volts. People tend to miss this point. Remember that higher frequency harmonics are present in any nonsine wave. These can be reduced or eliminated because of a limited bandwidth. For example, a 100 MHz square wave will look very much like a 100 MHz sine wave on a 100 MHz oscilloscope. In the past, if all you ever wanted to do was audio or TV servicing, you could get by with a 5 to 10 MHz oscilloscope. Well, that was yesterday. Today's TVs, stereos, CD players, and all the other electronics are using more and more digital circuits and higher-frequency switching power supplies. And, don't forget remote controls, embedded microprocessors, digital signal processor and so on. You will need a bandwidth of at least 50 MHz for any digital work. This will show 20 ns glitches. For high-speed digital work, you should multiply your speed by four or five for your bandwidth. So, for a clock speed of 50 MHz, get a 200 to 250 MHz oscilloscope.

The usable "one-shot bandwidth" (for nonrepetitive signals) of a digital oscilloscope is about 20% to 25% of the sampling rate. This value is most compatible to the analog oscilloscope's bandwidth. However, digital scopes have the capability of constructing an image of the signal by taking multiple samples and adding them together. So the "repetitive signal bandwidth" may be much greater. Know the difference.

You should always get better equipment than you currently need. This is because it is engineering's nature to evolve. Things always work faster or with more precision today than they did yesterday. Getting a piece of test equipment that just meets your needs today will make it inadequate tomorrow. Then you will either have to buy a new instrument or try to use something that cannot provide you with the precision necessary. It makes more sense to budget a little extra now instead of paying a lot more in the near future. Good instruments are expensive. You want it to last awhile. When purchasing an oscilloscope, try to get a multitrace unit with dual time base (not just delayed sweep). It's almost like getting two oscilloscopes in one. The extra time base can be set to trigger on an event sometime after the first trigger. It makes seeing parts of complex waveforms easy. For example, suppose you want to look at the 10th scan line in a video signal. Just set up the main trigger on the retrace signal and adjust the delay for the second time-base trigger, which is on the sync signal. The trace will step from line to line with the turn of a dial.

If you buy a scope and intend to service it yourself, be aware that all scopes that use tubes or a CRT contain lethal voltages. Treat an oscilloscope with the same care you would use with a tube-type, high-power amplifier. The CRT should be handled carefully because if dropped, it will crack and implode, resulting in pieces of glass being blown around the room. Another concern when servicing an older scope is the availability of parts, which may become harder to find with increasing age. Be sure to purchase a used scope from a reputable vendor, try to get the operators manual and, if possible, try the scope out before you buy it.

The Function Generator

The function or waveform generator is one of the most versatile and useful pieces of test gear on your electronics workbench. It can generate a variety of precision waveshapes over a range of frequencies from hertz to megahertz. Function generators can provide a wide range of controlled amplitudes from a low-impedance source and maintain constant amplitude over a wide frequency range. Voltage control of frequency enables a source of swept frequency to be generated for frequency response testing. AM and FM modulation analyses can also be performed. Function generators are available in many different styles with many different features and prices, and although they are very important to a test bench, they can be very expensive. A neophyte to electronics often cannot afford to purchase one immediately, but in this chapter you will learn how you can build your own function generator at a fraction of the cost of a commercial function generator.

The function generator is generally regarded as the second most important piece of electronics test equipment after the oscilloscope. In this chapter, you will learn about applications and features of function generators and how they work, as well as what to look for when purchasing one and how to build your own low-cost generator. The term function generator is a relatively new term that was derived from what was called a signal generator. A signal generator in past decades was a name given to a generator of audio and often radio frequencies or RF. Older signal generators were classified as audio/RF generators. A distinction is now made between a signal generator, which is now called an RF generator and which produces RF frequencies from a few kilohertz to a few megahertz, and an audio generator, which is now incorporated into a function generator.

Audio generators generally produce audio frequencies in the audio range from 20 to 20,000 Hz, but some audio generators operate at higher frequencies. They modulation output is often just sine wave but sometimes square wave signals as well. Audio generators evolved into function generators, which produce sine, square, and triangle waves as well.

Basic models have seven or eight frequency ranges from 1 Hz to 1 mHz. Medium-priced function generators usually have a dc offset control, which adjusts the dc component of the output signal; an amplitude control, which adjusts the signal output level from the function generator; and a slope control, which controls the polarity of the signal. More expensive function generators also have sweep generator functions built in, which allows the output signal to be "swept" between a range of frequencies. Newer and more expensive function generators also have means to produce various standard pulses and the ability to control the rise and fall time of the pulses.

Function generators are now often called arbitrary function generators. This new term is used to define a new type of function generator, which not only produces various frequencies with different forms of modulation output, but allows you to specify or generate a particular or arbitrary signal function. This means that you can now create you own tailored waveform, using the internal microprocessor of the function generator. For instance, if you want to create a 1000 Hz square wave, you can specify the length of the pulse or wave and the starting and ending shape of the waveforms as well. Modern generators usually have digital readout of the frequency and they usually have a computer interface for data logging and control, a far cry from the old single-modulation-output audio generator of earlier days

All About Signal Generators and Waveform Generators

Signal generators and waveform generators are used to test and align all types of transmitters and receivers, and to measure frequency. Signal generators can use ac energy, audio frequency (af), and radio frequency (rf) to function. They are also used to troubleshoot various electronic devices and to measure frequency. The function of a signal generator is to produce alternating current (ac) of the desired frequencies and amplitudes with the necessary modulation for testing or measuring circuits. It is important that the amplitude of the signal generated by the signal generator be correct. In many signal generators, output meters are included in the equipment to adjust and maintain the output at standard levels over wide ranges of frequencies. When using the signal generator, the output test signal is connected to the circuit being tested. The progress of the test signal can then be tracked through the equipment by using electronic voltmeters or oscilloscopes. In many signal generators, calibrated networks of resistors, called attenuators, are provided. Attenuators are used in signal generators to regulate the voltage of the output signal. Only accurately calibrated attenuators can be used because the signal strength of the generators must be regulated to avoid overloading the circuit receiving the signal.

Signal generators and waveform generators typically come as a portable or benchtop instrument, a fixed instrument, or a PC-based instrument or module. Common generator types include continuous wave, function, pulse, signal, and sweep. A source that can produce a sine wave is referred to as a CW source. The frequency and amplitude of the sine wave can be set in most CW sources. Function generators create square waves by applying a bipolar sine wave to a comparator's input. A pulse is a short burst of signal(s) generated by the instrument. Signal generators output signals or sine waves that carry information. There are numerous methods for adding information to sine waves. Basic signal generators have frequency, amplitude, and phase modulation capabilities. More advanced signal generators have pulse and IQ modulation capabilities. A swept source adds the ability to automatically vary the output frequency or amplitude of a sine wave over a range of frequencies or amplitudes in a controlled manner.

Frequency characteristics that are important to consider when searching for signal generators and waveform generators include maximum input channels, frequency range, frequency resolution, frequency accuracy, and switching speed. Maximum input channels refers to the maximum number of all analog input channels, general and specific. The frequency range specifies the range of output frequencies the generator can produce. The frequency resolution is the smallest frequency increment the generator can produce. The generator's internal clock determines the frequency accuracy, which is a measure of how accurately the source frequency can be set. Operation features to consider include onboard reference, onboard oscillator, reverse power protection, and battery power. An onboard reference is a source of information, usually referring to the clock, which supplies timing information. An oscillator creates the basic electrical fluctuation (ac) that is used to create the waveforms. Reverse power protection prevents signals from traveling the wrong direction from damaging the source.

Additional specifications to consider when searching for signal generators and waveform generators include user interface, connections to the host, memory and storage, computer bus, display options, and environmental parameters.

Function generators usually fall into two basic categories: analog and digital. Analog generators use a voltage-controlled oscillator (VCO) to generate a triangular waveforms of variable frequency. Sinusoidal waveforms and square waves are generated from this. Digital generators use a digital-to-analog converter (DAC) to generate a waveshape from values stored in memory. Normally, such generators only offer sine and square waves up to the maximum generator frequency. Triangular waves and other waveforms are limited to a much lower frequency. A third type of generator uses digital techniques to control an analog VCO.

Analog function generators offer several advantages over their digital counterparts. They provide simple and instantaneous control of frequency and amplitude, and they do not have the high-frequency limitations on nonsinusoidal waveforms such as triangles and ramps that digital generators do. Starting prices for an analog generators are considerably lower than for digital generators.

Digital function generators normally derive waveform frequency from a crystal clock using a digital technique. Consequently, the frequency accuracy and stability will usually be higher than can be obtained from an analog generator. Digital generators may be able to generate a much greater number of standard waveforms than analog generators.

A variety of techniques can be used, one of the most versatile of which is called direct digital synthesis or DDS. What is DDS? Take a look at the DDS generator block diagram shown in Figure 4-1. Over the last few years, improvements in LSI logic, fast random access memories (RAM), and digital-to-analog converters (DACs) have made DDS the technology of choice for this application. There are three major components to a direct digital synthesizer or DDS: a phase accumulator, a sine look-up table, and a DAC. The phase accumulator computes an address for the sine table (which is stored in RAM). The sine value is converted to an analog value by the DAC. To generate a fixed-frequency sine wave, a constant value (called the phase increment) is added to the phase accumulator with each clock value. If the phase increment is large, the phase accumulator will step quickly through the sine look-up table, and so generate a high-frequency sine wave.

The frequency resolution of the DDS is given by the number of bits in the phase increment and phase accumulator. Lots of digital bits provide a very high frequency resolution. Some arbitrary waveform generators (AWBs) use a 48 bit phase accumulator for a frequency resolution of one part in 1014. This provides 1 Hz resolution at all frequencies from 1 Hz to 30 MHz. The maximum frequency depends on how fast you can add the 48 bit phase increment to the phase accumulator. Using a highly pipelined architecture, these additions can be performed at 40 MHz. This allows direct digital synthesis to 15 MHz. A frequency doubler is used to reach 30 MHz. For agile frequency and phase modulation, it is necessary to change the phase increment values quickly. To do this, the phase accumulator may switch between two 48

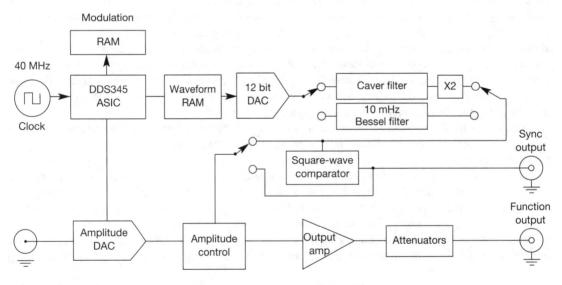

Figure 4-1　DDS generator block diagram.

bit phase increment values in 25 ns, and each of these 48 bit registers may be loaded in less than 1 s. During frequency modulation, one register is used while loading the other. A gate array does the 48 bit additions at 40 MHz and operates as a processor, modifying its control registers up to 10 million bytes per second, eliminating the traditional bottleneck that prevents rapid modulation of conventional direct digital synthesizers. The gate array also handles the trigger and counting logic for the arbitrary waveform functions.

The DDS generator offers not only exceptional accuracy and stability but also high spectral purity, low phase noise, and excellent frequency agility. A DDS generator can be swept over a much wider frequency range than an analog generator and can perform continuous-phase frequency hopping.

One of the disadvantages of digital function generators is that the maximum frequency for triangles and other nonsinusoidal waveforms is limited to a small fraction of the upper frequency for sine waves. This is related to the maximum clock rate combined with the filter characteristics. Rectangular waveforms can be generated from the sine wave using analog comparators and can therefore avoid this restriction, but performance limits will apply to pulse waveforms. Digital generators are more complex to use. This can be a drawback in simple or traditional test environments.

Arbitrary Waveform Generator

Since arbitrary waveform generators, commonly known as AWGs, are the future of waveform generators, we take a look to see what all the fuss is about. The AWG is definitely not your granddaddy's signal generator! It is a tool used to recreate real-world signals of all types, generally capable of generating standard periodic and unique complex waveforms with equal agility. Many types of single events can be precisely repeated using an AWG.

Arbitrary waveform generators can be used to create complex digital modulation base-band signals used in CDMA, W-CDMA, TDMA, and GSM cellular phone systems. They can also be used to generate digital TV signals. Signaling waveforms can be used for simulation of power disturbances to activate distribution relays or circuit breakers. Arbitrary waveform generators can also be used to test pacemaker operation, as well as testing airbag sensor signals that are initially recorded from actual automobile collisions.

How an Arbitrary Waveform Generator Works

In an arbitrary waveform generator, you define the waveform, using either the standard functions or custom profile data files to load into waveform memory. A set of start and stop addresses that corresponds to a group of waveform memory locations is assigned a waveform number. The address generator sequentially presents data values from each memory location to the digital-to-analog converter (DAC). The precision DAC converts the data into analog voltage values. This series of sequential voltage levels describes the output waveform with the frequency determined by the sample clock rate divided by the number of samples in the waveform. Changing the sample clock rate causes the address generator to change the speed at which the data is presented to the DAC, thereby changing the output frequency. Figure 4-2 is a block diagram of an arbitrary waveform generator

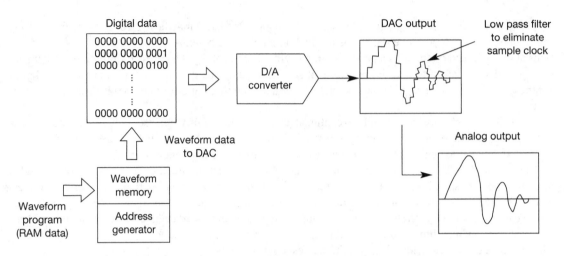

Figure 4-2 *Arbitrary waveform generator block diagram.*

Defining A Waveform Cycle's Frequency

All the data points in the specified waveform memory location make up one waveform cycle. The waveform generator will output all the points in the waveform at the sample clock rate specified. The resulting frequency is equal to the sample clock rate divided by the number of data points in the waveform. If multiple cycles of the waveform are entered into the same waveform memory location, the output frequency will increase proportionally to the number of cycles in memory. For example, if you create a triangular wave having three cycles using the same number of data points and the same sample clock, the frequency will be three times higher. The diagrams in Figures 4-3 and 4-4 show a formula for determining output frequency and an example frequency calculation, respectively.

The Sequence Generator

One of the more powerful options in an AWG is the sequence generator. A sequence is a stored program that sequentially selects (links) stored waveforms and repeats (loops) the waveform from 1one to over a million times. The process begins by selecting a waveform and repeating it a specified number of times. Then, the process continues synchronously, generating the next waveform a specified number of times. Each step in the sequence specifies a waveform and the number of repetitions. This process continues until all steps have been completed. At this point, the sequence generator may terminate the operation or continue it, depending on the mode, which may be single cycle or continuous.

Waveform generation can also be accomplished by utilizing a personal computer and a good sound card. A number of different types of waveforms can be generated using this technique and this method of waveform generation works reasonably well, but, unfortunately, the waveforms generated are limited to the audio range of a sound card, which is generally from 20 to 15,000 Hz.

$$f_o = \frac{f_{sc}}{n_s}$$

f_o = output waveform frequency
f_{sc} = sample clock frequency
n_s = number of samples in the selected waveform

Waveform frequency is dependent on both sample clock rate and number of samples.

Figure 4-3 Formula for determining output frequency.

Function Generator Applications

The function generator can perform many tasks around your electronics workbench. It can be used to:

1. Provide a signal to test an audio amplifier and its various stages of amplification.
2. Provide a test signal that can be inserted at various points in the radio circuit to determine whether a particular stage is working.
3. Provide a stable signal to align the intermediate frequency amplifier of a superheterodyne receiver and the radio frequency section of any receiver.
4. Generate complex waveforms and pulses.
5. In some instances, the function generator can serve as a wireless broadcaster.
6. Test pulse-shaping circuits to see if they are working.
7. Test all types of filter circuits.

Function Generator Terms

There are two basic types of alternating-current signals or waveforms: periodic and nonperiodic. Periodic signals, such as sine or triangular waves, behave in a uniform manner and repeat themselves over a given

Example

f_{sc} = 20 MHz
n_s = 32 samples

$$f_o = \frac{20\text{ MHz}}{32} = 625\text{ kHz}$$

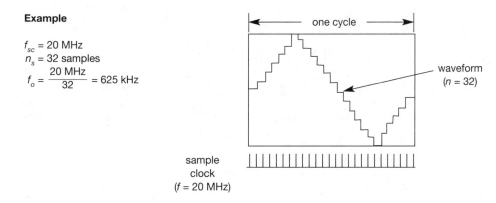

Figure 4-4 Example frequency calculation.

length of time. Each repetition of a repeating signal is called a period or a cycle. Nonperiodic signals, such as those generated by analog devices, behave in a nonuniform manner and do not repeat themselves over any given length of time. When working with ac signals, there are three properties that most people are concerned with: amplitude, period, and frequency.

The amplitude of a wave is defined as the maximum magnitude of the wave. The amplitude is the vertical component of the signal and is measured in units of volts (V). Since we are dealing with an ac signal, the voltage will change over a period of time. The maximum voltage of a signal during its cycle is commonly referred to as the peak voltage. The amplitude can be measured from the reference line to the peak (V_p) or from peak to peak (V_{pp}) (see Figure 4-5). With a periodic signals that is symmetrical (equidistant above and below the reference point), the peak-to-peak voltage is equal to twice the peak voltage.

The period (T) of the signal is defined as the time it takes for a signal to complete one full cycle. The period is the horizontal component of the signal, measured in units of seconds (s). In Figure 4-5, the period of the signal is measured as 250 milliseconds (250.0×10^{-3} s). The frequency (f) of a signal is defined as the rate at which a periodic signal repeats. It is usually measured in units of hertz (Hz), where 1 Hz = 1 cycle per second. In Figure 4-6, you can see four (4) cycles occurring within one second; therefore, the signal has a frequency of 4 Hz. The frequency, f, of a wave is inversely related to its period (T). Table 4-1 is a scientific-prefix conversion chart.

Example. The period of a signal is 250 milliseconds; therefore, the frequency of that signal is:

$$f = 1/T$$

$$f = 1/250 \text{ milliseconds}$$

$$f = 4 \text{ Hz}$$

The duty cycle of a square waveform signal is the ratio of the high signal value to the low signal value time the intervals within the signal's period. The dc offset of a periodic voltage/current signal is the quan-

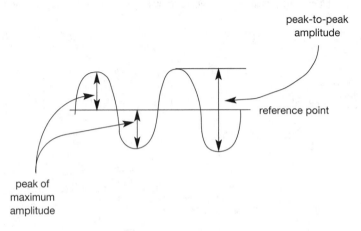

Figure 4-5 *Amplitude measurement.*

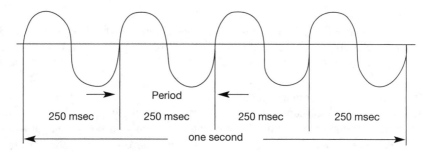

Figure 4-6   *Period measurement.*

tity representing the average signal value calculated over one period of the signal (constant voltage/current component of a voltage/current waveform).

Typical Function Generator Controls

Prior to connecting your function generator to a circuit or oscilloscope, you will need to properly adjust the controls. The controls listed below are typical and represent controls found on most function generators or signal generators. More expensive function generators will often have pulse output capabilities. Function generators with pulse outputs may have additional controls such as trigger/slope, sweep rate, and sweep width controls.

To operate a basic function generator,

1. Turn on the power switch.
2. Set the dc offset, which is used to regulate the amount of dc offset added to the signal, to 0 or no offset.
3. Set amplitude, attenuator, or output level (the signal output control) initially low, to 0 dB or –20 dB.
4. Select the function or signal waveform: sine wave, square wave, or triangular wave.

Table 4-1   *Scientific-prefix conversions*

Abbreviation	Prefix name	Factor
T	Tera	10^{12}
G	Giga	10^{9}
M	Mega	10^{6}
K	Kilo	10^{3}
m	milli	10^{-3}
μ	micro	10^{-6}
n	nano	10^{-9}
p	pico	10^{-12}

5. Set the frequency to the one that you want.
6. Select the Multiplier or Range switch. this determines if you want 10 Hz or 100 Hz, and so on.
7. Set the TTL/CMOS control to the logic type that your are using. The TTL/CMOS control may not be on all Function generators.

Most function generators can produce an output from 0 to 10 volts. Make sure that when you initially set up the generator, the amplitude control is first set to zero and ramp it up slowly so that you do not destroy the input to the circuit under test. Function generators generally have BNC output jacks and have a 50 ohm output impedance, which should me maintained. Inexpensive function generators may have only one single output jack, whereas more expensive generators may have a number of inputs and outputs such as time base, sweep, trigger, and Sync.

Waveform Generator Purchase Considerations

If you are thinking of purchasing a new waveform or function generator, decide what features are important to you, and how much money you can afford to spend. It is often not wise to purchase the least expensive function generator you can find. Try instead to look for a waveform generator that will serve you over many years. Likewise, it is nice to have the best or the latest and greatest model but who has the money to purchase a $4000 generator. Buy the best you can afford with the most features that you think you will need for a reasonable amount of money. At minimum, be sure that the function generator can generate sine, square, and triangular waveforms and has attenuation or amplitude and dc offset controls. A function generator that can produce some types of pulses would also be desirable as well. You can also buy or build a separate pulse generator if your function generator does not have pulse capability (see Chapter 16). Look at different specification sheets and compare features and functions. Consult with an experienced repair technician or a ham radio operator.

Consider the following when looking to purchase a new or used function generator:

Purity—does the generator produce harmonics in addition to the desired output signal?
Linearity—refers to the quality of delivering identical and accurate signals throughout a range of frequencies.
Precision—how many digits or places does the digital display have?
Accuracy—when checked against a standard, how close is your generator to the standard?
Calibration—if used, when was the unit calibrated? How was it calibrated?
Drift—does the generator remain on frequency at all times and from onset of power-up?
Display—does the generator have a frequency display? This is very useful if you do not have a frequency counter available.
Condition—how does the generator look? is it rusty, dirty, and so on.

If you are looking at the possibility of purchasing a used function generator, be aware of a few caveats. Look for one in good cosmetic condition. If the case is not rusty, chances are the inside will be clean and not badly corroded and the switches will still work. Get one with a manual: having the detailed instructions and a schematic can make using and maintaining it much easier. If it does not have a manual, advertise in *Radio Age* or *Antique Radio Classified* for a photocopy. Older tube-type generators will tend to drift quite a bit as they are warming up, so it is recommended to avoid equipment that is too old. Transistor and integrated circuit function generators will not have those types of problems. See Chapter 15 for a discussion on buying new test equipment versus used equipment.

Build Your Own Function Generator

The function generator presented in this section will not produce arbitrary waveforms but it will produce the three basic waveforms used most commonly in testing equipment. Our function generator will produce sine, square, and triangular waveforms from 20 Hz to 200 kHz in four stepped ranges, and with a little modification can be used to produce a swept-frequency output.

The heart of our function generator is the special purpose Exar XR-8038A generator chip, illustrated in the block diagram shown in Figure 4-7. The XR-8038A chip is a precision waveform generator integrated circuit capable of producing sine, square, triangular, sawtooth, and pulse waveforms, with a minimum number of external components and adjustments. Table 4-2 is the pin-out chart for the XR-8038A function generator chip. The XR-8038A allows the elimination of the external distortion adjusting resistor, which greatly improves the temperature drift of distortion, as well as lowers external parts count. Its operating frequency can be selected over eight decades of frequency, from 0.001 Hz to 200 kHz, by the choice of external R-C components. The frequency of oscillation is highly stable over a wide range of temperature and supply-voltage changes. Both full frequency sweeping as well as smaller frequency variations (FM) can be accomplished with an external voltage control. Each of the three basic waveform outputs, (sine, triangular, and square) are simultaneously available from independent output terminals. The XR-8038A monolithic waveform generator uses advanced processing technology and Schottky-barrier diodes to enhance its frequency performance.

The precision waveform generator produces highly stable and sweepable square, sine, and triangular waves across eight frequency decades. The time base of the chip uses resistors and a capacitor for fre-

Table 4-2 XR-8083 pin-out table

Pin #	Symbol	Type	Description
1	SA2	I	Waveform adjust input 1
2	SWO	O	Sine wave output
3	TWO	I	Triangular wave output
4	DCA1	I	Duty cycle adjustment input 1
5	DCA2	I	Duty cycle adjustment input 2
6	Vcc		Positive power supply
7	FMBI	I	Frequency modulation
8	FMSI	I	Frequency sweep input
9	SQO	O	Square wave output
10	TC	I	Timing capacitor
11	Vee		Negative power supply
12	SA2	I	Waveform adjust input 2
13	NC		No connect
14	NC		No connect

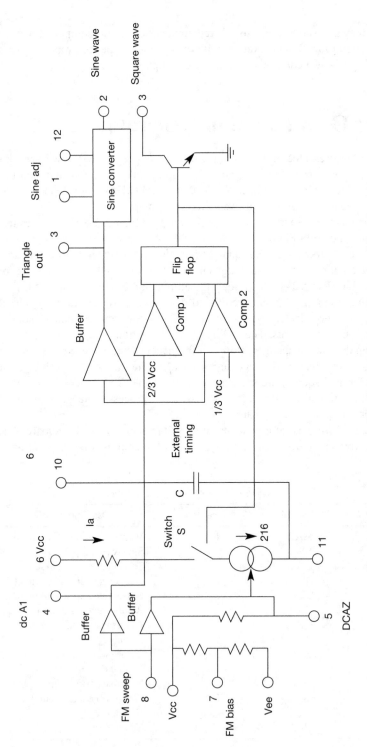

Figure 4-7 Block diagram of Exar XR-8038A chip. (Courtesy of Exar Corp.)

quency and duty cycle determination. The generator contains dual comparators, a flip-flop driving switch current sources, buffers, and a sine wave converter. The three frequency outputs are simultaneously available. Supply voltage can range from 10 to 30 volts or ± 5 to ± 15 volts. Small frequency deviations or FM is accomplished by applying modulation voltage between pins 7 and 8, large frequency deviations or sweeping is accomplished by applying voltage to pin 8 only. Sweep range is typically 1000:1.

The basic function generator circuit is depicted in Figure 4-8. The function generator revolves around U1, the Exar XR-8038A function generator chip, and an LF351 output amplifier chip. All of the waveforms are generated by U1. The frequency of the internal square wave generator is controlled by timing capacitors C2, C3, C4, and C5, the 10 k potentiometer at R6 and the range switch at S1, a four-position, single-pole rotary switch. The square wave output is differentiated to produce triangular waves, which in turn are used to shape the sine wave output. The purity of the sine wave output is then controlled by the two 100 kohm resistors at R1 and R10.

The wave shape selection is controlled by the shape switch at S2, a three-position rotary switch. The triangular wave is selected at the "a" position of switch S2, and the sine wave output is selected at the "b" position. The square wave function is selected at the "c" position of S2. Resistor R11 controls the overall waveform amplitude.

Integrated circuit U2 is an LF351 op-amp that is configured as a direct-coupled, noninverting buffer, thus providing isolation between the waveform generator while increasing the output current. Resistors R13 and R14 form the output attenuator. At the high output at J1, the maximum output is about 8 volts peak to peak with the square wave. The maximum for the triangular waveform output is about 4 to 6 volts. The low output at J2 is approximately 20 mV to 50 mV.

Power to the function generator is supplied through the two LM7805, 5 V, fixed-voltage regulators. The output of the two regulators are filtered by 100 μf capacitors. The input voltage to the regulators is provided from transformer T1 through two general-purpose silicon diodes, which convert the ac voltage from the transformer to the dc voltage that feeds the regulators. Transformer T1 is a 100 V primary to 6 V secondary, capable of producing 500 mA of current.

The function generator prototype was constructed on a small printed glass epoxy circuit board, which measured 5″ by 6″. When constructing the function generator circuit, be careful to observe the polarity of the capacitors when installing them on your circuit board, to avoid damaging the circuit when it is first powered up. When installing the diodes, be aware of the correct orientation before soldering them into the circuit board. The use of integrated circuit (IC) sockets for the two integrated circuits is a prudent choice, in the event of a circuit failure at a later date. When installing the ICs into their respective sockets, be sure to orient them correctly. There is often a notch or indentation at the top of the IC, near pin one, or a sometimes there is an indented circle at pin one. If the notch is at one end of the IC, then usually pin one is to the left of the center notch. The two regulator integrated circuits that supply power to the generator must be installed with the correct orientation to avoid damage to the circuit. The power transformer is mounted on the bottom-left section of the chassis box. Next, you will need to wire the circuit board to the controls, which will be mounted on the front panel of the chassis box. A parts list for the function generator follows:

R1, R10	100 kohm potentiometer (PC) mount
R2, R12	10 kohm ¼ W resistor
R3, R4, R13	2.2 kohm ¼ W resistor
R5	10 megohm ¼ W resistor
R6, R11	10 kohm potentiometer (PC) mount

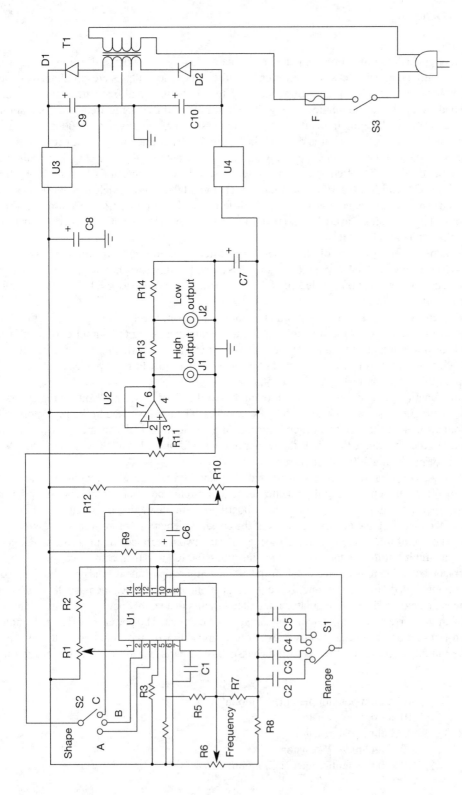

Figure 4-8 Basic function generator circuit.

R7	4.7 megohm ¼ W resistor
R8	22 kohm ¼ W resistor
R9	2.7 kohm ¼ W resistor
R14	47 kohm ¼ W resistor
C1	1 μf 25 V electrolytic capacitor
C2	220 nF polyester 30 V capacitor
C3	22 nF polyester 30 V capacitor
C4	2200 pF polystyrene 30 V capacitor
C5	220 pF polystyrene 30 V capacitor
C6 -	47 μF, 30 V electrolytic capacitor
C7, C8	100 μF 30 V electrolytic capacitor
C9, C10	1000 μF 30 V electrolytic capacitor
D1, D2	1N4001 silicon diodes
U1	EXAR XR-8038A function generator
U3	LF 351 op-amp
U3, U4	LM 7805 regulator
T1	110 V ac–6 V ac, 500 mA transformer
S1	Four-position rotary switch
S2	Three-position rotary switch
S3	SPST power switch
J1, J2	RCA jacks
F1	0.5 amp fuse
Miscellaneous	PC board, chassis, wire, and so on

The finished board is shown in Figure 4-9. The diagram in Figure 4-10 depicts the front panel layout of the function generator controls. The on–off switch, frequency, range switch, attenuator, and so on were all

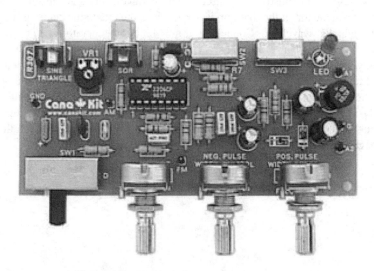

Figure 4-9   Finished function generator board.

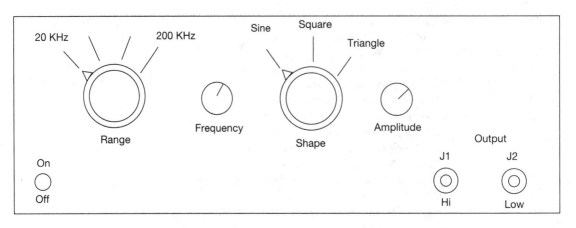

Figure 4-10 Function generator front panel.

mounted on the front panel of the chassis box. The power cord and a fuse holder were mounted on the rear of the chassis panel. This is an excellent time to look over your circuit board for any "shorts" caused by stray component leads. Look for solder bridges between circuit lands, and so on. Your function generator circuit is now ready to take its place on your electronics workbench. Now is a good time to test it. First, make sure that the power switch is turned off, then plug the power cord into the ac outlet. Connect a cable from the generator's output jack on the front panel to an oscilloscope. Turn on the function generator circuit with switch S3, and adjust the controls to product a 1–10 kHz sine wave output signal. Adjust the attenuator control and then look for a waveform on your oscilloscope. If all is well, you should see a sine wave on the scope. You may want to locate an inexpensive X1 oscilloscope probe to act as an output device for your new function generator.

The diagrams in Figure 4-11 illustrate the three different waveforms that the function generator will produce. The waveform at the top of the diagram is a sine wave, the one in the center is a square wave, and the bottom one is a triangular waveform. Connect your new function generator to an oscilloscope and you should be able to duplicate these waveforms.

When setting up your new function generator to inject a test signal into an unknown circuit for the first time, you will want to be sure to have the output signal or attenuator control on the function generator set to minimum. Use the minimum output from the function generator to inject a signal into the circuit "under test." Gradually increase the signal if needed as you watch the signal propagate through the circuit "under test" on your oscilloscope. You do not want to inject a distorted signal through the circuit "under test" or cause damage to the circuit that you are trying to test.

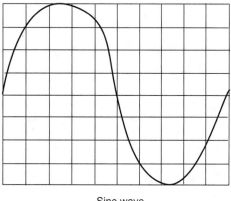

Sine wave

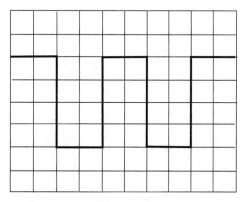

Square wave

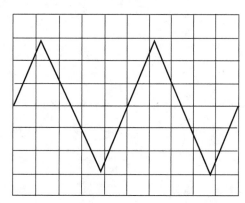

Triangular wave

Figure 4-11 Function generator output waveforms.

Frequency Counter

Inexpensive frequency counters can often measure frequencies between 500 to 1000 MHz or better, and are available to the hobbyist today for as little as $100. As you build your electronics workbench and as funds become available, you should definitely consider purchasing a good frequency counter. It will become one of the most indispensable pieces of test equipment on your bench.

A frequency counter is an excellent means of accurately determining the frequency of unknown signals, to test for frequency response, to see if an oscillator or a multiplier stage in a receiver or transmitter is working, or for testing frequency response of equipment.

Modern integrated circuit technology and microprocessors, as well as quality high-frequency components, have permitted the increased use of VHF, UHF, and SHF spectrum use for commercial radio, TV, and amateur radio use. With present technological demands, there is a much greater need for both accuracy and stability. This trend in increasing need for precision has increased over the years. As an approximation, the level of frequency precision required for modern electronics and radio gear has increased almost one power of 10 per decade since the 1950s!

Older frequency counter technology was based on high-chip-count, synchronous-decade counting techniques. The new frequency counter technologies have much lower chip counts, more accurate frequency counting, are more reliable, and cost considerably less than the older technologies. A basic frequency counter block diagram is shown in Figure 5-1.

In this chapter, we will look at frequency counter architecture, methods of counting, frequency counter terms, accessories, and, finally, you will also learn how to build your own low-cost frequency counter.

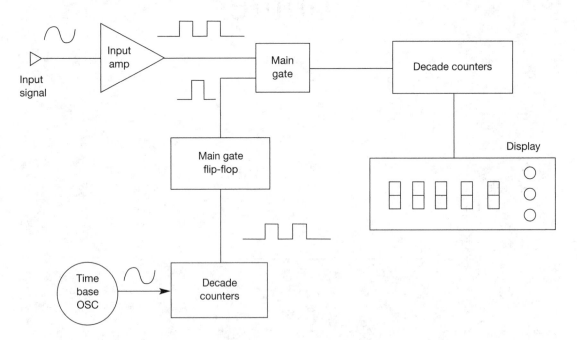

Figure 5-1 Basic frequency counter block diagram.

Frequency Counter Architecture

Radio frequency counters fall into two basic categories, and understanding the two different approaches will help you choose the correct counter and use it correctly. Direct counters simply count the number of times the input signal crosses zero during a specific gate time. The resulting count is sent directly to the counter's readout for display. This method is simple and inexpensive, but it means that the direct counter's resolution is fixed in hertz. For example, with a 1 second gate time, the lowest frequency the counter can detect is 1 Hz (since 1 zero crossing in 1 second is 1 Hz, by definition). Thus, if you are measuring a 10 Hz signal, the best resolution you can expect for a 1 second gate time is 1 Hz, or two digits in the display. For a 1 kHz signal and a 1 second gate, you get four digits, for a 100 kHz signal, six digits, and so on. Figure 5-2 illustrates this relationship. It is also interesting to note that a direct counter's gate times are selectable only as multiples and sub-multiples of 1 second, which could limit your measurement flexibility.

Reciprocal counters, in contrast, measure the input signal's period, then reciprocate it to get frequency. Thanks to the measurement architecture involved, the resulting resolution is fixed in the number of digits displayed (not Hertz) for a given gate time. In other words, a reciprocal counter will always display the same number of digits of resolution regardless of the input frequency, as shown in Figure 5-3. Note that you will see the resolution of a reciprocal counter specified in terms of the number of digits for a particular gate time, such as "10 digits per second."

Using the counter industry's benchmark of a 1 second gate time, Figure 5-4 compares the resolution of direct and reciprocal counters. You can readily see that the reciprocal counter has a considerable advantage over the direct counter in the lower frequencies. As an example, at 1 kHz, a direct counter gives

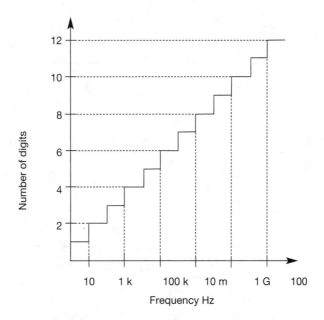

Figure 5-2 Direct count method. Number of digits versus frequency.

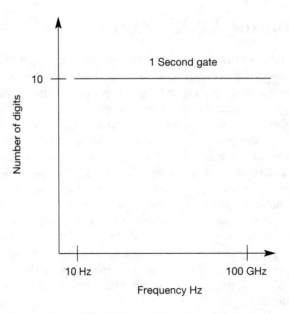

Figure 5-3   Reciprocal counter resolution.

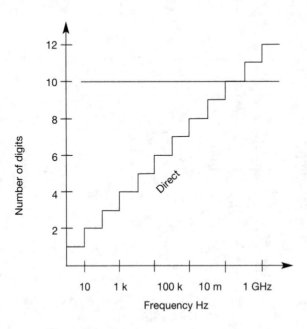

Figure 5-4   Direct versus reciprocal counters.

a resolution of 1 Hz (four digits). A 10 digit/second reciprocal counter gives a resolution of 1 μHz (10 digits).

Even if you don't need microhertz resolution, the reciprocal counter still offers a significant speed advantage: the reciprocal counter will give 1 mHz resolution in 1 ms, whereas a direct counter needs a full second to give you just 1 Hz resolution (see Figure 5-5). Reciprocal counters also offer continuously adjustable gate times (not just decade steps), so you can get the resolution you need in the minimum amount of time. The choice comes down to cost versus performance. If your resolution requirements are flexible and you aren't too concerned with speed, a direct counter can be an economical choice. For fast, high-resolution measurements, though, a reciprocal counter is the way to go.

You can figure out whether a counter is direct or reciprocal simply by looking at the frequency resolution specification. If it specifies resolution in hertz, it's a direct counter. If it specifies resolution in digits-per-second, it is a reciprocal counter.

The flexibility of microprocessor-based design permits a frequency counter to measure using both the direct and reciprocal methods of counting, as shown in the block diagram of Figure 5-6.

Choosing the Appropriate Time Base

Measurement accuracy in frequency counters begins with the time base because it establishes the reference against which your input signal is measured. The better the time base, the better your measurements can be. (Notice the "can be" part here; you still have to calibrate and take care of your counter to maxi-

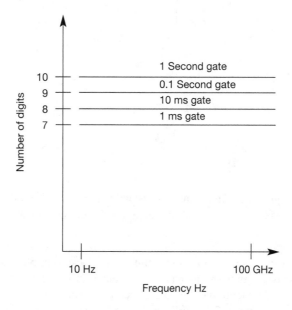

Figure 5-5 *Gate times versus resolution.*

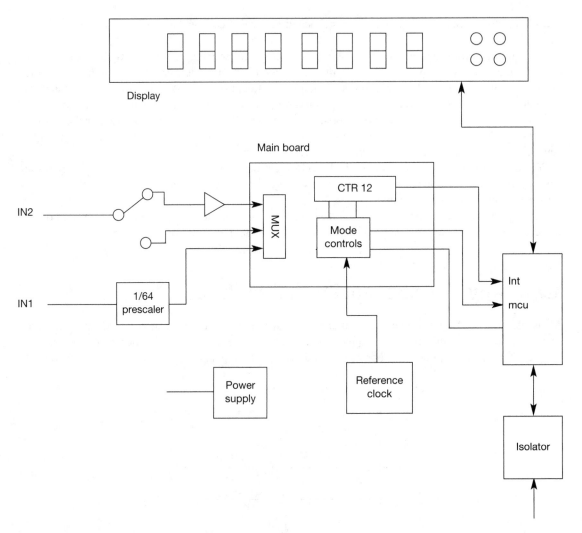

Figure 5-6 Microprocessor frequency counter.

mize performance.) The frequency at which quartz crystals vibrate is heavily influenced by ambient temperature, and time-base technologies fall into three categories based on the way they address this thermal behavior:

1. Standard. A standard or "room temperature" time base doesn't employ any kind of temperature compensation or control. Although this has the advantage of being inexpensive, it also allows the most frequency errors. As the ambient temperature varies, the frequency output can change by 5 parts per million (ppm) or more. This works out to ± 5 Hz on a 1 MHz signal, so it can be a significant factor in your measurements.
2. Temperature-compensated. One way to deal with the crystal's thermal variation is to make sure that

the other electronic components in the oscillator circuits have complementary thermal responses. This approach can stabilize the thermal behavior enough to reduce time-base errors to around 1 ppm (± 1 Hz on a 1 MHz signal).

3. Oven-controlled. The most effective way to stabilize the oscillator output is to simply take the crystal off this thermal roller coaster. Counter designers do this by isolating the crystal in an oven that holds its temperature at a specific point in the thermal response curve. The result is much better time-base stability, with typical errors as small as 0.0025 ppm (± 0.0025 Hz on a 1 MHz signal). There is more to this story than just temperature-related accuracy, however. Oven-controlled time bases also help with the effects of crystal aging, which means you don't have to take your counter out of service for calibration as often. The optional high-stability oven reduces this to < 0.015 ppm (± 0.015 Hz on a 1 MHz signal) per month. In other words, the standard time base ages 20 times faster than the high-stability model, and will therefore require calibration more frequently to maintain your required measurement accuracy.

Differences Between Resolution and Accuracy

Equating resolution with accuracy is a common mistake. They are related, but they are distinctly different concepts. Resolution can be defined as the counter's ability to distinguish closely spaced frequencies. All other things being equal (such as measurement time and product cost), more digits are better but the digits you see on the display need to be supported by accuracy. Digits can be deceptive when other errors push the counter's resolving ability away from the actual frequency. In other words, it is possible for a counter to give you a very precise reading of an incorrect frequency. True measurement accuracy is a function of both random and systematic errors. Random errors, which are the source of these resolution uncertainties, include quantization error (the uncertainty surrounding the final count in the gate time window), trigger error (such as triggering on noise spikes), and short-term instabilities in the time base. Systematic errors are biases in the measurement system that push its readings away from the actual frequency of the signal. This group includes effects on the time-base crystal, such as aging, and temperature and line voltage variations. Compare the two counters in Figure 5-7. Counter A has good resolution but a serious bias error, so its displayed result in most cases will be less accurate than those of Counter B, which has poorer resolution but a smaller systematic bias error.

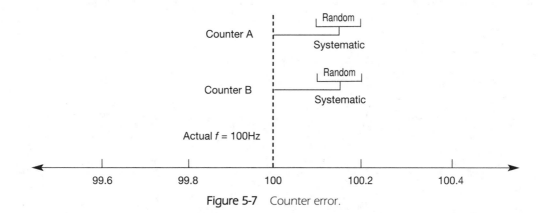

Figure 5-7 Counter error.

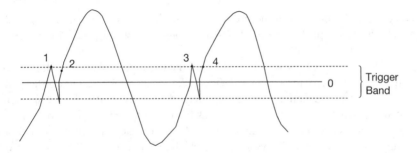

Figure 5-8 Spurious frequency count.

How to Adjust Sensitivity to Avoid Noise Triggering

The good news is that high-quality counters are broadband instruments with sensitive input circuits. That's also the bad news. To a counter, all signals basically look the same. Sine waves, square waves, harmonics, and random noise all just look like a series of zero crossings as far as the counter is concerned. A counter figures out the signal's frequency by triggering on these zero crossings to measure frequency. If your signal is clean and uncluttered, the process works quite well. Noisy signals, however, can trick the counter into triggering on spurious zero crossings. When this happens, the counter doesn't count what you think it is counting. Fortunately, all good counters offer a way around this problem. First, they require the signal to pass through both lower and upper hysteresis thresholds before they register a zero crossing. The gap between these two levels is called trigger sensitivity, the hysteresis band, the trigger band, or something similar. Second, good counters let you adjust this band to minimize unwanted triggering. Figure 5-8 shows a signal with some spurious components that are causing trouble with the count. The trigger band is fairly narrow, so both the unwanted noise (at points 1 and 3) and the real signal (at points 2 and 4) cause the counter to trigger. What is really just two cycles of the signal get counted as four. By adjusting the trigger band to make the counter less sensitive, you can avoid these spurious triggers. In Figure 5-9, the trigger band is wide enough (which is to say the sensitivity is low enough) that the spurs don't get counted as zero crossings. The counter registers two valid zero crossings and goes on to compute the appropriate frequency. If you think your signal might have some noise problems, try switching your counter into low-sensitivity mode. If the displayed frequency changes, chances are you were triggering on noise.

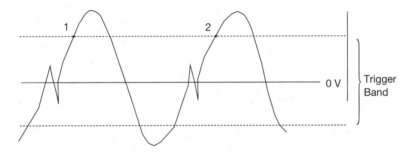

Figure 5-9 Desired frequency count.

The Problem of Low-Frequency Measurements

The problem of triggering on unwanted frequency components in the signal you are trying to measure can be even more acute with low-frequency signals (roughly 100 Hz and below), since the chance of spurious triggering on irrelevant high-frequency components is that much greater. In addition, the signal's slew rate affects trigger accuracy; the lower the slew rate, the more chance there is for error. Here are three quick steps you can take to help improve the quality of counter measurements on low-frequency signals.

1. Use the counter's low-pass filter, assuming that your counter has one. This reduces the chance of triggering on harmonics and high-frequency noise.
2. Use manual triggering. When a counter is set to use autotriggering, it estimates the peak-to- peak level of the signal and computes the midpoint to establish a trigger level. This approach generally leads to good results but can cause trouble on low-frequency signals. The problem is that the autotrigger algorithm can take less time to run than the signal takes to transition between its minimum and maximum values. As a result, the autotrigger can wind up following the signal level up and down, rather than setting a single trigger level based on a consistent estimate of the minimum and maximum values. The solution is to turn off autotrigger and set the trigger level manually.
3. Use dc coupling Many counters offer a choice between ac and dc coupling on their primary input channel. This setting works the same way on a counter as it does on an oscilloscope: ac coupling removes any dc offset from the signal, whereas dc coupling admits the entire signal, offset and all. The trouble with ac coupling is that it also attenuates lower frequencies. In fact, your counter's performance probably isn't even specified below a certain frequency when ac coupling is used. To ensure better results all the way down to fractions of a Hertz, use dc coupling instead.

Generally, reciprocal counters are not plagued by the problems of low-frequency counting and are often chosen by persons working exclusively with low-frequency signals.

Frequency Counter Measurement Techniques

Naturally, as with all types of measurement, you must be careful when using a frequency counter as well as aware as to what is really being measured and exactly what the counter is truly "seeing." Often, a counter will give a false reading that is not generally the fault of the counter. After all, a frequency counter is a device that ideally reads events per unit time. For sine or square waves, or pulse train signals encountered in RF and digital work, ideally the counter is reading the number of cycles per second, or level transitions in a particular direction. However, for complex waveforms, especially if random signals such as noise pickup, video or audio program components, or other extraneous frequencies are present at the counter input, the counter may produce a reading entirely different from what was expected. Furthermore, the reading may jump wildly around, since a frequency counter is counting transitions in level about a reference point (generally, zero crossings) and a waveform containing several frequency components, especially if nonharmonically related or not phase coherent, will have many possible zero crossings depending on the amplitudes of all the frequency components present at the counter input. Due to the counter frequency response, it may favor some of these components over others as the counter may have much higher sensitivity at the favored frequencies. A counter connected to a circuit point containing both RF and audio signal may not see the audio component if the counter preamplifier cuts off below, say, 1 MHz, as some of the small hand-held counters do.

So if you observe wildly varying counter reading that are way off the "mark" of what you are expecting, this is a possible sign of the presence of interfering signals. To improve the reading, you can also reduce the number of displayed digits in one of three ways. The first option is simply to reduce the number of digits displayed using the "fewer digits" button or whatever that function is called on your counter. Although this can quiet the display, it might hide information you need to make decisions about circuit behavior. Note that this is strictly a display function that doesn't have any effect on the actual measurement. The second method is to use limit testing with a visual indicator, if you just need to know whether the signal is within a certain band of frequencies. The catch here, of course, is that your counter needs to have this feature before you can use it. The third method is to use signal averaging. Averaging (labeled "mean" on many counters) is a good option to consider any time your signal is jumping around. Unlike simply reducing the number of displayed digits, of course, averaging actually improves the quality of your measurements. By reducing the effects of random variations in the signal, it reduces the number of display changes. Also use low input signals levels.

A proper reading is generally steady and repeatable. Note that especially at VHF and UHF, normal frequency instability as encountered in free-running oscillator circuits may cause a slow, steady drifting in the last few digits. These will be somewhat predictable after watching the change for a minute or so. Random jumping is a sign of interference or insufficient signal, or the presence of excessive phase noise. However, in crystal-controlled circuits this should not be occurring, except in the last digit or so, due to normal uncertainty in the least significant bit.

Frequency Counter Terms

Stability

Stability is a measure of how well the equipment stays where you set it. It is defined for crystals and oscillators as the change in operating frequency over the quoted temperature range. Modern synthesized rigs rely on one or more crystal oscillators, so (in comparison to free-running oscillators) they are reasonably stable. The best equipment refers all frequency generation to a single high-stability oscillator, chosen for the best possible results.

Accuracy

Accuracy describes how well the equipment is calibrated. Accuracy is of lesser importance than stability, as there is no point in calibrating equipment carefully if it won't be at the same setting next time you check! If the equipment is stable but inaccurate, you can at least write down the calibration error and subtract it from your readings.

Aging

All oscillators change frequency slowly with time and this is known as aging. Crystals that vibrate mechanically will undergo molecular-level changes that minutely affect the oscillation frequency, so even if nothing else changes (oscillator supply voltage, capacitance, temperature), the crystal will slowly change in frequency, usually increasing in frequency. Old oscillators are frequently better in this regard as aging is fastest with new oscillators. Cesium and rubidium Standards are used commercially, in expensive equipment because they have very low aging rates.

Calibration

The process of referring (or transferring) a reference to a standard, for the purpose of determining the offset in the standard from the reference, is called calibration. The standard might be adjusted to higher accuracies, or, since adjustments can become quite difficult, the offset might simply be noted and taken into account during subsequent measurements. A "transfer standard" is a calibrated standard that is portable and is taken to other standards for calibration. The alternative method is to receive the reference (typically by radio), and perform the calibration remotely.

Oven-Controlled Xtal Oscillator

A crystal oscillator in an oven is called an oven-controlled Xtal oscillator or OCXO. Crystals designed for use in ovens have a thermal characteristic that has a plateau at the oven temperature; hence, they are very stable, provided the oven temperature is stable. Operating temperatures are typically 70–80°C, above even the highest ambient temperatures. The best OCXOs are the "double oven" type, with the oscillator and crystal in an inner oven having proportional control, and the whole inside an outer oven. Even the cheaper OCXOs can benefit from good thermal lagging provided by a polystyrene cover. OCXOs have slow warmup and use more power than other types. OCXO stabilities can be as high as 1 in 108, with aging rates as low as 1 in 109 per day.

Temperature-Compensated Crystal Oscillator

A temperature-compensated crystal oscillator or TCXO is a type of oscillator with performance superior to a normal crystal oscillator, since the temperature variations are corrected by a thermistor and varicap (variable capacitance) diode. Some of the better units use quite sophisticated control methods. TCXOs use much less power than OCXOs, so they are good for portable equipment, but the stability is more modest—1 in 106 to about 2 in107.

Voltage-Controlled Crystal Oscillators

Voltage-controlled crystal oscillators or VCXOs are often used in phase-locked devices and other devices where a crystal oscillator is "steered" to follow another standard (the technical term is "disciplined"). A varicap (variable capacitance) diode is used to alter the crystal frequency very slightly. Some TCXOs and OCXOs have voltage control, and so offer better performance.

Offset

The difference between a standard (or equipment being measured) and a reference is called the offset. This is also usually quoted in ppm and represents the current working error correction due to all error sources. You need to work this out just before and just after making any critical frequency measurements, and use the mean to correct the readings.

Parts per Million

Frequency accuracy, stability and aging are typically quoted in "ppm " (parts per million). This is because invariably these factors are dependent not on any error in hertz, but in the amount of error per hertz. Although this could be expressed as a percentage, the very small numbers that would result are not convenient. Hence, the error is typically expressed in parts per million. An error of 1 ppm is the same as 0.0001% or 1 part in 106. Reference sources are quoted with errors as small as 0.000001 ppm, or 1 part in 1012. Typically, the accuracy, stability, and aging rate per week are numerically similar. For example, a 0.1 ppm stability TCXO will have a calibrated accuracy of around 0.1 ppm and an aging rate of less than 0.1 ppm/week.

Reference

A device with higher (and traceable) inherent accuracy than the standards it calibrates is called a reference. Invariably, the reference will be traceable to a national or international standard, and must have at least one power of 10 better performance (accuracy, stability) than the devices it calibrates. If the reference is received by radio, its received version must have the necessary stability.

Standard

A calibrated oscillator used to measure other oscillators is known as a standard. Normally, the standard will be on a "round" frequency such as 1, 5, or 10 MHz. The standard may be built into equipment such as a frequency counter or signal generator, or be in a separate box simply used as a signal source. The standard must, of course, have suitable stability, low aging, and be regularly calibrated.

Frequency Counter Accessories

The features and accessories described in this section are items to look for when choosing a used counter or when purchasing a new frequency counter. These features aid in the measurement process and make the job of frequency determination much smoother and more accurate. Keeping these facts in mind may save hours of frustrating labor wasted in the efforts to find circuit problems that do not exist. The correct coupling technique is the one that produces a steady, reliable reading in the expected frequency range you are trying to measure.

Whip Antenna

A whip antenna is useful for sampling the near field of a transmitter antenna. Although supplied as an included accessory on many counters, the antenna is almost useless for any other purpose. Whips are only useful in cases where it is definitely known that the antenna is in an area that is dominated by one large, strong signal that is at least 10 to 20 dB stronger than anything else. Most counters have short whips and these act as high-pass filters that cut off below 30 to 50 MHz; hence, the whip is best used for VHF and UHF measurements, as weaker VHF and UHF signals may dominate strong signals in the lower HF range (below 10 MHz)

Probe or Clip Leads

Probe or clip leads are good for general work in the HF range up to about 50 MHz or so. If connections are kept short, they are fairly reliable and foolproof, especially at audio and lower RF frequencies. They may cause loading down and detuning of sensitive circuits such as oscillators, and the frequency read with the counter may not be the actual frequency due to this detuning. They are poor to useless when several frequencies may be present. Also remember that many counters have low input impedances at the higher frequencies (above 50 MHz) or separate inputs that are 50 ohms for high and very high frequencies.

Active Preamp Probes

Active preamp probes are used so as not to cause excessive loading at high frequencies or in high-impedance circuits, but have same limitations as a scope probe—the input impedance will still be a few picofarads, which can cause severe detuning of VHF and UHF circuitry.

Coupling Loop

A coupling loop is very good for RF work. It consists of one to several turns of preferably insulated wire large enough to hold its shape (#18 AWG or larger) connected to a BNC female connector. It provides a low impedance to the counter, and a length of 50 or 75 ohm coaxial cable fitted with suitable connectors at each end can be used to connect it to the counter. If made small (1–2 cm in diameter) it has a small pickup area and is useful in RF work in which several frequencies are present in a small area, as the loop can be positioned near the circuit to be measured. It can be held near a tuned circuit and the coupling adjusted by positioning the loop. The tuned circuit acts as a bandpass filter and mainly whatever energy is present in the tuned circuit is sampled. This method of coupling is very useful and reliable. No direct connection to the circuit is needed, and only enough coupling sufficient to get a reading is needed. Properly used, this technique has much merit and causes almost negligible loading of the circuit being measured. It is unwieldy at lower frequencies (<5 MHz) unless a large loop with a number of turns is used. It is not generally useful for audio or very low (<100 KHz) frequency work.

Tuned Antenna and Preselectors

These act as variable-frequency filters and are useful where several frequencies are present, as in off-the-air measurements. They are excellent as long as you know the approximate frequency to be measured in advance. There are a few older counters on the surplus market such as the HP 5245 models, available to experimenters at reasonable prices, that have plug-ins with this feature.

1 GHz Frequency Counter Project

The frequency counter is one of the more useful pieces of test gear that you can have on your electronics workbench. A frequency counter can be used for checking radio transmitters and as a frequency display for radio equipment. It can be used to determine the frequency of oscillators, the timing speed of circuits, and so on. The 1 gigahertz frequency counter described in this project (see Figure 5-10) is relatively simple and straightforward and consists of only three integrated circuits. The heart of the frequency counter is the high-frequency prescaler IC (a Phillips SAB6456), the preprogrammed PIC16F84 microprocessor, and a voltage regulator chip. The frequency counter can display a range of frequencies from 0 to 999.9 MHz with a resolution of 0.1 MHz. This frequency counter is very fast, with a short measuring period, has overrange indication, and can be used directly with a 10.7 MHz I.F. offset for receiver connection.

The 1 Ghz frequency counter circuit shown schematic form in Figure 5-11 begins with the input conditioner circuit formed by the network of capacitors C2 and C3, and the input protection diodes D1 and D2. The input network is then fed to the input of the SAB6456 prescaler IC on pin 2, one of the differential input pins. The prescaler is a divide-by-64 or divide-by-256 switchable device controlled by the mode selector on pin 5. The SAB6456, shown in Figure 5-12, is a 70-to-1 Ghz prescaler chip designed for VHF/UHF TV tuners. The prescaler chip is comprised of an input amplifier, a divider stage and an output stage. Power is applied to the prescaler on pin 8 and ground is connected via pin 4. For our frequency counter application, the prescaler chip mode selector is programmed by connecting pin 5 to ground. The output of the prescaler chip is next coupled to resistors R1 and R2 and transistor Q1, which

Figure 5-10   One gigahertz frequency counter.

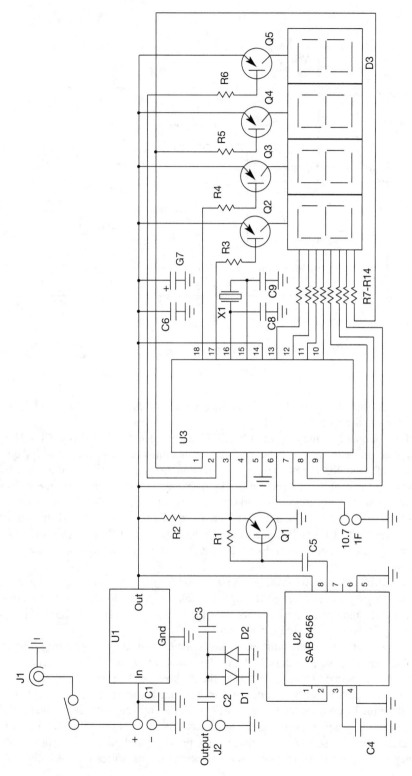

Figure 5-11 Schematic of 1 GHz frequency counter.

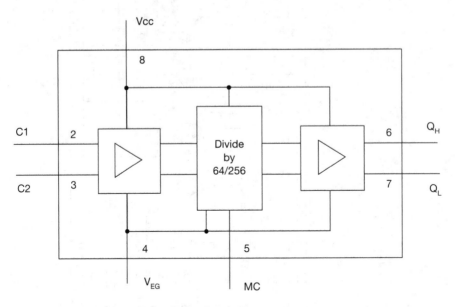

Figure 5-12 *SAB6456 prescaler chip.*

form the interface between the prescaler and the microprocessor chip at U3. Table 5-1 lists the pin-outs for the SAB6456 prescaler chip.

The brain of the frequency counter is the PIC16F84 microprocessor from MicroChip Corp (Figure 5-13). The 18 pin microprocessor is an off-the-shelf device that must be programmed with the included program (see Table 5-2 for the PIC 16F84 pin-out specifications). The microprocessor is powered from the 5 volt regulator at U1. The input to the regulator can be from any dc power supply from 8 to 20 volts. A 9 to 12 volt "wall wart" power supply module would work fine for this application. The microprocessor is controlled by the 4 MHz crystal at X1. The outputs of the of the microprocessor at RB1 through RB6 are used to drive the seven-segment LED display. The LED displays are either a Hewlett Packard HD-M514RD or HD-512RD four-digit LED display with decimal point. The HD514 is a red display and the HD512 is a green display unit.

The microprocessor outputs RA0 through RA3 are used for the display multiplexing. Each of the microprocessor multiplexing outputs are fed through 2.2 kohm resistors to transistors T2 through T5. The collectors of each of the multiplexing transistors drive the display, while each of the transistor emitters are connected to the 5 volt output of the regulator at U1.

The frequency counter circuit was built on a glass epoxy circuit board measuring 5 by 6 inches (see Figure 5-14). The bottom of the four-digit LED display panel was epoxied to the circuit board and wires from the display were extended to mate the holes on the PC board. When assembling the circuit board, pay careful attention to the polarity of the diodes, electrolytic capacitors, and transistors. Integrated circuit sockets were and should be used for the U2 and U3, in the event of a any later IC problems. When placing U2 and U3 in their respective IC sockets, be sure to observe the cutout or the alignment dot on both the IC socket and integrated circuits, so you do not damage the prescaler and microprocessor.

The frequency counter is placed in a plastic enclosure which measures 6 by 8 inches. The "on–off"

Table 5-1   SAB6456T Prescaler pin-outs

	Pin-out	
1	N.C.	No connection
2	C1	Differential input
3	C2	Differential input
4	V_{EE}	Ground (0 V)
5	MC	Mode control
6	Q_H	Complementary output
7	Q_L	Complementary output
8	V_{CC}	Positive supply voltage

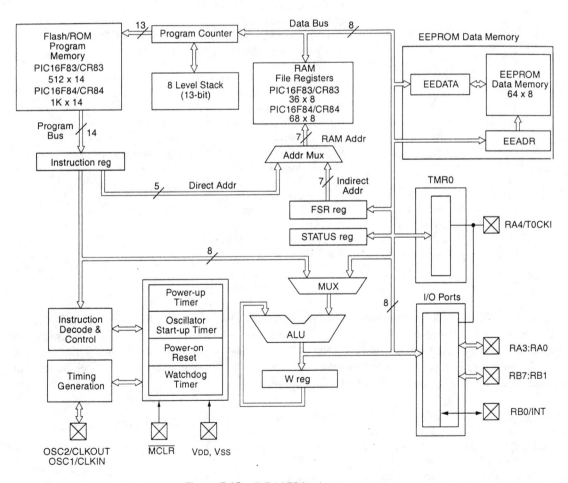

Figure 5-13   PIC 16F84 microprocessor.

Table 5-2   PIC 16F84 Pin-out chart

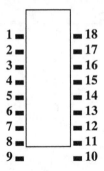

Pinouts		
1	RA2	PORTA is a bidirectional I/O port
2	RA3	PORTA is a bidirectional I/O port
3	RA4/T0CKI	PORTA. Can be selected to be clock input to TMR0 timer
4	MCLR	Master clear (reset). Also active low-reset pin
5	V_{SS}	Ground reference pin
6	RB0/INT	PORTB bidirectional I/O port. Can be selected as interrupt
7	BR1	PORTB bidirectional I/O port
8	RB2	PORTB bidirectional I/O port
9	RB3	PORTB bidirectional I/O port
10	RB4	PORTB bidirectional I/O port. Interrupt on change pin
11	RB5	PORTB bidirectional I/O port. Interrupt on change pin
12	RB6	PORTB. Interrupt on change pin, serial programming clock
13	RB7	PORTB. Interrupt on change pin, serial programming data
14	V_{DD}	Positive power pin
15	OSC2/clkout	Oscillator crystal output, 1/4 freq of OSC1
16	OSC1/clkin	Oscillator crystal input, clock source input
17	RA0	PORTA is a bidirectional I/O port
18	RA1	PORTA is a bidirectional I/O port

power switch and a BNC jack are mounted on the front of the box along with the four-digit LED display. The front panel of the plastic chassis box is replaced by a 1/8 inch red plastic sheet, so the display can show through. A coaxial power jack is placed on the rear panel of the plastic enclosure. A 12 volt "wall wart" power supply cube is fitted with a matching coaxial plug to mate with the power jack on the rear panel. The following is a complete parts list for the unit:

R1	39 kohm ¼ watt resistor
R2	1 kohm ¼ watt resistor
R3, R4, R5, R6	2.2 kohm¼ watt resistor
R7–R14	220 ohm ¼ watt resistor
C1, C5, C6	100 nF mini 35 volt capacitor

Figure 5-14   Frequency counter circuit board.

C2, C3, C4	1 nF 35 volt capacitor
C7	100 μf 35 volt electrolytic capacitor
C8, C9	22 pF 35 volt mylar capacitor
U1	LM7805 5-volt regulator IC
U2	SAB6456 prescaler IC (Phillips Corp)
U3	PIC 16F84 microprocessor (Microchip Corp)
Q1	BC546B transistor or equivalent
Q2, Q3, Q4, Q5	BC556B transistor or equivalent
D1, D2	BAT41 diode
D3	HD-M514D or HD-M512D (see text)
X1	4 MHz crystal
J1	coaxial power jack
J2	BNC jack
S1	SPST toggle switch
Misc	PC board, connectors, chassis box, wire, and so on

The microprocessor is "dumb" at this point and needs to be programmed for it to act as a frequency counter. You will need to borrow or purchase a PIC programmer to load the HEX program from the listing in Table 5-3. Once the COUNTER.HEX program is loaded into the PIC16F84 chip your frequency counter is now ready to operate. Place the microprocessor back into the frequency counter circuit and check all your connections and apply power to the circuit and attach a function generator or oscillator to the input of the counter circuit and you see a frequency displayed on the LED display.

The 1 GHz frequency counter can also be used as frequency display unit for any radio receiver with a

Table 5-3   1 Ghz Frequency counter program listing

---

counter.hex*
:10000000831610308500013086007330810083122 2
:1000100097018F01910191158F0397030F088500B8
:10002000111594018F108F12910A9415940A971349
:10003000971081018C010B118E01D720D720D7207A
:10004000D72001088D008D1F262828280B198E0A1D
:100050000C088E078D0A03198E0A06183728D63029
:10006000 8D02031837288E080319CF288E039901B3
:100070009A019B019C010E1F452808309900013010
:100080009A0009309B0002309C008E1E4D280430DF
:10009000990709309B0706309C070E1E5528023031
:1000A000990704309B0708309C078E1D5D2801309E
:1000B000990702309B0704309C070E1D6528053008
:1000C0009A0701309B0702309C078E1C6D28023076
:1000D0009A0705309B0706309C070E1C75280130D7
:1000E0009A0702309B0708309C078D1F7B2806303B
:1000F0009B0704309C070D1F812803309B070230AB
:100100009C078D1E872801309B0706309C0708300E
:100110000D1A9C0704308D199C0702300D199C079D
:100120000013088D189C070A301C02031C9A289C0081
:100130009B0A93280A301B02031CA1289B009A0AE1
:100140009A280A301A02031CA8289A00990AA128A2
:100150001908093C031CCF28FF3091008F001908B3
:100160000319B828C42091001A08C4208F00BD28A4
:100170001A080319BD28C4208F001B08C42097004B
:100180001C08C4209400192882078134BB3425340C
:1001900023341B34433441348B340134033445302D
:1001A00091007D308F009700FF309400192885154D
:1001B000110886000510EF2005140F088600851031
:1001C000EF208514170886000511EF200515140887
:1001D0008600851 1EF2000000000000008006F304D
:1001E00095000000000000000B1DF9280B118C0A7F
:1001F000FC280000000000000000000000000000DB
:10020000000000000000000000000000000000000EE
:10021000000000000000000000000000000000000DE
:10022000000000000000000000000000000000000CE
:10023000000000000000000000000000000000000BE
:080240000000950BF4280800F2
:00000001FF

---

*The counter.hex listing is the hex code program which is loaded into the PIC16F84 microprocessor chip using a programmer.

10.7 mHz I.F. Simply connect the output of the receiver's IF output to the 10.7 MHz input on the counter circuit which is provided at pin 6 on the microprocessor.

Your new low cost frequency counter is now ready for operation and will serve you for many years to come. In a test bench setup, your, new frequency counter could be connected directly to the output of a function generator if your signal source doesn't have a display device. The function generator project in elsewhere in this book has no display device, so this circuit could be used in conjunction with the function generator to display the actual frequency output of the generator if desired. The frequency counter could also be used in conjunction along with your oscilloscope to observe signals on both devices at once.

CHAPTER 6

Test
Bench
Equipment

In this chapter, we will look at a variety of test instruments and their applications. Some pieces of test equipment can serve several purposes and may substituted for others. This chapter does not cover all equipment available; only the more common and most useful instruments were chosen.

Whether you are repairing or building an electronic circuit, you will generally need to have some way to determine whether or not it is working correctly, and if so, how well. If it is not working, you will need to try to locate the problem so you can repair it. A wide range of test equipment and instrumentation has been developed for testing and repair work. Compare the equipment presented and determine which pieces of equipment you think you will need for your electronics workshop and work toward obtaining them. Prices for test equipment can vary greatly between different manufacturers and models. Chapter 15 will help you determine whether you should buy new or used equipment or whether you should build your own test equipment.

Multimeter

The multimeter is one the most often used pieces of test equipment on the electronics test bench (see Figure 6-1). There are basically two major types of multimeters: the analog meter, which uses an analog needle that moves along a calibrated scale in response to an input signal, and the modern digital multimeter. The two major meter types are again broken down into subcategories. In the analog category you will find the vacuum-tube voltmeters or VTVMs, the volt-ohm-milliammeter (VOMs), and field-effect transistor VOMs (FET-VOMs). In the digital category you will find the digital multimeter or DMM. Multimeters are presented in more detail in Chapters 2 and 10. Chapter 2 presents the differences between analog and digital multimeters and also discusses how multimeters work. Chapter 10 discusses how to use your multimeter to test electronic components.

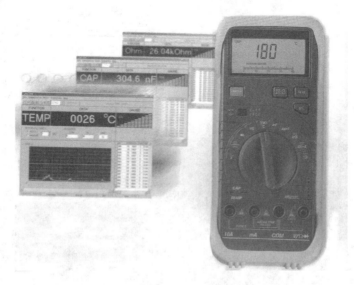

Figure 6-1   Multimeter.

Many low-cost multimeters, both analog and digital, often do not measure current. A meter that measures current requires more circuitry and thus will cost more money than a meter that will only measure voltages. If you can only afford a single good multimeter then it would be prudent to purchase a meter that will measure current. You will undoubtedly at some point need to measure the current consumption of a circuit or a load, such as a lamp or motor. Look for a multimeter that will measure both dc and ac current. In the long run, this is a wise investment. Note that many extremely low cost analog and digital meters may only measure dc voltage and not be able to measure ac voltage. If you can afford two meters and do a lot of dc-voltage-only testing or automobile voltage testing, or want to carry a meter in your car or truck, you could consider a dc-only voltage tester. If you could only afford a single meter at a moderate price, then look for less "bells and whistles" but make sure that your new meter will measure ac and dc voltage and current.

Multimeters are used to read voltages, current, resistance, and signal levels (with an appropriate probe), as well as to test components. Modern low-cost digital multimeters can also be used test resistors, capacitors (within certain limitations), diodes, and transistors. Some digital multimeters can also measure audio frequency and some are capable of interfacing with a personal computer. These meters can display their reading on a computer as well as log readings over time. Digital multimeters or DMMs have a relatively high input impedance, good accuracy, and flexibility, and offer many features for low to moderate cost. DMMs have become quite inexpensive in recent years. Some DMMs are affected by RF, so many technicians keep an analog-display VOM on hand for use near RF equipment.

When buying an analog VOM meter, look for one with an input impedance of 20 kohm/V or better. Reasonably priced models are available with 30 kohm/V ($35) and 50 kohm/V ($40) impedances. The 10 Mohm or better input impedance of DMMs, FET-VOMs, VTVMs, and other electronic voltmeters makes them the preferred instruments for voltage measurements.

Vacuum Tube Voltmeter (VTVM)

Vacuum tube voltmeters are used to extensively by repair technicians to measure ac volts (RMS), ac volts (peak-to-peak), dc volts, resistance, and decibels. The main advantage of a VTVM is its high input impedance, which means that the loading effect of the instrument is negligible and the circuit under test is not disturbed nor "loaded" by the test instrument. The VTVM also has additional advantages and is especially valuable in all electronic applications. A typical VTVM is shown in Figure 6-2. VTVMs can be used to measure all operating voltages and potentials such as B+ voltage in ac–dc, straight ac power supplies, filament voltage, bias voltage, AVC voltage, line voltage. They are ideal for measurements in all types of AM, FM, and TV circuits. Radio technicians use VTVMs to check discriminator or detector operation and AVC or AGC performance, as well as use the ohmmeter to measure circuit continuity and circuit resistances, test individual components with resistance measurements, or trace out circuit wiring through cables or chassis openings. These features make the VTM the perfect all-around instrument for either laboratory or service shop activities.

The most commonly used VTVM probes are the low-capacitance ($\times$ 10) probes. This type of probe isolates the meter from the circuit under test, preventing the test-probe capacitance from affecting the circuit and changing the reading. A network in the probe serves as a 10:1 divider and compensates for frequency distortion in the cable and test instrument. Demodulator probes are used to demodulate or detect RF signals, converting modulated RF signals to audio that can be heard in a signal tracer or seen on a low-band-

Figure 6-2   A typical vacuum tube voltmeter.

width oscilloscope. Older VTVMs used vacuum tubes whereas later models utilized field effect transistors (FETs). VTVMs are fairly difficult to locate these days, since modern DMMs can perform most of the tasks that the VTVM could perform, and their input impedance is usually high enough to prevent circuit interaction when measuring parameters.

Capacitance Meter

A capacitance meter is generally a small electronic measuring device that looks similar to a modern digital multimeter (see Figure 6-3). Capacitance meters have become low-cost necessities on the electronic test bench. They are used for measuring unknown capacitor values or to compare known values against their markings. Capacitance meters can be used when building new projects or kits to check component markings and values before installing them into a new circuit and can be used to test "junk box" capacitors before installing them into your new circuit designs or when you repairing older circuits and want to know their capacitance value and whether they are good or defective. Capacitance meters can also be used when repairing electronics circuits, to check the actual known capacitor values to see if they are correct. They usually are low-current-consumption devices that can operate from a 9 volt battery for quite some time. Most capacitance meter are generally used to perform "out-of-circuit" testing. Digital capacitance meter prices are very reasonable and are generally available from many electronics mail-order suppliers. Quality capacitance meters can measure values to 200 μ range and are priced from $40 and up.

Figure 6-3   Capacitance meter

ESR Capacitance Meter

The equivalent series resistance (ESR) capacitor meter would be a most useful addition to your electronics workbench or test bench. Conventional capacitor testers test the capacitor in question for opens and shorts and to make sure that the capacitors have the correct value. The one thing overlooked by most general-purpose capacitor testers is the "leakage" values that a capacitor may exhibit. Often, a capacitor may look good to a conventional tester but may in fact be defective. The ESR capacitance tester tests for the value series resistance of the capacitor (see Figure 6-4). Filter capacitors usually fail because of the increasing ESR values. The ESR acts like a resistor in series with a capacitor. This includes the resistance of the dielectric, plate material, electrolytic solution, and the terminal lead at specific frequencies. As the capacitor ESR value increases, so does the ripple voltage across a filter capacitor, eventually causing it to fail. The ESR meter is a good investment for electronics technicians and hobbyists who are actively involved in repairing old radios and televisions that contain electrolytic capacitors. ESR capacitance meters are generally more expensive than conventional capacitance meters, but well worth their price. A good ESR meter will cost from $150 to $200 or more and are available from a number of major suppliers.

Inductance/Capacitance (L/C) Meter

An inductance/capacitance or L/C meter is a very handy tester to have around your electronics test bench. This type of meter can measure both the capacitance values as well as inductor values and give

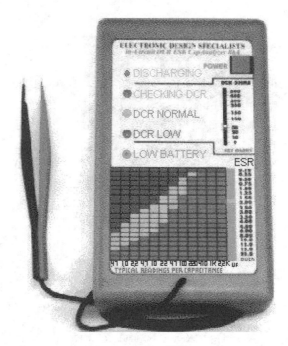

Figure 6-4 ESR capacitance meter. (Courtesy of Electronic Design Specialists.)

you a readout on an LCD display instantaneously (see Figure 6-5). Inductance/capacitance meters are usually about the size of a common multimeter and are often powered from a single 9 volt battery. Most modern L/C meters perform out-of-circuit testing on both capacitors and coils or inductors. You can also purchase L/C/R meters that also test resistance as well as inductance and capacitance, but these are a bit harder to find as portable instruments, cost more, and are often only available as larger bench-top test units. It is very convenient to have a single instrument that can perform all three tests in one piece of test gear. Remember that L/C and L/C/R instruments usually do not test for ESR, so this type of meter should also be used in conjunction with an ESR meter for serious equipment repair work. Quality L/C meters will check capacitors with values up to 200 μF and inductors up to 2 henries. Hand-held L/C meters are priced from $50, and laboratory L/C/R meters can be found for between $100 to $150.

Resistance/Capacitance Decade Boxes

Resistance decade or substitution boxes contain numerous resistors from 10 ohms to 20 megohms. Capacitance substitution boxes usually contain a number of capacitors from 50 to 100 pF and up to 0.5/10 μF in switched banks. These devices are very helpful when repairing or designing electronic circuits. The substitution boxes allow you to quickly insert a component value to test it in the circuit you are designing or repairing, so that you will quickly know if it represents the correct circuit value. Substitution boxes are

Figure 6-5 Inductance/capacitance meter.

very helpful when building and repairing electronics circuits, especially if your "junk" box or parts supply is limited. Substitution boxes will let you quickly determine the correct values and whether you need to purchase a new part. Capacitor and resistor substitution boxes are often combined into one box for convenience, but often the combined boxes do not include as many values of resistance and capacitance as do the separate substitution boxes. Substitution box prices start at about $30.00.

Wheatstone Bridge

The Wheatstone bridge is a very simple but amazingly accurate device for making measurements (see Figure 6-6). Wheatstone bridge circuits can be used to measure resistance and capacitance or inductor values. The Wheatstone bridge concept is used in a wide range of circuits and instruments but is unknown to many people. The Wheatstone bridge has been around for many years and is the granddaddy of test gear. Wheatstone bridges can be configured for ac or dc measurements. They are balance measurement devices. Essentially the Wheatstone bridge compares a known value of a component, such as a resistor or inductor, to an unknown value for a match. For example, a known value of resistance is dialed up on a potentiometer and compared against an unknown resistor. As the comparison between the known and unknown values of resistance takes place, the balance bridge gets closer or farther away from the balance and the meter display indicates whether you are getting closer or farther from the balance point. The Wheatstone bridge and its variations are great when you have "standards" or known values with which to

Figure 6-6 *Wheatstone bridge.*

compare. Stand-alone Wheatstone bridges can often be found on eBay or via classified ads in electronics and radio magazines.

Frequency Counter

A frequency counter is an instrument that measures and displays the frequency of the signal presented at its input (Figure 6-7). A frequency counter operates by amplifying the input signal and clipping it into a clean rectangular waveform, then counting the number of pulses that occur within a precise time interval, usually 1 second or less. Frequency counters can be used to measure the frequency of oscillators, receiver I.F. frequencies, digital circuits, and so on. Frequency counters work with both analog or digital signals, over the frequency range for which they are specified. Modern frequency counters come in a variety of architectures, are available for frequency counting into the gigahertz range, and have become quite inexpensive in recent years.

If you are thinking of purchasing a frequency counter, look for one with a high-stability time-base option, [i.e., one part per million (PPM) or better] Also look closely at "universal" counters. These allow you to take ratio measurements between two different frequencies. This saves time and produces fewer errors in counting.

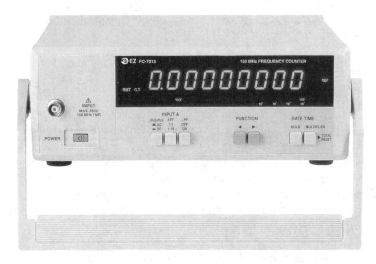

Figure 6-7   Frequency counter.

If you plan to do much work in the electronics field, you may want to build or purchase your own frequency counter. Low-cost frequency counters are available for about $100.00. A low-cost 1 gHz frequency counter construction project is featured in Chapter 5.

Audio-Signal Generator

An audio-signal generator is a device that can provide one or more waveforms at frequencies ranging from approximately 0.1 hertz (one cycle every 10 seconds) to perhaps up to100 kHz. Typical waveforms are sine, square, and triangular. In many cases, a dc offset can be imposed on the signal, or the signal can be coupled through a suitable capacitor to eliminate any dc bias. This allows a known signal to be inserted into any analog circuit being tested, so that the progress of the signal through the circuit can be traced. Chapter 4 shows how to construct a practical, working function generator.

It is a good idea to have a known signal source so you can test and troubleshoot other circuits. Although signal generators have many uses, in troubleshooting they are most often used for signal injection. When buying a signal generator, look for one that can generate sine wave and square wave signals. A good signal generator is double or triple shielded against leakage. Fixed-frequency audio should be available for modulation of the RF signal and for injection into audio stages. The most versatile generators can generate amplitude- and frequency-modulated signals. Good generators have stable frequency controls with no backlash. They also have multiposition switches to control signal level. A switch marked in dBm is a good indication that you have located a high-quality instrument. The output jack should be a coaxial connector (usually a BNC or N connector), not the kind used for microphone connections.

Some older, high-quality units are easy to find. Look for World War II surplus units of the URM series, such as Boonton, GenRad, Hewlett-Packard, Tektronix, or other well-known brands. Some home-built signal generators may be quite good, but make sure to check the construction techniques, level control, and shielding quality.

Function Generator

A function generator is an advanced or modern type of signal generator that provides the basic sine, square, and triangular waveforms. Many modern function generators can also provide additional waveforms, such as sawtooth waves, as well as various types of pulses, including level-triggered and edge-triggered pulse waveforms. General-purpose function generators can generate signals from 0.1 Hz to 1 mHz and above (see Figure 6-8). Higher-end function generators can generate signals from 0.001 to 2 mHz and also have additional features available, depending on how much money you want to spend. Additional features can include variable sweep rates, internal versus external triggering as well as slope control, inverted outputs, and various types of triggered outputs. Basic function generators can be purchased for between $160 to $200. A low-cost, three-waveform, basic function generator building project is featured in Chapter 4.

Arbitrary-Waveform Generator

An arbitrary-waveform generator is a signal generator and function generator all rolled into one with the newest features available. Arbitrary-waveform generators add the feature of generating your own specific waveforms as needed. Arbitrary-waveform generators incorporate a microprocessor that can be used to control the timing functions needed to program specific waveforms with complex repetition rates. Most arbitrary-waveform generators have a built-in table of fixed waveform types as well as the ability to gen-

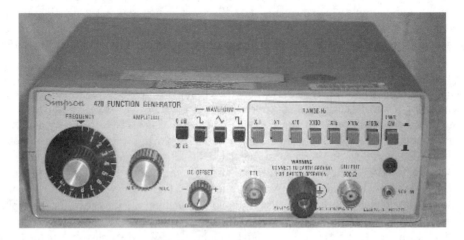

Figure 6-8   Function generator.

erate your own specific waveforms. Additional features include front panel LCD displays as well as a keyboard for direct waveform programming functions. Arbitrary-waveform generators are more expensive than function generators; prices are generally in the $500 to $1000 dollar range.

Pulse Generators

Pulse generators are designed to specifically create precise pulses with low rise/fall times and high repetition rates up to the megahertz range. Pulse widths are controllable from 0.1 Hz to 10 mHz. Many pulse generators feature pulse delay control, with normal inverted and, special sync outputs. Pulse generators can be used as missing-pulse detectors and for tracing digital logic flow and analyzing microprocessor programs. Pulse generators are often used for testing audio/video as well as remote-control devices. Stand-alone pulse generators often have many additional features not found in combination devices such as function generators with pulse generation capabilities. Precision stand-alone pulse generators are available for about $250 to $300. Chapter 16 features a stand-alone pulse generator construction project.

RF Signal Generator

RF signal generators produce sine waves at frequencies ranging from 100 kHz to 50 MHz or higher (the higher the top frequency, the more expensive the instrument). This is a higher-frequency version of the audio-function generator. The RF generator emits a radio frequency signal that can be adjusted via front panel control over a wide range of frequencies. It often includes a simple 1 kHz audio oscillator that can be used to modulate either the frequency or the amplitude of the main carrier wave signal. The RF generator is primarily used to test and troubleshoot radio receivers, tuners, intermediate frequency (I.F.) amplifiers, and detector circuits, and to determine band edges. RF signal generators are primarily utilized by radio or RF engineers, radio technicians, TV/radio repair personnel, and ham radio operators, and are generally available as bench-type instruments that cost from $150 and up. Used RF signal generators can often be found in classified ads, radio magazines, or at local "hamfests."

Tone Test Set

A tone test set is a handy device for tracing wires of all sorts. Telephone installers use them, as do twisted-pair Ethernet cable installers. Tone test sets are an electrician's best friend as well. Basically, the test set consists of two hand-held units—a sender (transmitter) and a receiver unit (see Figure 6-9). The transmitter unit produces a fixed-tone frequency, a wide-band RF signal modulated by a fixed 1 kHz tone that is fed to a small inductive coil used to send the signal to the receiver unit, which is used remotely. The receiver unit consists of a pickup coil front end that is fed to a detector/amplifier circuit that reproduces the 1 kHz tone on a small speaker. The transmitter unit is placed on the end of the wire to be traced and the receiver unit is used to "hunt" for the wire that you are looking for.

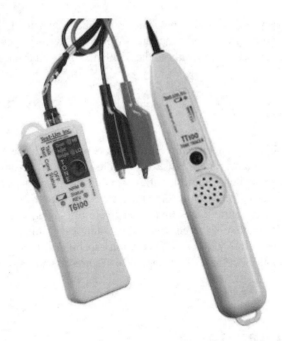

Figure 6-9   Tone test set.

Signal Injector

Signal Injectors are signal source devices that are usually small circuit boards with small, low-power RF (radio frequency) generator, producing an audio signal superimposed on the RF signal. The signal injector usually operates on a single frequency in the 50 to 100 kHz range with a 1000 Hz audio signal impressed on the RF signal. These devices are generally small circuit boards mounted in a rectangular plastic case, often in the shape of a pen or cylinder. Signal injectors are usually powered by internal batteries and have a probe tip at one end, from which the signal is sent. A ground lug is often found on the side of the enclosure; it is connected to the circuit under test's ground plane. The signal injector can be used for testing both receivers' RF stages and audio amplifiers. Since the signal generator emits both RF and audio signals, it is versatile and can be used for multipurpose applications. With the signal injector and a circuit schematic, you can follow the various stages of an audio amplifier or RF circuit in a receiver from its input through the successive stages, in order to locate the stage that is not passing the signal from the signal injector. Tone test sets are available for about $75.00.

Signal Tracer

Signals can be traced with a voltmeter and an RF probe, grid-a dip meter with headphones, or an oscilloscope but there are some devices made especially for signal tracing. A signal tracer is primarily a high-gain audio amplifier. Some signal tracers may have a built-in RF detector and probe as well. Most convert the traced signal to audio through a speaker. The signal tracer functions as a detector and/or receiver and amplifier. A signal tracer is usually a high-impedance input device designed to prevent circuit loading. Signal tracers are especially useful for testing receivers and amplifiers. They are available in a number of different packages and usually cost less than $50 or $60.

Note that a general-coverage receiver can be used to trace RF or IF signals if the receiver covers the necessary frequency range. Most receivers, however, have a low-impedance input that severely loads the test circuit. To minimize loading, a capacitive probe or loop pickup is commonly used at the input of the receiver. When the probe is held near the circuit, signal will be picked up and carried to the receiver. It may also pick up stray RF, so make sure you are listening to the correct signal by switching the circuit under test on and off while listening.

Clamp-on Ammeter

The clamp-on ammeter is a dedicated instrument that is used to measure and display current passing through a wire. Clamp-on ammeters are generally used to measure higher currents through a single wire (see Figure 6-10). Clamp-on ammeters can have either analog or digital displays. They are usually about the size of a multimeter. A lever on the side of the meter allows the clamp to open up, allowing a wire to be placed inside the clamp circle for measurement. Clamp-on ammeters are often used by electricians to

Figure 6-10   *Clamp-on ammeter.*

quickly measure high-current wires when checking heaters, furnaces, motors, compressors, or refrigeration units. Analog meters are generally less expensive than digital meters. Analog clamp-on ammeters can cost $40 and up depending on the accuracy. Digital clamp-on ammeter prices begin about $75.

Insulation Tester

The insulation tester is designed to perform a number of insulation tests on wire and cables. The tester applies a voltage of 1000 volts or more for reading up to 10 Gohm. These testers are sometimes combined with other functions such as continuity testing, resistance, and/or voltage measurements. Newer meters are often analog bargraph or LCD bargraph. These meters are usually portable and are rather expensive—over $400. Insulation testers are often used by electricians and field power engineers to locate cable faults and line noise problems.

Cable Testers

Cables testers are available for many different types of cables, and range from simple continuity testers to graphic TDR testers. Also in this category are Ethernet cable testers, which are available to test shielded coaxial and twisted-pair Ethernet cables (10-BASE-T or 10 BASE-2 cables). These testers perform a number of tests such as testing opens, shorts, and high/low resistance. TDR or time domain reflectometry testers can detect a variety of defects and display how far down the line the fault is in graphic form. Ethernet cable testers are generally expensive and can cost from $900 to $3000 and above depending upon their features.

TV Pattern Generator

TV pattern generators are available as bench-top or handheld instruments. These instruments generate NTSC composite video test graphics such as color bars, crosshatch patterns, dots, staircase patterns, and color raster patterns. The pattern generators often have multiple output configurations such as S-video, mini-DIN, and RCA outputs. Output signals can be produced as either progressive scan or interlaced modes. Handheld generators are usually powered from a 9 volt battery. Pattern generators are used to repair televisions and computer monitors. This instrument will allow you to repair a PC monitor without the computer attached. TV pattern generators are relatively low cost and can be purchased for about $80 to $150.

Transistor Tester

Transistor testers are similar to transconductance tube testers. Device current is measured while the device is conducting or while an ac signal is applied at the control terminal. Commercial surplus units are often seen at ham radio flea markets. Some DMMs being sold today also include a built-in, simple transistor tester. Most transistor failures appear as either an open or shorted junction. Opens and shorts can be found easily with an ohmmeter; a special tester is not required. Transistor gain characteristics vary widely, however, even between units with the same device number. Testers can be used to measure the gain of a transistor. A tester that uses dc signals measures only transistor dc alpha and beta. Testers that apply an ac signal show the ac alpha or beta. Better testers also test for leakage. In addition to telling you whether a transistor is good or bad, a transistor tester can help you decide if a particular transistor has sufficient gain for use as a replacement. It may also help when matched transistors are required, but the final test is the repair circuit.

A quality transistor tester is able to test many different types of transistors, such as bipolar, power, and FETs, for their overall operating condition—whether they are good or bad. Additionally, a good transistor tester will also test for hfe or gain of the transistor. Additional transistor tests include beta, leakage, and conductance. Transistor testers are available in many configurations from simple bad/good LED testers to transistor curve tracers using built-in oscilloscopes. Transistor testers are available as high-quality bench testers that are powered by ac power, and as portable units that run on batteries. Transistor testers are often found on some multimeters as well as stand-alone testers, and are available for under $100.

Transistor Curve Tracer

A curve tracer can be an expensive dedicated piece of test equipment or an add-on to an oscilloscope. It provides a graphical display of the voltage/current characteristics of an electronic component (see Figure 6-11). The design of a curve tracer is simple in principle. For the horizontal input to the oscilloscope or

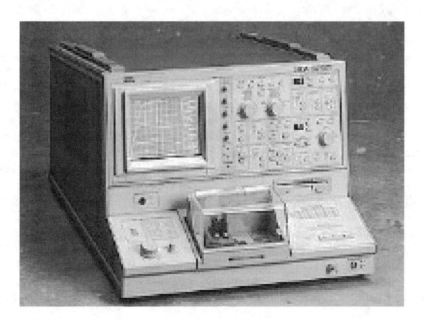

Figure 6-11   Transistor curve tracer.

the transistor's collector supply, you need a variable ramp generator. If your scope has a sweep output, then you can derive it from this. For the base drive of the transistor, you need a programmable current source capable of putting out a series of constant currents for the base drive, such as a counter driving a D/A set up for a current output mode. Use the trigger output or sweep output of the scope to increment the counter so that it sequences through a set of, say, 10 current settings. Finally, you need some way of sensing collector current to drive the vertical channel—a small series resistor in the emitter circuit, for example. For simple diode tests, you can use a variable ac voltage source like a variable isolation transformer (with a current-limiting resistor) across the diode. The X (horizontal) input of the scope goes across the device under test. The Y (vertical) input of the scope goes across the current-limiting resistor or a separate series-current-sense resistor. Commercial stand-alone curve tracers are usually quite expensive and can cost over $5000.

An old issue of *Popular Electronics* (May 1999) has complete plans for a "Semiconductor Tester" that can handle NPN and PNP bipolar transistors, JFETs and MOSFETs, all sorts of diodes including zeners, and a variety of other devices. It is basically a curve tracer adapter for an oscilloscope.

Tube Tester

There are two basic categories of tube testers. The general "drug store" variety was found in back of drugstores, where people could come in and test their radio tubes by themselves. These testers measure a few parameters, including open and shorted filaments.

Most simple tube testers measure the cathode emission of a vacuum tube. Each grid is shorted to the plate through a switch and the current is observed while the tube operates as a diode. By opening the switches from each grid to the plate (one at a time), it is possible to check for opens and shorts. If the plate current does not drop slightly as a switch is opened, the element connected to that switch is either open or shorted to another element. The emission tester does not necessarily indicate the ability of a tube to amplify.

The "real" tube testers are of the transconductance type (see Figure 6-12). The transconductance tube testers measure a number of useful tube parameters, including gain and the relationship between tube elements. These were the tube testers used by the professional electronics and radio technicians. Some transconductance testers read plate current with a fixed bias network. Others use an ac signal to drive the tube while measuring plate current. Most tube testers also check interelement leakage. Contamination inside the tube envelope may result in current leakage between elements. The paths can have high resistance and may be caused by gas or deposits inside the tube. Tube testers use a moderate voltage to check for leakage. Leakage can also be checked with an ohmmeter using the × 1 M range, depending on the actual spacing of tube elements.

There are many aficionados of vintage gear who enjoy working with old vacuum tube equipment. The tube tester is now a very difficult piece of test equipment to find, but it is still very useful if you are involved in repairing or restoring old radios, TVs, and amplifiers. They are scarce because tubes are no longer used in modern consumer equipment. You can often find good tube testers through buy/sell bulletin boards, classified ads, auction sites such as eBay, at hamfests, or from TV repair shops going out of business.

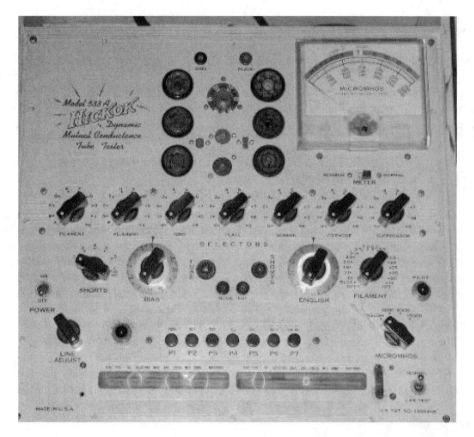

Figure 6-12   Tube tester.

RF Impedance Bridge

Like most bridges, the design of the RF impedance bridge laces reference components in one leg of the circuit and unknown components in the opposite leg. RF is applied at the generator input, at the desired frequency, and the detector is tuned to the same frequency. The reference components are adjusted for a balance, which implies a null. The null causes the detector (receiver) to indicate a minimum signal level.

The concept of an impedance bridge is simple. An impedance bridge is generally used in conjunction with a signal generator and detector, combined into one package. The generator is usually an RF signal generator, although some newer impedance bridge systems use a wideband noise source. The detector should be a narrowband receiver. The impedance bridge reactance of the unknown impedance is then measured in ohms; finding the null, however, takes skill. The latest impedance bridges are digital LCD display devices that can directly read off the correct impedance without the interpretation needed with the older instruments.

Field-Strength Meters

Field-strength meters are usually passive RF detectors and are used to detect nearby transmitters or indicate how much RF strength is radiated from the transmitter. The field-strength meter is a low-cost device consisting of a broadband RF detector, a diode detector, and a display device. Some RF field-strength meters have an internal amplifier built in. They allow you to see the strength of an RF field, which can be measured at various distances from a transmitter. Field-strength meters costs begin at $50.00.

SWR or VSWR Bridge

An VSWR or SWR meter is a an invaluable aid for RF technicians and amateur radio operators who work on various types of radio frequency transmitters. SWR stands for standing wave ratio, whereas VSWR stands for voltage standing wave ratio and may be illustrated by considering the voltage at various points along a cable or coaxial cable driving a poorly matched antenna. A mismatched antenna reflects some of the transmitted power back toward the transmitter, and since this reflected wave is traveling in the opposite direction as the incident wave, there will be some points along the cable where the two waves are in phase and other points where the waves are out of phase. If you attached an RF voltmeter at these two points, the two voltages could be measured and their ratio would be the SWR.

In most cases, it is most desirable to match every component of a system to the chosen system impedance so that device matching is not frequency sensitive and critically dependent upon the cable lengths.

The SWR is a useful number for evaluating the actual voltages and currents present along transmission lines. SWR readings can generally be directly measured and presented on a digital meter display. SWR bridges can cost from $100 to $300. Good used SWR meters can be obtained on eBay or via ham radio classifieds or from local hamfests.

Oscilloscope

An oscilloscope is a very effective measuring instrument that provides much more information about a signal than can conventional multimeters. The oscilloscope "paints" a "real-time" picture of a particular waveform's characteristics, Oscilloscopes can measure and display voltage relative to time as well as frequency. Using an oscilloscope, engineers and technicians can effectively isolate faults as well as optimize circuit performance of faulty circuits. Circuits adjustments and "fine-tuning" often require an oscilloscope for efficient operation.

Oscilloscopes are available in two basic classifications: analog and digital. The analog scope was the original instrument, designed to measure real-time signals and display them on a CRT, as shown in Figure 6-13. More recently, the digital scope has become the preferred measurement device. Digital scopes have many more features than analog scopes, such as data logging, delay, and storage. Newer digital oscilloscopes are much more accurate, smaller, and lighter, and will often run on batteries and utilize LCD displays.

A general-purpose scope is usually adequate for most basic measurements; however, specialized waveform analysis requires the use of advanced digital models that provide added capabilities such as delayed time-base measurements, high-frequency operation, and simultaneous multiple channel displays.

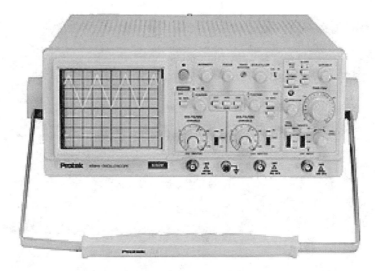

Figure 6-13   Oscilloscope.

Recently, oscilloscope PC cards have become more popular and are becoming less expensive as time passes. They are an interesting alternative but you need to be very careful in comparing specifications between conventional oscilloscopes and the new digital scope cards. Older oscilloscope cards often have serious high-end frequency response limitations. The latest PC card oscilloscopes are approaching the response of bench-top digital oscilloscopes. Oscilloscopes are covered in more detail in Chapter 3. Analog oscilloscopes can be purchased for as little as $250 for a dual-channel 20 MHz model. Digital oscilloscope are more expensive and start at about $950. Good oscilloscope cards start around $250 to $300, and high-end models can cost up to $1000.

Spectrum Analyzer

A typical spectrum analyzer (Figure 6-14) can display the RF spectrum from 2 to 2150 MHz and act as a continuous-tuning AM/FM receiver. It can be used to measure the amplitude and frequency of RFI generated by your computer or electrical appliances, and instantly evaluate the results of filtering or shielding. You can also use the spectrum analyzer to examine satellite TV signals and their subcarriers; identify modulation modes such as AM, FM, SSB, FSK, PCM; sweep an area for illegal transmitter "bugs"; signal-trace transmitters and receivers; check "gain-per-stage" when building or troubleshooting; and test for harmonic or intermodulation distortion. If you are a ham radio buff, you can use the spectrum analyzer to check transmitter outputs for "spurs" or to receive "on-carrier" or "subcarrier" ATV sound. Many hams use it to see if the band is "open" at a glance, to find a quiet spot on the band as well as to monitor all the local repeaters simultaneously! Making field-strength measurements or orienting and tuning antennas for maximum results across a band of frequencies, tuning antenna duplexers or diplexers, and making VSWR measurements can all be made using a spectrum analyzer.

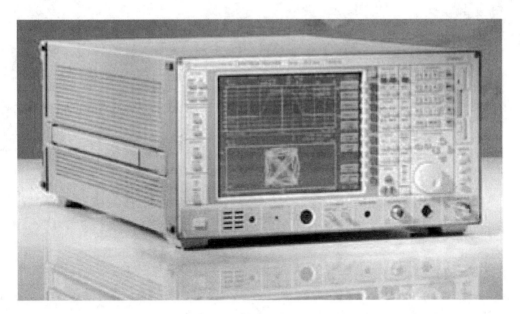

Figure 6-14   Spectrum analyzer.

Basically, a spectrum analyzer sweeps a voltage-tuned front-end over a range of frequencies in synchronism with the horizontal sweep of a scope. The received signal is passed through a narrowband filter and the detected signal is applied to the vertical amplifier of the scope. If there is no signal, there will be no vertical deflection. The deflection produced by the signal is proportional to the received signal's strength. Resolution is approximately 250 kHz and is determined by the bandwidth of the filter. The output of the spectrum analyzer is audio; therefore, it can use any oscilloscope for the display! The analyzer functions as a tunable RF voltmeter with "eyes" and "ears." A spectrum analyzer is a necessity if you are an RF engineer or RF technician involved in the design and repair of commercial radio systems and equipment. Spectrum analyzer prices can range from $1500 to $20,000 depending upon upper-frequency ranges and features.

Logic Pulser

The logic pulser is an inexpensive electronic piece of test gear (see Figure 6-15). It is usually used and paired with a logic tracer or logic probe. Essentially, the logic pulser is the sender or transmitter type device and the logic probe or logic tracer is the receiver unit. A logic pulser is used to send a logic pulse through a circuit under test, in order test a logic gate. The logic pulser is a small solid state circuit housed in a small plastic marker-sized package. It has a metal tip probe protruding from the housing. A ground terminal on the logic pulser is used to connect to the ground the circuit under test. The logic pulser can be internally powered by a small battery or powered via an external battery or alligator clips. Some logic pulsers have one or more frequency or pulse repetition rates available for possible use.

Figure 6-15   Logic pulser.

The logic pulser is great for testing many types of individual logic gates or a series of gates connected together that form complex systems. Logic pulsers are generally low cost (often $40 or less) and simple to operate.

Logic Probe

A logic probe or tracer is a small inexpensive logic latch circuit that is used to test logic circuits under test (Figure 6-16). A logic probe looks for state changes in a logic circuit under test. A logic pulser is used to send pulses to a circuit under test and the logic probe looks for the pulse propagated through the test circuit. The logic probe circuit board is usually housed in a small plastic enclosure or housing that looks much like a marker and is usually powered by AAA or button sized batteries. Some logic probes are externally powered from that which exit the probe and connect to an external battery. Logic probes often have two or three small LEDs that are used as logic state indicators to aid in troubleshooting the circuit under test. Usually, one LED is for state changes indications. Sometimes, an LED is used for showing negative- or positive-going states and often there is an LED that is used as the gate latch indicator. Logic probes are generally available for less than $30 or $40.

Logic Analyzers

The logic analyzer is an invaluable tool for digital circuit designers working at high speeds. Most logic analyzers are high-performance multichannel devices. Many logic analyzers have 24 to 32 channels. A typi-

Figure 6-16   Logic probe.

cal capture rate for a 32 channel analyzer is 25 mHz or up to 100 mHz for eight channels, and it can display the captured results. Most analyzers display in both timing and list formats on an LCD screen or computer monitor. Most logic analyzers feature multilevel triggering with event counting and restart, and most feature glitch capture and glitch triggering. Most logic analyzers are comprised of two or more clocks with independent qualifiers. Clock inputs can usually be fixed or set up as variable threshold. Logic analyzers are generally expensive and range from $1000 to $1500 or more.

Microprocessor Programmers

Many electronic workbenches now have EEPROM and microprocessor programmers on them, and many electronic enthusiasts now experiment with microprocessors. Microprocessor Programmers come in many types and capabilities. Many universal programmers can program a wide variety of EEPROMs and microprocessors and generally use a computer's parallel or serial port to communicate between the programmer and the personal computer. Most programmers now come with extensive software libraries and numerous adaptors for different-sized microprocessors. Good microprocessor programmers are also supported via their manufacturers with technical support and web tools. When looking for a microprocessor programmer, read the literature very carefully as to the microprocessors that it will support before purchasing it. Microprocessor programmers can cost as little as $30.00 for the most basic models. There are a number of expandable/flexible and moderately priced microprocessor programmers in the $150 to $200 range.

Variac

The variac, also known as the autotransformer, is a variable line current transformer, (Figure 6-17). The variac can be thought of as a piece of test equipment, a valuable diagnostic tool that can be very useful on your electronics workbench. The variac is an adjustable line current or ac variable transformer. Variacs come in many different sizes and current handling capabilities. They often have a 120 volt line cord at the input end of the transformer and an ac outlet on the housing for the output connection. Variacs usually have a large dial calibrated from 0 to 100 that is connected to a shaft that can be used to vary the output of the autotransformer from 0 to the maximum line voltage. Variacs can be used to test power devices or circuits by varying the voltage to the circuit under test. They are ideal for gradually applying voltage to circuits under test. Sometimes, the voltage can be raised up to point where a fuse blows in a circuit under test to help determine a fault condition. Variacs can be rather expensive and range from $25 for a 2 ampere unit to $100 or more for a 10 ampere unit. Variac can often be purchased on the surplus market for a fraction of the cost of a new one, and there is basically nothing to go wrong, so they will last a lifetime. Check classified ads in radio or electronic magazines or auction sites such as Ebay.

Isolation Transformer

Isolation transformers are generally 1:1 ratio 115 volt ac line transformers with a line cord and power plug at one end (primary) and a 115 volt standard outlet at the other end (secondary). The isolation

Figure 6-17   Variac.

transformer is very helpful when troubleshooting circuits on the bench. Many old radios and TVs have a "live" chassis, which puts a service technician in possible danger, since many of these circuits do not have transformers in them and, therefore, no isolation from line voltages. Often, touching the chassis and any ground causes a very bad 115 volt ac shock to the person servicing the equipment. Isolation transformers come in many current capacities, so they should be chosen appropriately for the current consumption of the devices they will be used on. Isolation transformers are often used on pieces of test equipment such as oscilloscopes to protect the scope from "hot" chassis equipment since many low-cost oscilloscopes do not have isolation protection. It is wise to ensure that your expensive scope will not be damaged by "hot" chassis equipment. Isolation transformers can be expensive; cost varies with power handling capabilities.

Battery Tester

Battery testers are available in many forms from a low-cost Radio Shack multibattery tester to commercial battery testers for automobile or deep-cycle batteries. True battery testers are not just voltmeters, but voltmeters with resistors or resistive devices that simulate a battery under a load condition. A specific resistor is switched into the measuring circuit for each particular type of battery being tested. Small household D, C, AA, AAA, or button-type batteries can be easily tested using one of these inexpensive battery testers. The problem of battery testing becomes more difficult when trying to test automotive batteries or other large lead–acid batteries. It is more difficult to test larger batteries under load. Often, large batteries are tested by a hygrometer, which tests the chemical reaction of the battery. You can also get an idea of a battery's condition if you have a good battery charger with a panel meter display. When applying the charging leads from a battery charger, you can get a rough indication of the battery's condition via the battery charger's front panel meter. When the battery charger cables are applied to the battery, you will see the condition of the battery based on how the meter responds. If the meter displays a large movement when the battery is connected, then the battery is in need of a charge and the battery voltage is low. Household battery testers for testing AA, AAA, C, and D batteries can be readily obtained for less than $20.00.

Batteries and Chargers

There are many different types of battery chargers available, from the small desk-type nickel–cadmium and nickel–metal hydride battery chargers to the deep-cycle marine and automotive battery chargers. There are fast chargers, trickle chargers, and stepped-voltage chargers. Chargers are usually designed for specific types of batteries. It is usually not prudent to use a Ni–Cd charger to charge nickel–metal hydride batteries or any other type of battery. Certain rechargeable batteries are designed for fast charging, whereas other batteries are designed for slow charging. So-called smart chargers can often charge more than one type of battery, and do so at different rates as well. Smart chargers often can charge batteries at a faster rate, then sense when the battery is near full charge so it can slow the rate of charging down to a trickle charge until the battery is fully charged, and then turn the charger off so as to not damage the battery being charged. Rechargeable batteries have become very efficient and cost-effective as well very "green" for our environment. The capacities of small AA batteries, for example, have reached 2000 mA/h or better in the last few years. The discharge rate of nickel–cadmium rechargeable batteries generally drops off gradually and the batteries tend to develop a "memory" effect if not fully discharged. Nickel–metal hydride batteries have an output that usually remains high and drops off a little less quickly. These types of batteries are not as susceptible to the "memory" effect. These batteries, like the Ni–Cd types, should be fully discharged at every cycle for the longest life. The newest rechargeable lithium-ion batteries discharge at a constant rate and fall off near the end very rapidly. These batteries do not have a "memory" effect and do no have to be fully discharged at every cycle, but periodically they should be fully discharged for optimum performance.

Many engineers and technicians prefer to power their new circuit designs with batteries rather than from ac-to-dc power supplies. Some designers prefer clean power when perfecting their new amplifier or radio receiver design to avoid any power line noise, spikes, or ac hum problems that can occur with older power supplies. Many times, battery power supplies are fabricated from AA batteries and plastic battery holders mounted to a block of wood or mounted to a chassis box. Multiple battery taps are often provided from 3 to 12 volts, and sometimes adjustable regulators are designed into these portable power sources. If you choose to build your own dc battery prototype power source, then you will want to locate a supply of nickel–metal hydride or lithium-ion batteries and a smart charger, to keep your rechargeable batteries ready. Twelve volt deep-cycle batteries are available in large capacities if desired, and many people also hook up an inverter for emergency use.

Power Supplies

Power supplies come in all shapes and sizes, from microcurrent dc-to-dc power supplies to monster 5 volt, 80 ampere computer power supplies. Lab or bench power supplies are available in single output, dual output, triple output, and even quadruple output variations. A good high-quality laboratory power supply is an invaluable piece of equipment to have on your electronics workbench or test bench and will last many useful years if treated well.

A good power supply is more important than most people realize. A good, current-limited, adjustable, linear supply is the ideal power supply for your new electronics workshop. Switching power supplies are electrically noisy and really only acceptable for digital circuits. Current limiting is important because it often stops components from failing, that is, exploding or burning up, when there is a short circuit. Note that poor power supply regulation can lead to noise and all sorts of strange or intermittent problems.

A good choice for a workhorse lab or bench supply is a dual-voltage 0–30 or 0–50 volt power supply. A laboratory power supply with both adjustable voltage and current controls is a great choice for your workbench power supply. Voltage and current metering is also very desirable as well. Look for at least a 0–2 ampere and preferably a 0–5 ampere current capacity for your lab supply. Dual-voltage supplies (Figure 6-18) are ideal if you plan on working with linear op-amps or filters, since these types of integrated cir-

Figure 6-18   Dual-voltage power supply.

cuits need both a plus and minus voltage power supply to power them. A good general-purpose, dual-voltage power supply is ideal for designing both analog and digital circuits. Power supplies can also be used for bench testing of components from your "junk" box such as switch, relays, motors, and lamps, found at hamfests or from surplus dealers. A good general-purpose, dual-voltage lab power supply can be purchased for about $150 to $200. Used power supplies can be purchased through classified ads in electronics and radio magazines as well as online through eBay for less than $100.

Power supplies, as mentioned earlier, are available in other configurations for other purposes. There are two major categories of dc-to-dc power supplies that can used in electronics work. One type of dc-to-dc power supply is the down-converter or step-down type supply that takes a voltage and reduces the output to a lower voltage level. The second type of dc-to-dc power supply is the up-converter or step-up style that has a low-voltage primary voltage and converts the input voltage to a higher voltage. For example, a 6 volt dc input to a step-up supply could produce an output of 50 volts. Other power supplies include high-voltage power supplies with a 110 ac primary voltage that produce a high voltage of 1000 volts dc. These power supplies are used in copy machines, fax machines, and so on.

Temperature-Controlled Soldering Station

A temperature-controlled soldering station is a highly desirable accessory for your electronics workbench. It allows the soldering iron to be adjusted to a precise temperature for soldering critical circuit components. The temperature-controlled-soldering station is ideal for soldering static-sensitive electronic components, and many types feature ceramic soldering tips to eliminate static problems. The temperature control is usually calibrated for fine control of the tip temperature.

Desoldering Stations

Another type or soldering station is called a desoldering station or extraction station. This is a special-purpose soldering station designed to remove the solder from circuit boards in order to remove integrated circuits from them. Desoldering stations usually employ a temperature-controlled desoldering iron with a special assortment of tips specially designed for solder removal. Many of the desoldering tips are shaped like integrated circuits with 8, 14, and 16 heat pads to allow heating of all the circuit pads under an IC all at once. Desoldering stations usually have suction tools attached to the desoldering tips so that the solder can be removed at the same time as it is heated. Some of the newer desoldering stations also have a fume removal hood or carbon-based filter with a fan to remove the solder fumes, all in one package. The desoldering station is ideal for technicians or repair personnel who have to perform a great deal of parts removal or substitution every day. Temperature-controlled soldering stations start at about $40 and desoldering stations begin at $100. Combined stations are also available starting at about $300.

Power
Supplies

Power supplies are very important components of any electronics workbench. They are available in a number of types, shapes, and sizes. We will discuss a number of different types of power supplies that you will come across as an electronics enthusiast. Power supplies can be constructed to produce many different fixed or variable voltages and can be fabricated to generate all types of current, depending on your needs.

There are two main categories of power supplies: the linear type or the newer switching type found in computers and small portable electronics equipment. The linear power supply has been around for a long time and represents the bulk of all power supplies up until the 1980s and early 1990s, when integrated circuits and computers hit the marketplace. The linear supply usually consists of a transformer, a diode or set of diodes for rectification, and a number of filter capacitors to smooth out the power supply ripple that results from the rectification process. Linear supplies are simple to implement but are only about 50% efficient as far as power transfer, and usually get warm or hot. They are very quiet, economical, and easy to build and troubleshoot, and they usually incorporate a bulky steel-plate and copper-coil power transformer. These types of power supplies have been around for many years and have been the backbone of power supply industry until fairly recently.

Switching power supplies, also called switch mode power supplies (SMPSs) have been around for a much shorter time than linear power supplies. They entered mainstream electronics in the late 1980s. Switching power supplies have much more complex circuits than linear power supplies. The advantage of switching power supplies is that they provide high current and can supply multiple voltages, as in the power supply from a personal computer. Switching power supplies can provide higher current per size/weight than linear power supplies and are easily obtainable at a low cost due to the competition between computer suppliers. They are very efficient, sometimes reaching 85% or more, and they run much cooler because of this. They are lightweight because they do not use a power/isolation transformer. They are sometimes noisy because they actually switch power at frequencies between 60 and 300 kilohertz. Although some early switchers produced objectionable amounts of RF noise, today you can build very quiet switchers using proper design techniques and careful EMI filtering. Proper design must be implemented to keep noise out of the equipment and isolate the mains from your equipment. The disadvantages of switching power supplies are that they have a high parts count, are not as straightforward to repair, and the supplies must often be loaded in order for them to provide and output voltage.

How to Choose a Power Supply

There a number of methods and considerations for choosing power supplies for your electronics workbench. The first consideration when looking a for a power supply is to decide if you want to purchase or build one. Do you want a single, dual, or triple voltage power supply? Do you want a high-voltage and current supply? Do you want a variable power supply with metering? There are many questions to ask yourself, and there are many answers, based on your budget and power supply needs.

When selecting a general-purpose power supply for your electronics workshop, a good place to start is to purchase or build a an adjustable-voltage, linear power supply, as shown in Figure 7-1. A 30 to 50 volt adjustable 3 to 4 ampere power supply with circuit breaker protection and a voltage and current meter is an ideal bench supply. Adjustable current output control and/or current limiting is desirable. You can purchase a new power supply for about $100 to $150 from Jameco Electronics, Circuit Specialists, and a number of other electronic suppliers.

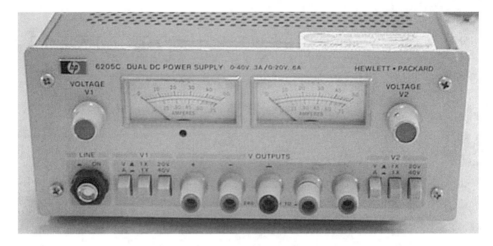

Figure 7-1 Adjustable-voltage, linear power supply.

Another approach to obtaining a power supply is to look to surplus electronics suppliers. You can locate good used laboratory power supplies through both mail-order dealers or local surplus suppliers. A local supplier is preferable, since the cost of shipping heavy power supplies may be high. You can purchase a name-brand surplus adjustable voltage power supply for about $50–60. Spending some extra money could provide a used calibrated power supply from a certified dealer/reseller of used test equipment for about $80.

The third approach is to build a variable-output voltage power supply, which is ideal for the electronic enthusiast who doesn't have much money and is just starting out. If you enjoy building circuits and want to get some building experience "under your belt," then you might want to consider building your own power supply. An adjustable-output voltage power supply that you can build yourself is featured at the end of this chapter.

If you have considered purchasing a power supply and have looked at a catalog of power supplies, then you may have noticed that dual-voltage and even triple-voltage power supplies are available. If you can afford a dual- or a triple-voltage power supply, it would be money well spent. In your electronics pursuits, at one time or another, you may get involved with integrated circuits. Eventually, you may have to deal with analog integrated circuits, which often require both plus and minus voltages to operate. Integrated circuits such as op-amps, amplifiers, and filters often require a plus and minus supply from 9 to 15 volts. This is the primary reason for purchasing a dual-voltage adjustable power supply. If you can afford a dual power supply with adjustable current outputs and voltage as well as current metering, your money well be well spent. A triple-voltage output supply is also very useful for working on both analog and digital circuits at the same time. For example, you may be designing an analog filter that requires both a plus and minus 12 volt supply. The filter circuit might then be fed into a microprocessor operating at 5 volts, hence the need for three power supplies at the same time.

The diagram in Figure 7-2, illustrates two single-voltage power supplies. When these two power supplies are connected as shown, with the minus connection of the first power supply connected to the plus voltage output of the second power supply, it is now possible to have a dual-output power supply capable

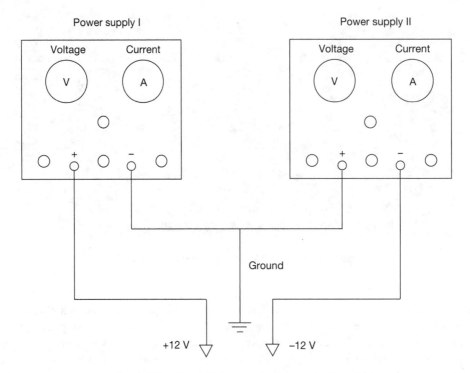

Figure 7-2 Two single-voltage power supplies.

of providing both a plus and minus voltage for analog or linear integrated circuits. Looking through a surplus catalog, you might be able to locate two identical low-cost power supplies that could be connected together to form a dual supply.

Dual- or triple-voltage power supplies are, of course, more expensive than a single-output power supply, so another option might be to purchase a good single-output, 30 to 50 volt dc 5 ampere adjustable laboratory power supply, a workhorse power supply for your electronics workbench output, with good filtering, metering, and output protection. Then, at a later date you could purchase or build a compact low-current power supply for op-amp circuitry and prototypes. Later in this chapter, we will also present a low-current, dual-voltage power supply that would be ideal for experimenting with op-amps that require plus and minus power supplies.

Power Supply Components

The Half-Wave Rectifier

Power supplies generally consist of at least a power transformer, a half-wave or full-wave rectifier, and a filter capacitor. Usually, a step-down transformer converts line voltage of 115 volts or 220 ac down to a lower ac voltage at the secondary of the power. The stepped-down ac voltage at the output of the transformer is then coupled to a rectifier circuit, which in turn is fed to a filter capacitor.

The diagram in Figure 7-3 illustrates a simple half-wave rectifier circuit. The semiconductor diode conducts current in one direction but not the other. During half of the ac cycle, the rectifier conducts and current flows through the rectifier to the load. During the other half cycle, the rectifier is reverse biased and there is no current to the load. As shown, the output is in the form of pulsed dc, and the current always flows in the same direction. A filter capacitor is then used to smooth out these variations and provide a higher average dc voltage from the circuit. The average output voltage read by a dc voltmeter with no filter at the output is $0.45 \times E_{rms}$ of the ac voltage delivered by the secondary or the transformer. E_{rms} is RMS or root mean square voltage. Since the frequency of the pulses is low, considerable filtering is required to provide a good smoothed output. For this reason, the circuit is usually limited to applications in which the required current is small. The peak inverse voltage (PIV) that the rectifier must withstand when it isn't conducting varies with the load. The capacitor connected across the circuit stores the peak positive voltage when the diode conducts on the positive pulse.

Full-Wave Rectifier I

The diagram in Figure 7-4 depicts the commonly used full-wave, center-tap rectifier circuit. In this circuit arrangement, the outputs of two half-wave rectifiers are combined. This type of rectifier circuit makes use of both halves of the ac cycle. Note that this circuit utilizes a transformer with a center-tapped secondary. Each outer leg of the transformer's secondary is connected to a silicon diode, shown at D1 and D2. The center-tap of the transformer is used as the common-point connection or ground reference.

The average output voltage is $0.9 \times E_{rms}$ of half of the transformer secondary; this is the maximum that can be obtained using a choke-input filter arrangement. The peak output voltage is $1.4 \times E_{rms}$ of half the transformer secondary; this is the maximum voltage that can be obtained from a capacitor-input filter.

The peak inverse voltage (PIV) presented to each diode is independent of the type of load at the output. This is because the peak inverse voltage condition occurs when diode D1 conducts and diode D2 does not

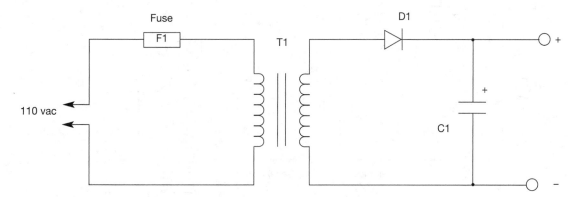

Figure 7-3 Half-wave rectifier circuit.

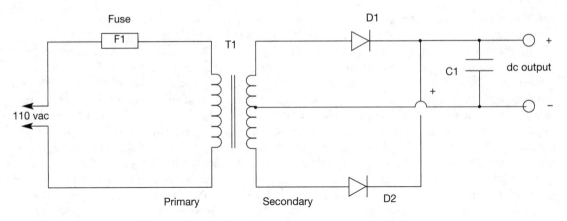

Figure 7-4   Full-wave, center-tap rectifier circuit.

conduct. The positive and negative voltage peaks occur at precisely the same time, a condition different from that in the half-wave circuit. As the cathodes of diodes D1 and D2 reach a positive peak ($1.4\ E_{rms}$), the anode of diode D2 is at a negative peak, also at $1.4\ E_{rms}$, but in the opposite direction. The total peak inverse voltage is then $2.8\ E_{rms}$.

Full-Wave Rectifier II

Another commonly used full-wave rectifier circuit is shown in Figure 7-5. In this circuit, two rectifiers operate in series during each half of the cycle, one rectifier being in the lead to the load, and the other being in the return lead. As shown, when the top lead of the transformer's secondary is positive with respect

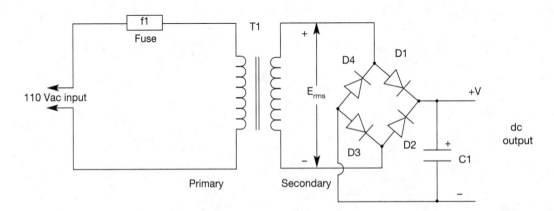

Figure 7-5   Full-wave bridge rectifier circuit

to the bottom lead, diodes D1 and D3 will conduct while diodes D2 and D4 are reversed biased. During the next half of the cycle, when the top lead of the transformer is negative with respect to the bottom, diodes D2 and D4 will conduct while diodes D1 and D3 are reversed biased.

The shape of the output voltage, is the same as in the center-tapped, full-wave bridge shown in Figure 7-4. The average dc output voltage into a resistive load or choke-input filter is 0.9 times the RMS voltage delivered by the transformer secondary. With a capacitor filter and a light load, the maximum output voltage is 1.4 times the secondary RMS voltage. The circuit shows the inverse voltage to be 1.4 E_{rms} for each diode. When an alternate pair of diodes is conducting, the other diodes are essentially connected in parallel in a reverse-biased direction. The reverse stress is then 1.4 E_{rms}. Each pair of diodes conducts on alternate half cycles, with the full load current through each diode during its conducting half cycle. Since each diode is not conducting during the other half cycle, the average current is one-half the total load current drawn from the supply.

Filters

The pulsating dc waves from the rectifiers are not sufficiently constant in amplitude to prevent hum corresponding to the pulsations. Filters are required between the rectifier and the load to smooth out the pulsations and provide an essentially constant dc voltage. The design of the filter depends to a large extent on the dc voltage output, the voltage regulation of the power supply, and the maximum load current ratings of the rectifier. Power supply filters are low-pass devices using series inductors and shunt capacitors, as depicted in Figure 7-6.

Voltage Multipliers

Voltage multipliers operate by the process of charging one or more capacitors on one-half cycle of an ac waveform, then connecting the capacitor or capacitors in series with the opposite polarity of the ac waveform on the alternate half cycle. Note that with full-wave multipliers, this cycle occurs during both half cycles. Voltage multipliers are often used in high-voltage power supplies, to double, triple, or quadruple voltages.

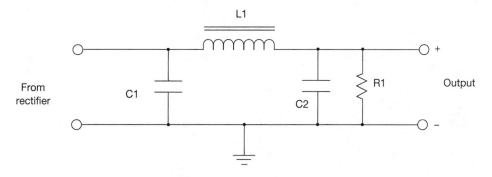

Figure 7-6 Capacitor-inductor filter.

Half-Wave Doubler

Figure 7-7, illustrates a half-wave voltage-doubler circuit. During the first negative half cycle, diode D1 conducts, charging C1 to the peak rectified voltage (1.4 E_{rms}). Capacitor C1 is charged with the polarity. During the positive half cycle of the secondary voltage, diode D1 is cut off and D2 conducts, thus charging capacitor C2. The amount of voltage supplied to C2 is then the sum of the transformer's peak secondary voltage plus the voltage stored in C1. On the next negative half cycle, diode D2 is non-conducting and C2 will discharge the load. With a load connected to the doubler output, the voltage across C2 drops during the negative half cycle and is recharged to 2.8 E_{rms} during the positive half cycle.

Voltage Tripler

Figure 7-8 presents a half-wave voltage-tripler circuit. On one-half of the ac cycle, capacitors C1 and C3 are charged to the source voltage through diodes D1, D2, and D3. On the opposite half of the cycle, D2 conducts and C2 is charged to twice the source voltage, since it sees the transformer plus the charge in C1 as its source. At the same time, diode D3 conducts, and with the transformer and the charge in C2 as the source, C3 is charged to three times the transformer voltage.

Voltage Quadrupler

The voltage-quadrupler circuit shown in Figure 7-9 operates very similar to the voltage-tripler circuit. The output voltage in this circuit approaches an exact multiple of the peak ac voltage when the output current drain is low and the capacitors' values are high.

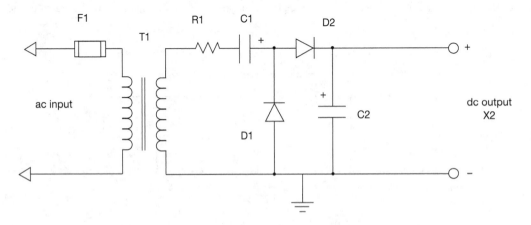

Figure 7-7   Half-wave voltage-doubler circuit.

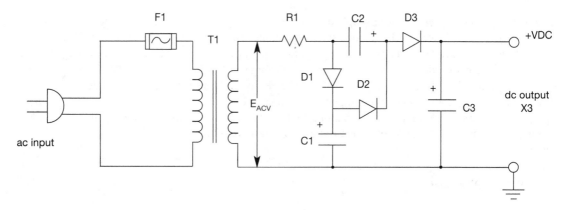

Figure 7-8   Half-wave voltage-tripler circuit.

Power Supply Regulation

Rectifier and filter circuits by themselves are unable to protect equipment from problems associated with input power line fluctuations, load-current variations, and residual ripple voltages. Further measures may be necessary to provide sufficiently clean and stable power. Regulators can eliminate these problems, but not without costs in circuit complexity and power-conversion efficiency. Voltage regulators are often used to provide an additional level of conditioning.

Zener Diode Regulation

One of the most basic forms of low-cost voltage regulation can be accomplished by using a zener diode. The zener diode can be used to maintain the voltage applied to a circuit at a practically constant value, regardless

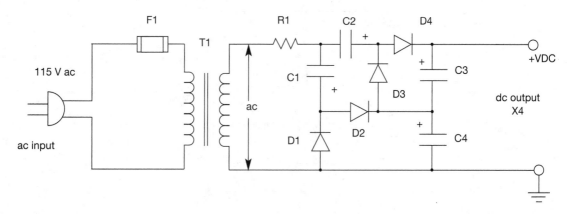

Figure 7-9   Voltage-quadrupler circuit.

of the voltage regulation of the power supply or variations in load current. A zener diode regulation circuit is shown in Figure 7-10. Note that the cathode side of the diode is connected to the positive side of the supply.

Zener diodes are available in a wide variety voltages and power ratings. The minimum voltage regulation range of a zener diode is about 1.2 volts and extends up to a few hundred volts. Power ratings of zener diodes run from less than 0.25 W to over 50 W. The ability of the zener diode to stabilize a voltage depends on the series diode's conducting impedance. This can be as low as 1 ohm or less in a low-voltage, high-power diode or as high as 1000 ohms in a high-voltage, low-power diode. Zener diodes provide a simple low-cost means of regulation where critical regulation is not required.

Linear Regulators

Linear regulators come in two varieties: series and shunt. The shunt regulator is simply an electronic version of the zener diode. For the most part, the active shunt regulator is rarely used since the series regulator is a superior choice for most applications. The series regulator consists of a stable voltage reference, which is usually established by a zener diode, a pass transistor in series with the power source and the load, and an error amplifier, as shown in Figure 7-11. In critical applications, a temperature-compensated reference diode is used instead of the zener diode. The output voltage is sampled by the error amplifier, which compares the output, which is usually scaled down by a voltage divider and fed to the reference. If the scaled-down output voltage becomes higher than the reference voltage, the error amplifier reduces the drive current to the pass transistor, thereby allowing the output voltage to drop slightly. Conversely, if the load pulls the output voltage below the desired value, the amplifier drives the pass transistor into increased conduction. The "stiffness" or tightness of regulation of a linear regulator depends on the gain of the error amplifier and the ratio of the output scaling resistors. In any regulator, the output is cleanest and regulation stiffest at the point where the sampling network or error amplifier is connected. If heavy load current is drawn through long leads, the voltage drop can degrade the regulation at the load. To combat this effect, the feedback connection to the error amplifier can be made directly to the load. This technique, called remote sensing, moves the point of best regulation to the load by bringing the connecting loads inside the feedback loop.

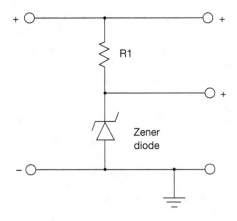

Figure 7-10   Zener diode circuit.

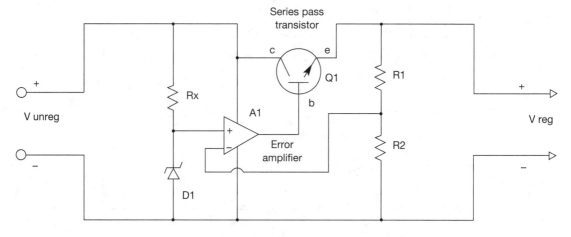

Figure 7-11 *Series regulator circuit.*

In a series regulator, the pass-transistor power dissipation is directly proportional to the load current and input–output volt differential. The series pass element can be located in either leg of the supply. Either NPN or PNP devices can be used, depending on the ground polarity of the unregulated input. A simple zener diode reference or IC op-amp error amplifier may not be able to source enough current to a pass transistor that must conduct heavy load current. The Darlington pass transistor multiplies the pass transistor beta, thus extending the control range of the error amplifier. If the Darlington arrangement is implemented with discrete transistors, resistors across the base–emitter junctions may be necessary to prevent collector base leakage currents in Q1 from being amplified and turning on the transistor pair. These resistors are contained in the envelope of a monolithic Darlington device.

When a single pass transistor is not available to handle the current required from a regulator, the current-handling capability may be increased by connecting two or more pass transistors in parallel. The differential between the input and output voltages is a design trade-off. If the input voltage from the rectifiers and filter is only slightly higher than the required output voltage, there will be minimal voltage drop across the series pass transistor, resulting in minimal thermal dissipation and high power-supply efficiency. The supply will have less capability to provide regulator power in the event of power line brownout and other reduced line voltage conditions, however. Conversely, a higher input voltage will provide operation over a wider range of input voltage, but at the expense of increased heat dissipation.

Overcurrent Protection

Damage to a pass transistor can occur when the load current exceeds the safe mount. Figure 7-12, illustrates a simple current-limiter circuit that will protect Q1. All of the load current is routed through R1. A voltage difference will exist across the R1; the value will depend on the exact load current at a given time. When the load current is determined to be at safe value, the voltage drop across R1 will forward bias Q2 and cause it to conduct. Because Q2 is a silicon transistor, the voltage drop across R1 must exceed 0.6 V to turn Q2 on. This being the case, R1 is chosen for a value that provides a drop of 0.6 V when the maximum safe load cur-

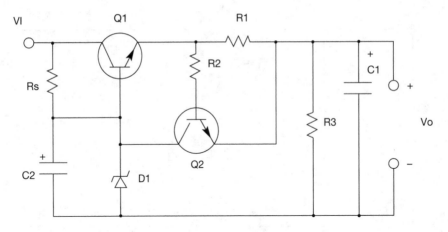

Figure 7-12 Regulator with overload protection.

rent is drawn. In this instance, the drop will be 0.6 V when I_L reaches 0.5 A. R2 protects the base emitter junction of Q2 from current spikes, or from destruction in the event Q1 fails short-circuit conditions.

When Q2 turns on, some of the current through Rs flows through Q2, thereby depriving Q1 of some of its base current. This action, depending upon the amount of Q1 base current at a precise moment, cuts off Q1 conduction to some degree, thus limiting the current through it.

"Crowbar" Circuits

In a regulated power supply, the only component standing between an elevated dc source voltage and an electronic circuit is one transistor, or a group of transistors wired in parallel. If the transistor, or one of the transistors in the group, happens to short internally, your equipment could suffer damage.

To safeguard load equipment against possible overvoltage, some power supply manufacturers include a circuit known as a crowbar. This circuit usually consists of a silicon-controlled rectifier (SCR) connected directly across the output of the power supply, with a voltage-sensing trigger circuit tied to its gate. In the event that the output voltage exceeds the trigger set point, the SCR will fire and the output will be short circuited. The resulting high current in the power supply (shorted out in series with a series pass transistor failed short) will blow the power supply's line fuses. This is both a protection for the supply as well as an indicator that something has malfunctioned internally. For these reasons, never replace blown fuses with ones that have a higher current rating.

IC Voltage Regulation

The modern trend in regulators is toward the use of three-terminal devices commonly referred to as three-terminal regulators. Figure 7-13 illustrates a three-terminal positive voltage, integrated circuit regulator. Integrated circuit regulators are generally connected directly to the output of a filter capacitor, which is connected to the line-current input transformer. Inside each regulator is a voltage reference, high-gain er-

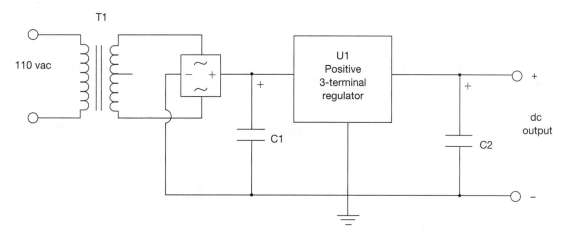

Figure 7-13   *Basic three-terminal regulator.*

ror amplifier, temperature-compensated voltage-sensing resistors, and a pass element. Many currently available units have thermal shutdown, overvoltage protection, and current fold-back, making them virtually destruction-proof.

Three-terminal regulators have three connections, one for unregulated dc input, regulated dc output, and ground. Fixed-voltage integrated circuit regulators comes in a number of different voltage and current outputs. The most commonly available are the 78L05, a 5 volt regulator; the 78L12, a 12 volt regulator; and the LM78L15, a 15 volt regulator capable of handling from 100 mA to 5 amperes. Three-terminal regulators are also available as negative voltage regulators, such as the LM7905, the LM7912, and the LM7915 series regulators. The LM78 and LM79 series regulators teamed together form a plus/minus dual-voltage power supply.

Three terminal regulators are also available in adjustable-voltage regulators, as shown in Figure 7-14. The LM117 regulator, for example, accepts a voltage range of 1 to 28 volts and can provide an adjustable output with a range of 1.2 volts to 25 volts. An adjustable IC regulator requires only a few more external components than the fixed-voltage regulators, and can provide a simple and low-cost, continuously adjustable power supply.

It is easy to see why regulators of this sort are so popular when you consider the low price and the number of individual components they can replace. The IC regulators are available in several different package styles, depending on current ratings. Low-current (100 mA) devices frequently use the plastic TO-92 and DIP-style cases. TO-220 packages are popular in the 1.5 A range, and TO-3 cases house the larger 3 A and 5 A devices.

Dual-Voltage Power Supply

Figure 7-15 shows a dual-voltage power supply that produces both a positive 12 volt output and a minus 12 volt output. The circuit begins with a 115 to 28 volt ac, 1 ampere center-tapped transformer. The stepped-down ac voltage from T1 is fed to the input of the bridge rectifier at BR. The pulsed dc output from the bridge

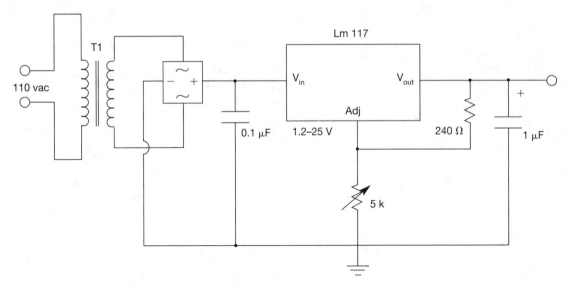

Figure 7-14 Adjustable three-terminal regulator.

rectifier is then filtered by capacitors C1 and C3 in the positive supply and by C2 and C4 in the negative sup-ply. The filtered dc voltage is next passed on to the IC regulators U1 and U2. On the positive supply side, a LM7812 drops the positive input voltage down to 12 volts dc and capacitor C5 provides a smooth output on the positive output of the power supply. The LM7912 regulator in the minus side of the power supply drops the output to minus 12 volts and C6 provides a clean and smooth output at the negative supply output pins.

If you already own a single-output lab/bench power supply, then you may want to consider building the dual-voltage power supply. This straightforward and easy to construct dual power supply is ideal for power-ing linear op-amp circuits used in filter and amplifier circuits. Note that you can use individual silicon diodes

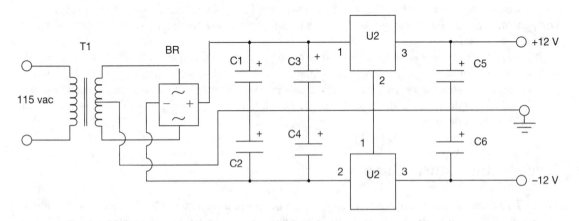

Figure 7-15 Dual-voltage power supply.

for the bridge rectifier or the four-pin package as shown . This circuit can be constructed either on a small breadboard or on a printed circuit board, and will provide a low-cost reliable dual-voltage power supply.

dc-to-dc Power Supply

dc-to-dc power supplies come in a variety of types. Sometimes they supply higher voltages than the input voltage and operate as multipliers, and they often take the form of voltage converters, which take a positive voltage and convert it to a negative voltage.

The heart or building block of many dc-to-dc converter circuits is the Maxim constant-frequency pulse-width-modulation (PWM) controller, which operates in the current mode. A block diagram of the MAX 1846/1847 is shown in Figure 7-16, and list of the device pinouts is given in Table 7-1. The MAX

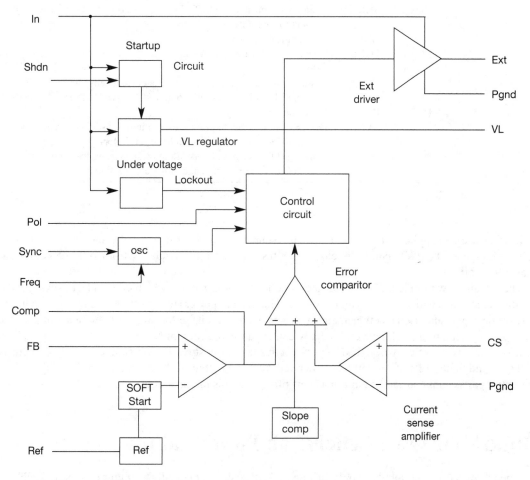

Figure 7-16   Maxim MAX 1846 constant-frequency pulse-width-modulation controller.

Table 7-1 *MAX 1846/1847 PWM controller pinouts*

Pin, Max1846	Pin, Max1847	Name	Function
	1	POL	Sets polarity of EXT pin, ground for PMOS FET, set to V_L for NMOS FET in transformer applications
1	2	V_L	VL low dropout regulator, connect 0.47 µF capacitor from V_L to ground
2	3	FREQ	Oscillator frequency set input, resistor from FREQ to ground
3	4	COMP	Compensation node, connect a resistor/capacitor from COMP to ground
4	5	REF	1.25 V reference output; REF can source to 500 µA; bypass with 0.1 µF capacitor from REF to ground
5	6	FB	Feedback input; connect FB to center of resistor/divider between output and REF
	7,9	N.C.	No connection
	8	SHDN	Shutdown control; drive SHDN low to turn off controller; drive high or connect to IN for normal
6	10,11	GND	Analog ground; connect to PGND
7	12	PGND	Negative rail for EXT, driver and negative current sense input; connect to ground
8	13	CS	Positive current sense input; connect a current sense resistor
9	14	EXT	External MOSFET gate-driver output; EXT swings from IN to PGND
10	15	IN	Power supply input
	16	SYNC	Drive low or connect to ground to set internal oscillator frequency with Rfreq. Drive SYNC with logic level clock input signal to externally set the converters' operating frequency. dc-to-dc conversion cycles initiate on rising edge of the clock's signal

1846/1847 can be supplied with an input voltage source of +3 volts to +16 volts and produces a –2 to –200 volt output. The PWM controller clock operates at a switching frequency from 100 to 500 kHz and is over 90% efficient.

The circuit shown in Figure 7-17 illustrates a dc-to-dc power supply, which converts a positive 12 volt input voltage to a negative 12 volt supply voltage. This type of power supply can be used when you have a positive dc supply and need a complementary negative supply voltage for op-amp integrated circuits. The plus 12 volt to minus 12 volt converter/supply centers around a Maxim MAX 1846/1847 integrated circuit. This chip is specifically designed to perform this task. A positive 12 volt source is applied to pin 10 of the IC and minus 12 volts at 250 mA is derived from the output of the circuit at C6. The circuit has a minimum parts count and can be built in a short time using this modular approach.

Build Your Own Bench/Lab Power Supply

The bench or lab power supply shown in Figure 7-18 and schematically in Figure 7-19 is capable of providing a variable output voltage up to 30 volts with current output to 10 amperes. This is a heavy-

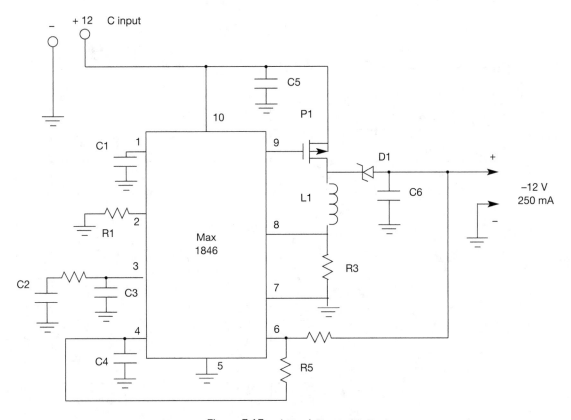

Figure 7-17   dc-to-dc converter.

Figure 7-18   Bench/lab power supply.

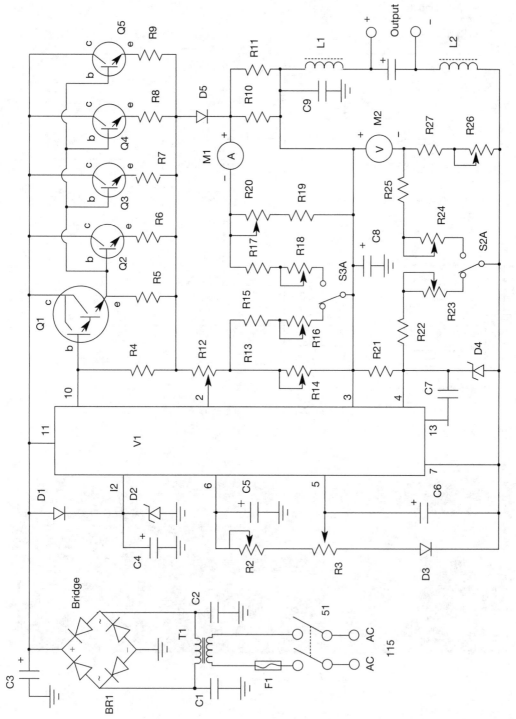

Figure 7-19 30 volt, 10 ampere lab power supply.

duty lab power supply that would make a great addition to your electronics workbench, and you can build it yourself. The power supply revolves around the 14 pin Motorola MC1723 regulator IC. The MC1723 chip is the workhorse of the power supply. It can be used to deliver a load current up to 150 mA. Using an external "pass" on the output of the regulator can increase the output current to 10 amperes. Pins 2 and 3 of the MC1723 are the current-limit and current-sensing terminals, respectively. The inverting input and the noninverting inputs are shown on pins 4 and 5. Pin 6 represents the voltage reference input, whereas pin 13 is the frequency compensation terminal. Power to the regulator is provided on pin 11, and power is available on pin 12. Table 7-2 lists the LM723 pinouts. A 6.2 volt zener diode is connected to pin 9 and the regulator output is provided at pin 10. The output of the regulator is fed to Q1, followed by power transistors Q2 through Q5, which boosts the current capability to 10 amperes. Power diode D3 is connected from the "pass" transistor network to the panel current meter. Potentiometer R12 is the current-limiting control. Potentiometers R14, R16, R18, and R20 are used for setting the current-range settings for the current panel meter at M2. Potentiometers R23, R24, and R26

Table 7-2   LM 723 voltage regulator IC pinout chart

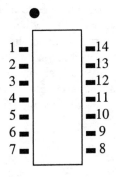

Pin	Function	Description
1	NC	No connection
2	CL	Current limit pin
3	CS	Current-sensing pin
4	Invert IN	Inverting input
5	Non-Invert	Noninverting input
6	V_{REF}	Voltage reference pin
7	$-V$	Minus voltage input
8	NC	No connection
9	Vz	Zener regulation input
10	V_{OUT}	Voltage output
11	Vc	Open collector output
12	$+V$	Plus voltage input
13	FC	Frequency compensation
14	NC	No connection

are utilized for adjusting the voltage-range setting for the voltmeter shown at M1. The power "on" indication is formed by the LED, zener diode D1, and resistor R1. The voltage reference input on pin 6 is adjusted via R2.

The power supply has two range selector switches that provides a 0.7 volt to 6 volt range and a 6 to 37 volt range via S2. The power supply also provides two current ranges; the low-current range is from 0 to 1 ampere and the high-current range covers from 1 ampere to 10 amperes, selected via S3. The 110 volt ac input is first fed to the primary of the transformer T1 through power switch S1 and the 3.15 ampere fuse. The secondary of T1 is coupled to a bridge rectifier. Capacitor C3, a 15,000 μf, is the "big boy" capacitor and is mounted on the chassis (see Figure 7-20).

Much of the lab/bench power supply circuit can be built on a printed circuit board. Output coils L1 and L2 each have two turns, hand wound on a ferrite bead as shown in Figure 7-21. Potentiometers R3 and R12 are mounted on the front panel of the lab/bench power supply. Potentiometer R3 is used to adjust current settings, whereas R12 is used to make voltage adjustments to the meter circuit. The voltmeter and ammeter are both mounted on the front panel along with the output terminal posts and the current and voltage range-selector switches S2 and S3. Power switch S1 is also mounted on the front panel. You can mount the 115 VAC power receptacle, the same type used on computer power supplies, as well as the fuse holder on the rear panel.

Before mounting the power transistors, you will need to look at Figure 7-22, which illustrates how to provide isolation when mounting the transistors so they will not short out. Transistors Q1, Q2, Q4, and Q5 and power diode D3 are all mounted on a large aluminum heat sink at the rear of the power supply (see Figure 7-23). If desired, you can mount two ventilation fans over the heat sink to remove the heat produced when the supply is under heavy load in 10 ampere mode. When mounting the transistors, be careful to isolate the individual pins from the heat sink. Use pin feed-through insulators to prevent the transistor pins from shorting to the heat sink. Be sure to observe the same precautions when mounting the power diode D3 as well.

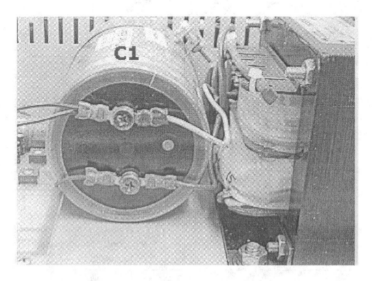

Figure 7-20 15,000 μf capacitor mounted on chassis.

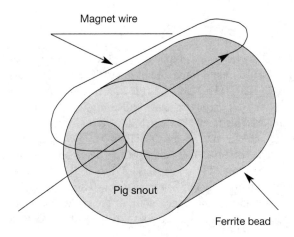

Magnet wire

Pig snout

Ferrite bead

Figure 7-21 Output coils.

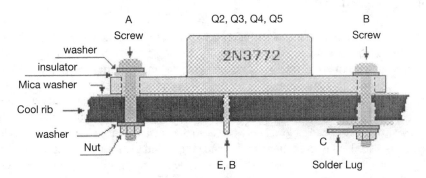

A
Screw

Q2, Q3, Q4, Q5

B
Screw

washer
insulator
Mica washer
Cool rib
washer
Nut

2N3772

E, B

C

Solder Lug

Figure 7-22 How to provide isolation when mounting transistors.

Figure 7-23 Aluminum heat sink.

Construction

Construction of the bench power supply is accomplished on three different platforms. Most of the small components are mounted on a 4-by-6 inch printed circuit board. The regulator chip as well as all trim potentiometers R2, R14, R16, R18, R20, R23, R24, and R26 are mounted on the PC board. Power resistors R10 and R11 and capacitor C4 as well as power resistors R6 through R9 are all board mounted. Potentiometers for current and voltage adjustment, R3 and R12, are both mounted on the front panel of the bench supply. Both the voltage and current meters as well as the power-binding posts and hi–low range switches are front panel mounted. The on–off switch, line cord jack, and fuse holder are all mounted on the rear panel of the chassis. Transformer T1 and capacitor C1 as well as the diode bridge package are all mounted on the bottom of the main chassis. Power transistors Q1 through Q5 as well as power diode D3 are all mounted on a large aluminum heat sink, the length of the entire rear of the power supply chassis. The power transistor and diode leads are all routed back to the circuit board, which is mounted on standoffs on the main chassis.

Adjustment Procedure

Begin by first getting familiar with all the potentiometers, and refer to Table 7-3 when adjusting the potentiometers. The two front panel potentiometers are R3 and R12; the remaining trim pots are all printed circuit board types. Be sure to "zero" the panel meters with the small plastic screw on the front of the meter. First turn all the potentiometers counterclockwise before applying power to the circuit. Insert a small-value fuse (0.3 A) into the fuse holder. Next, set R3 and R12 to the midway point, and set switches S2 and S3 to the "low" current and voltage settings. Connect a multimeter to the output of the power supply and observe the polarity of the test leads. Now, apply power to the power supply. Connect the plus lead of your multimeter to the top-side lead of potentiometer R3 and you should get a reading of 6 volts. If needed, this voltage can be adjusted by turning R2. Now, if you move the red meter probe to the power supply output terminal you will find you can adjust R3 for an output from 0.7 to 6 volts. Leave the multimeter probes connected to the output terminals and switch S2 to the "high" or 30 volt position. You will notice that the front panel meter will jump up immediately. Adjust R3 all the way to the right (clockwise) and adjust trimmer R23 until your multimeter shows 30 volts. Now, you can adjust trimmer R26 until the panel me-

Table 7-3 Trim-pot adjustment procedures

Trim-pot	Adjustment procedure	Measured result
R2	Top of R3	6 volt
R23	Output	30 volt
R26	Panel-meter adjust (M1)	Full scale, 30 volts
R24	Panel-meter adjust (M1)	Full scale, 6 volts
R20	Panel-meter adjust (M2)	Set to same value as multimeter reading
R18	Panel-meter adjust (M2)	Set to same value as multimeter reading
R14	Adjust until panel meter reads 1 ampere	Control with multimeter
R16	Adjust until panel meter reads 10 amperes	Control with multimeter

ter shows the same 30 volts. Switch back to the "low" or 6 volt position and the panel meter to the 6 volt full-scale with trim pot R24. Switch off the power supply, unplug the power cord, and remove the "temporary'" fuse and replace it with the 3.15 ampere fuse.

Switch the panel meters to the 6 volt and 1 ampere positions via S2 and S3. Now, turn the current limiter control R12 all the way to the left. Set the voltmeter to 4 volts via potentiometer R3. Select a setting on your multimeter of 100 to 300 mA dc. Take the red meter probe and insert a 39 ohm resistor between the probe lead and the plus power terminal on the output of the power supply. The panel meter will show a bit of current and at the same time the needle of the panel voltmeter will fall back to about 2 volts or lower. This lets you know that the current limiter is working. Remove the 39 ohm resistor, switch you multimeter to the highest current reading, and connect it to the power supply output terminals. Now adjust R12 until you see the current increase on both the panel meter and the multimeter. Switch the power supply off.

Now it is time to set up the 20 ampere adjustments. Set the front panel switch to 10 amperes, and turn R12 all the way to the left. Turn on the power supply with the multimeter leads connected to the power supply output terminals. Now, adjust R12 so that 5 amperes is displayed on the multimeter. Next, adjust the panel meter with R20 until it shows the same on the multimeter. Be sure that your multimeter has a 10 ampere position before you increase R12 to "high."

Next, turn R12 all the way to the left and switch to the 1 ampere current setting. Adjust R12 all the way to the right and with R18 adjust the values of the multimeter and the panel meter until they are both equal.

Remove the multimeter. Turn both R12 and R3 all the way to the left. Return the switches to the 1 ampere and 6 volt settings. Short out the output terminals on the power supply. Now, turn R12 all the way to the right and adjust R14 until the "current" meter indicates 1 ampere full scale.

Finally, turn R12 all the way back to the left and place the current switch to the 10 ampere position. Adjust the full scale of the panel meter with R16 until it shows exactly 10 amperes. All the adjustments are now correct and your power supply is ready to use. Unplug the power supply and remove the shorting wire. Once the construction of the main power supply is completed, you can build the automatic fan cooling switch circuit, which is used to cool the power transistors.

Cooling Circuit

The circuit diagram shown in Figure 7-24 illustrates an automatic temperature-controlled fan switch. The thermistor sensor shown at R4 is a 1.7 k at 70 degree unit available from Radio Shack. It is fed to the positive input of the comparator at pin 5 of U1:A. The thermistor is mounted on the large heat sink to sense the heat generated by the power transistors. A 47 k trim potentiometer forms the temperature threshold control, which is fed to the minus input of U1:A. An LM339 quad comparator acts as a comparator and switch used to drive transistor Q1. When transistor Q1 is activated, it in turn drives the relay at RY1, which can be used to turn on two 12 volt miniature muffin fans mounted on the heat sink over the power transistors. The entire automatic fan switch circuit is powered from the main power supply from a 12 volt fixed source at C3. Once the automatic fan cooling circuit has been completed, you can test it to make sure it works correctly. Connect fan or 12 volt load to the relay contacts and connect the circuit to a 12 volt dc power source. Grab a hair dryer and turn it on, facing it toward the thermistor sensor. Now adjust the potentiometer R1 to trigger the relay as the heat begins to blow on the thermistor. The automatic fan switch circuit is now complete and ready. The thermistor can be mounted on the heat sink. Two miniature

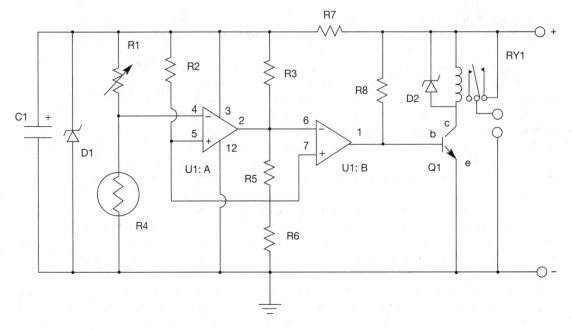

Figure 7-24 Circuit diagram of automatic temperature-controlled fan switch.

fans can be mounted on the heat sink, over the power transistors. Now you can mount the automatic fan cooling circuit inside the main power supply chassis. Your lab/bench power supply is now ready to serve you in your electronics workshop. Complete parts lists for the unit follow.

± 12 Volt Dual-Voltage Power Supply Parts List

C1, C2	2200 μf, 50 volt electrolytic capacitors
C3, C4	100 μf, 50 volt ceramic capacitors
C5, C6	220 μf, 50 volt electrolytic capacitors
BR1	2 ampere silicon diode bridge
T1	26–36 volt CT, 1 ampere transformer
U1	LM7812, 12 volt positive regulator
U2	LM7912, 12 volt negative regulator
Miscellaneous	PC board, wire, terminals, etc.

+12 to −12 Volt dc-to-dc Power Supply Parts List

R1	150 kohm, ¼ watt resistor
R2	10 kohm, ¼ watt resistor

R3	0.1 ohm, ¼ watt resistor
R4	97.6 kohm, 1 %, ¼ watt resistor
R5	10 kohm, 1 %, ¼ watt resistor
C1	0.47 µf, 25 volt ceramic capacitor
C2	27 nf, 25 volt capacitor
C3	220 pf, 25 volt capacitor
C4	0.1 µf, 25 volt capacitor
C5, C6	10 µf, 25 volt ceramic capacitor
P1	Si9435DY P-channel MOSFET (Siliconix)
L1	100 µh 1 ampere inductor (Coilcraft DS5022P-104)
D1	EP05/Q03L zener diode
U1	MAX 1846 IC regulator

Bench/Lab High-Current Power Supply Parts List

R1	470 ohm, ½ watt resistor
R2	2 kohm potentiometer
R3	5 kohm linear potentiometer
R4	560 ohm, ¼ watt resistor
R5	47 ohm, ¼ watt resistor
R6, R7, R8, R9	0.1 ohm 5%, 1 watt resistor
R10, R11	0.33 ohm 5%, 10 watt resistor
R12	470 ohm linear potentiometer
R13, R22	820 ohm, ¼ watt resistor
R14, R23	500 ohm potentiometer
R15	150 ohm, ¼ watt resistor
R16	100 ohm potentiometer
R17	1.2 kohm, ¼ watt resistor
R18	1 kohm potentiometer
R19	100 ohm, ¼ watt resistor
R20	100 ohm potentiometer
R21	5 k metal film resistor
R24	20 kohm potentiometer
R25	47 kohm, ¼ watt resistor
R26	5 kohm potentiometer
R27	8.2 kohm ¼ watt resistor
C1, C2	3.3 nf, 75 volt ceramic capacitor
C3	15,000 µf 45 volt electrolytic capacitor
C4	1000 µf, 63 volt electrolytic capacitor
C5, C6	4.7 µf, 25 volt tantalum capacitor
C7	4.7 nf, 63 volt ceramic capacitor
C8	10 µf, 63 volt electrolytic capacitor
C9	1 µf, 63 volt foil capacitor

C10	22 nf, 63 volt ceramic capacitor
D1	1N4004 silicon diode
D2	1N4148 silicon diode
D3	1N4389 20 ampere power diode
D5, D6, D7, D8	Red LEDs
ZD1	1N4754A 1 watt zener diode, 39 volt
ZD2	1N4636A 250 mW zener diode, 6.2 volt
Q1	2N6388, MJ2501, BD267A, or TIP140 transistor
Q2, Q3, Q4, Q5	2N3772 or NTE181 transistor
L1	Ferrite bead (see text)
L2	Ferrite bead (see text)
U1	MC1723 Motorola regulator IC, or NTE -923
BR1	Bridge rectifier
M1-	0–500 µA panel meter
M2	0–1 mA panel meter
T1	30–36 volt, 10 ampere Transformer
F1	Fuse, 3.15 ampere
S1	On–off toggle switch, DPDT
S2	On–off–on toggle switch, SPDT
S3	On–off–on toggle switch, SPDT
S4	On–off–on DPDT for meter
Miscellaneous	PC board, wire, connectors, knobs, chassis, etc.

Automatic Fan Cooling Switch Circuit Parts List

R1	47 kohm potentiometer
R2, R3, R6	1.2 kohm, ¼ watt resistor
R4	1.7 kohm, 70 degree thermistor
R5	15 kohm, ¼ watt resistor
R7	270 ohm, ¼ watt resistor
R8	2.2 kohm, ¼ watt resistor
C1	22 µF 25 volt electrolytic capacitor
D1	5.6 volt zener diode
D2	1N4148 silicon diode
Q1	2N2222 PNP transistor
U1	LM339 quad comparator
RY1	12 volt relay (Radio Shack)
Miscellaneous	PC board, wire, connectors, hardware

Battery Power

Batteries and Charging

The availability of solid-state equipment makes it practical to use battery power under portable or emergency conditions. Since almost all types of modern electronics, from toys to test equipment, operate from batteries, we should spend some time talking about batteries and their differences. Low-power equipment can be powered from two types of batteries. The "primary" battery is intended for one-time use and is then discarded; the "storage" (or "secondary") battery may be recharged many times.

A battery is a group of chemical cells, usually series-connected to give some desired multiple of the cell voltage. Each assortment of chemicals used in the cells gives a particular nominal voltage. This must be taken into account to make up a particular battery voltage. For example, four 1.5 V carbon–zinc cells make a 6 volt battery and six 2 volt lead–acid cells make a 12-volt battery.

One of the most common primary-cell types is the alkaline cell, in which chemical oxidation occurs during discharge. When there is no current drawn, the oxidation essentially stops until current is required. A slight amount of chemical action does continue, however, so stored batteries eventually will degrade to the point where the battery will no longer supply the desired current. The time taken for degradation without battery use is called shelf life.

The alkaline battery has a nominal voltage of 1.5 V. Larger cells are capable of producing more milliampere hours and less voltage drop than smaller cells. Heavy-duty and industrial batteries usually have a longer shelf life. Lithium primary batteries have a nominal voltage of about 3 V per cell and by far the best capacity, discharge, shelf life, and temperature characteristics. Their disadvantages are high cost and the fact that they cannot be readily replaced by other battery types in an emergency.

The lithium–thionyl–chloride battery is a primary cell, and should not be recharged under any circumstances. The charging process vents hydrogen, and a catastrophic hydrogen explosion can result. Even accidental charging caused by wiring errors or a short circuit should be avoided. Silver oxide (1.5 V) and mercury (1.4 V) batteries are very good where nearly constant voltage is desired at low currents for long periods. Their main use (in subminiature versions) is in hearing aids, though they may be found in other mass-produced devices such as household smoke alarms.

Secondary or Rechargeable Batteries

Many of the chemical reactions in primary batteries are theoretically reversible if current is passed through the battery in the reverse direction. Primary batteries should not be recharged for two reasons: (1) it may be dangerous because of heat generated within sealed cells, and (2) even in cases where there may be some success, both the charge and life are limited.

Lead–Acid Batteries

The most widely used high-capacity rechargeable battery is the lead–acid type. In automotive service, the battery is usually expected to discharge partially at a very high rate, and then to be recharged promptly while the alternator is also carrying the electrical load. If the conventional auto battery is allowed to discharge fully from its nominal 2 V per cell to 1.75 V per cell, fewer than 50 charge and discharge cycles may be expected, with reduced storage capacity. The most attractive battery for extended high-power electronic applications is the "deep-cycle" battery, which is intended for such uses as powering electric fishing motors and the accessories in recreational vehicles. Size 24 and 27 batteries furnish a nominal 12 V and are about the size of small and medium automotive batteries. These batteries may furnish between l000 and 1200 watt-hours per charge at room temperature. When properly cared for, they may be expected to last more than 200 cycles. They often have lifting handles and screw terminals, as well as the conventional truncated-cone automotive terminals. They may also be fitted with accessories such as carrying cases, with or without built-in chargers.

Lead–acid batteries are also available with gelled electrolyte. Commonly called gel cells, these may be mounted in any position if sealed, but some vented types are position sensitive.

Lead–acid batteries with liquid electrolyte usually fall into one of three classes:

1. The conventional type, with filling holes and vents to permit addition of distilled water lost from evaporation or during high-rate charge or discharge
2. Maintenance-free, from which gas may escape but water cannot be added
3. Sealed batteries.

Generally, the deep-cycle batteries have filling holes and vests.

Nickel–Cadmium Batteries

The most common type of small rechargeable battery is the nickel–cadmium (NiCad), with a nominal voltage of 1.2 volts per cell. Carefully used, these are capable of 500 or more charge and discharge cycles. For best life, the NiCad battery must not be fully discharged. Where there is more than one cell in the battery, the most discharged cell may suffer polarity reversal, resulting in a short circuit or seal rupture. All storage batteries have discharge limits, and NiCad types should not be discharged to less than 1.0 V per cell. There is a popular belief that it is necessary to completely discharge NiCad cells in order to recharge them to full capacity. Called the "memory effect," professional engineers have proved this to be a myth.

Nickel–cadmium cells are not limited to D cells and smaller sizes. They also are available in larger varieties ranging to mammoth 1000 amp-hour units having carrying handles on the sides and caps on the top for adding water, similar to lead–acid types. These large cells are sold to the aircraft industry for jet-engine starting, and to the railroads for starting locomotive diesel engines. They also are used extensively for uninterruptible power supplies. Although expensive, they have a very long life. Surplus cells are often available through surplus electronics dealers, and these cells often have close to their full rated capacity.

Advantages for the ham in these vented cell batteries lie in the availability of high discharge current to the point of full discharge. Also, cell reversal is not the problem that it is in the sealed cell, since water lost through gas evolution can easily be replaced. Simply remove the cap and add distilled water. By the way, tap water should never be added to either nickel–cadmium or lead–acid cells, since dissolved minerals in the water can hasten self-discharge and interfere with the electrochemical process.

Nickel–Metal Hydride

The nickel–metal hydride (NiMH) battery is quite similar to the NiCad, but the cadmium electrode is replaced by one made from a porous metal alloy that traps hydrogen. The voltage is nearly the same, they can be slow-charged from a constant source, and they can safely be deep cycled, but there are some important differences. The most attractive feature of NiMH batteries is their much higher capacity for the same cell size, often nearly twice as much as the NiCd types. The typical AA NiMH cell has a capacity between 1000 and 1300 mAh, compared 600 to 830 mAh for the same size NiCad. Another advantage of theses cells is complete freedom from memory effect. We can also find comfort in the fact that NiMH cells do not contain any dangerous substances, whereas both NiCad and lead–acid cells do contain quantities of toxic heavy metals.

The internal resistance of NiMH cells is somewhat higher than that of NiCad cells, resulting in reduced performance at very high discharge current. This can cause slightly reduced power output from a HT (handie talkie) powered by a NiMH pack, but the effect is barely noticeable, and the higher capacity resulting in longer run time far outweighs this. At least one manufacturer warns that the self-discharge of NiMH cells is higher than for NiCad, but again, in practice this can hardly be noticed. The fast-charge process is different for NiMH batteries. A fast charger designed for NiCad will not correctly charge NiMH batteries, but many commercial fast chargers are designed for both types of batteries. NiMH batteries outperform NiCad batteries whenever high capacity is desired, but NiCad batteries still have advantages when delivering very high peak currents. At the time of this writing, many cell phones and portable computers use NiMH batteries. Standard-sized NiMH cells are widely available from the major electronic parts suppliers.

Lithium Ion Cells

The lithium ion cell is another possible alternative to NiCad cells. It features, for the same amount of energy storage, about one-third the weight and one-half the volume of a NiCad. It also has a lower self-discharge rate. Typically, at room temperature, a NiCad cell will lose from 0.5 to 2% of its charge per day. The Lithium ion cell will lose less than 0.5% per day and even this loss rate decreases after about 10% of the charge has been lost. At higher temperatures, the difference is even greater. The result is that Lithium ion cells are a much better choice for standby operation where frequent recharge is not available. One major difference between NiCad and Li ion cells is the cell voltage. The nominal voltage for a NiCad cell is about 1.2 V. For the Li ion cell it is 3.6 V, with a maximum cell charging voltage of 4 V. You cannot substitute Li ion cells directly for NiCad cells. You will need one Li ion cell for three NiCad cells. Chargers intended for NiCad batteries must not be used with Li ion batteries, and vice versa.

Chemical and Other Hazards

In addition to the precautions given above, the following general precautions are recommended. When in doubt, always follow the manufacturer's advice.

Gas escaping from storage batteries may be explosive. Keep flames or lighted tobacco products away.

After adding electrolyte to dry-charged storage batteries, allow them to soak for at least half an hour. They should then be charged at about a 15 A rate for 15 minutes or so. The capacity of the battery will build up slightly for the first few cycles of charge and discharge, and then have fairly constant capacity for many cycles. Slow capacity decrease may then be noticed.

No battery should be subjected to unnecessary heat, vibration, or physical shock. The battery should be kept clean. Frequent inspection for leaks is a good idea. Electrolyte that has leaked or sprayed from the battery should be cleaned from all surfaces. The electrolyte is chemically active and electrically conductive and may ruin electrical equipment. Acid may be neutralized with sodium bicarbonate (baking soda), and alkalis may be neutralized with a weak acid such as vinegar. Both neutralizers will dissolve in water and should be quickly washed off. Do not let any of the neutralizer enter the battery.

Keep a record of the battery use and include the last output voltage and (for lead–acid storage batteries) the hydrometer reading. This allows you to predict the useful charge remaining, when recharging will be necessary, or when to buy extra batteries, thus minimizing failure of battery power during an excursion or emergency.

Internal Resistance

Cell internal resistance is a very important topic because the internal resistance is in series with the battery's output and, therefore, reduces the available battery voltage at the high discharge currents demanded by the transmitters and other high-current devices. The result is reduced transmitter output power and power wasted in the cell itself by internal heating. Because of different cell-construction techniques and battery chemistry, certain types of cells typically have lower internal resistance than others.

The NiCad cell is the undisputed king of cell types for high discharge current capability. Also, the NiCad maintains this low internal resistance throughout its discharge curve because the specific gravity of its potassium hydroxide electrolyte does not change. Next in line is probably the alkaline primary cell. When these cells are used in handheld radio transceivers, it is not uncommon to have lower output power, and often to have the low-battery indicator come on, even with fresh cells.

The lead–acid cell is pretty close to the alkaline cell in internal resistance, but only at full charge. Unlike the NiCad, the electrolyte in the lead–acid cell enters into the chemical reaction. During discharge, the specific gravity of the electrolyte drops as it approaches that of water, and the conductivity decreases. Therefore, as the lead–acid cell approaches a discharged state, the internal resistance increases.

Battery Capacity

The common rating unit of battery capacity is ampere hours (Ah), the product of current drain and time. The symbol "C" is commonly used; C/10, for example, is the current available for 10 hours of continuous use. The value of C changes with the discharge rate and might be 110 at 2 amperes but only 80 at 20 amperes.

Battery capacity may vary from 35 mAh for some of the small hearing-aid batteries to more than 100 Ah for a size 28 deep-cycle storage battery. Sealed primary cells usually benefit from intermittent (rather than continuous) use. The resting period allows completion of chemical reactions needed to dispose of by-products of the discharge.

The output voltage of all batteries drops as they discharge. "Discharged" condition for a 12 volt lead–acid battery, for instance, should not be less than 10.5 volts. It is also good to keep a running record of hydrometer readings, but the conventional readings of 1.265 charged and 1.100 discharged apply only to a long, low-rate discharge. Heavy loads may discharge the battery with little reduction in the hydrometer reading.

Batteries that become cold have less of their charge available, and some attempt to keep a battery warm before use is worthwhile. A battery may lose 70% or more of its capacity at cold temperatures, but it will recover with warmth. All batteries have some tendency to freeze, but those with full charges are less susceptible. A fully charged lead–acid battery is safe to –30°F (–34°C) or colder. Storage batteries may be warmed somewhat by charging. Blowtorches or other flames should never be used to heat any type of battery.

A practical discharge limit occurs when the load will no longer operate satisfactorily on the lower output voltage near the "discharged" point. Much gear intended for "mobile" use may be designed for an average of 13.6 V and a peak of perhaps 15 V, but will not operate well below 12 V. For full use of battery charge, the gear should operate well (if not at full power) on as little as 10.5 V with a nominal 12 to 13.6 V rating.

Somewhat the same condition may be seen in the replacement of carbon–zinc cells by NiCd storage cells. Eight carbon-zinc cells will give 12 V, whereas 10 of the same size NiCd cells are required for the same voltage. If a 10 cell battery holder is used, the equipment should be designed for 15 V in case the carbon–zinc units are plugged in.

Discharge Planning

In radio transceivers, power drain from a battery is determined as two or three rates: one for receiving, one for transmit standby, and one key-down or average voice transmit. Considering just the first and last of these (assuming the transmit standby is equal to receive), average two-way communication would require the low rate three-quarters of the time and the high rate one-quarter of the time. The ratio may vary somewhat with voice. The user may calculate the percentage of battery charge used in an hour by the combination (sum) of the rates. If, for example, 20% of the battery capacity is used in an hour, the battery will provide 5 hours of communications per charge. In most actual traffic, the time spent listening should be much greater than that spent transmitting.

Charging/Discharging Requirements

The rated full charge of a battery, C, is expressed in ampere-hours. No battery is perfect, so more charge than this must be offered to the battery for a full charge. If, for instance, the charge rate is 0.1 C (the 10 hour rate), 12 or more hours may be needed.

Basically, NiCad batteries differ from the lead–acid types in the methods of charging. It is important to note these differences, since improper charging can drastically shorten the life of a battery. NiCad cells have a flat voltage-versus-charge characteristic until full charge is reached; at this point the charge voltage rises abruptly. With further charging, the electrolyte begins to break down and oxygen gas is generated at the positive (nickel) electrode and hydrogen at the negative (cadmium) electrode.

Since the cell should be made capable of accepting an overcharge, battery manufacturers typically prevent the generation of hydrogen by increasing the capacity of the cadmium electrode. This allows the oxygen formed at the positive electrode to reach the metallic cadmium of the negative electrode and reoxidize it. During overcharge, therefore, the cell is in equilibrium. The positive electrode is fully charged and the negative electrode less than fully charged, so oxygen evolution recombination "wastes" the charging power being supplied

In order to ensure that all cells in a NiCad battery reach a fully charged condition, NiCad batteries should be charged by a constant current at about a 0.1 C current level. This level is about 50 mA for the AA-size cells used in most hand-held radios. This is the optimum rate for most NiCads since 0.1 C is high enough to provide a full charge, yet low enough to prevent overcharge damage and provide good charge efficiency. Although fast-charge-rate (3 to 5 hours, typically) chargers are available for handheld transceivers, they should be used with care. The current delivered by these units is capable of causing the generation of large quantities of oxygen in a fully charged cell. If the generation rate is greater than the oxygen recombination rate, pressure will build in the cell, forcing the vent to open and the oxygen to escape. This can eventually cause drying of the electrolyte, and then cell failure. The cell temperature can also rise, which can shorten cell life. To prevent overcharge from occurring, fast-rate chargers should have automatic charge-limiting circuitry that will switch or taper the charging current to a safe rate as the battery reaches a fully charged state.

Gelled-electrolyte lead–acid batteries provide 2.4 V/cell when fully charged. Damage results if they are overcharged. (Avoid constant-current or trickle charging unless battery voltage is monitored and charging is terminated when a full charge is reached.) Voltage-limited charging is best for these batteries. A proper charger maintains a safe charge current level until 2.3 V/cell is reached (13.8 V for a 12 V battery). Then, the charge current is taped off until 2.4 V/cell is reached. Once charged, the battery may be safely maintained at the "float" level, 2.3 V/cell. Thus, a 12 V gel-cell battery can be "floated" across a regulated 13.8 V system as a battery backup in the event of power failure.

Deep-cycle lead–acid cells are best charged at a slow rate, but automotive and some NiCad types may safely be given quick charges. This depends on the amount of heat generated within each cell, and cell venting to prevent pressure build-up. Some batteries have built-in temperature sensing, used to stop or reduce charging before the heat rise becomes a danger. Quick and fast charges do not usually allow gas recombination, so some of the battery water will escape in the form of gas. If the water level falls below a certain point, acid hydrometer readings are no longer reliable. If the water level falls to plate level, permanent battery damage may result.

Overcharging NiCads in moderation causes little loss of battery life. Continuous overcharge, however, may generate a voltage depression when the cells are later discharged. For best results, charging of NiCad cells should be terminated after 15 hours at the slow rate. Better yet, circuitry may be included in the charger to stop the charging or reduce the current to about 0.02 C when the 1.43 V per cell terminal voltage is reached. For lead–acid batteries, a timer may be used to run the charger to make up for the recorded discharge, plus perhaps 20%. Some chargers will switch over automatically to an acceptable standby charge.

Battery Charging

Charging batteries is an important topic. As we discussed earlier, there are a number of different types of batteries for our present-day electronic technologies. Probably the most ubiquitous type of battery is the 12 volt lead–acid battery or the 12 volt gel-cell battery.

For the most part, lead–acid batteries are made up of lead plates submerged in a sulfuric acid solution. The positive electrode plates are formed from lead dioxide (PbO_2) and the negative electrodes are made of sponge metallic lead (Pb). The porous nature of the lead plates allows the electrolyte, a dilute mixture of

35% sulfuric acid and 65% water, to efficiently contact the maximum surface area and obtain the most charge carriers. The electrolyte solution provides the sulfate ions formed during the discharge chemical reaction process, giving us the electrons needed for current flow into the load.

One of the by-products created during the discharge process of freeing sulfate ions is lead sulfate ($PbSO_4$). As the battery discharges, the lead sulfate attaches to the electrode plates, raising the internal resistance of the battery which in turn lowers its working terminal voltage.

Determining the SOC (state of charge) of a lead–acid battery using the classic voltmeter approach does not work very well. The terminal voltage will vary widely between batteries as a function of things like ambient temperature and the relative age of the battery. A full set of temperature profile tables would show big differences in the open-circuit terminal voltage over a wide temperature range. This is why a good charger must incorporate a temperature compensation network to avoid over- or undercharging the battery at different temperatures. To test a lead–acid battery's SOC, the best indicator is a hydrometer. When you test a battery's SOC with a hydrometer, you are actually measuring the amount of sulfuric acid left in the electrolyte solution. As more energy is drained from the battery, the ratio of sulfuric acid to water decreases and the created lead sulfate by-product begins forming on the electrode plates. A low hydrometer reading means the chemical makeup that generates the free electrons is diminished, so not as much energy is stored for use.

The term specific gravity is often used to benchmark a lead–acid battery's SOC. The specific gravity of a substance is a comparison of its density to that of water (1.000). imagine a one gallon bottle filled with water and a second filled with feathers. There are equal volumes of material present in both, but the bottle with the feathers will weigh less than that containing the water. The resultant specific gravity value of the bottle of feathers would be less than that of the bottle of water. With lead–acid batteries, the sulfur atoms break down and leach out of the electrolyte solution as it discharges. The breakdown of the electrolyte reduces its overall weight as the sulfur is removed from the solution, thus reducing the specific gravity measurement.

Great care should be taken to avoid discharging a battery beyond the 75% SOC point. Once the specific gravity drops below the 1.210 level, excessive sulfate deposits form on the electrode plates. This process is called sulfation and leads to the hardening of the electrode plates. If the battery is kept in a low charge state for long a period of time, the sulfation process will eventually reduce the ability of the battery to generate ion charges to the point that it no longer provides the needed power. This point is otherwise known as a dead battery.

When you recharge the battery, the process is reversed and the sulfur returns to the electrolyte solution. Proper cycling of the battery will ensure a long and functional life. If the battery is abused by allowing sulfation of the electrode plates on a regular basis or over an extended period of time, the charging process will not be able to restore the battery to its former full potential.

"Smart" Lead–Acid Battery Charger

The "smart" charger was designed as a dependable means to charge and hold your 12 volt lead–acid batteries at their peak level, ensuring a long life and maximum performance. The "smart" 12 volt lead–acid gel-cell battery charger is a very useful item to have around the electronics workbench.

The "smart" charger circuit diagram shown in Figure 8-1 depicts a 12 volt lead–acid battery-charging system that step charges a 12 volt battery. The battery charger applies the full current when the battery

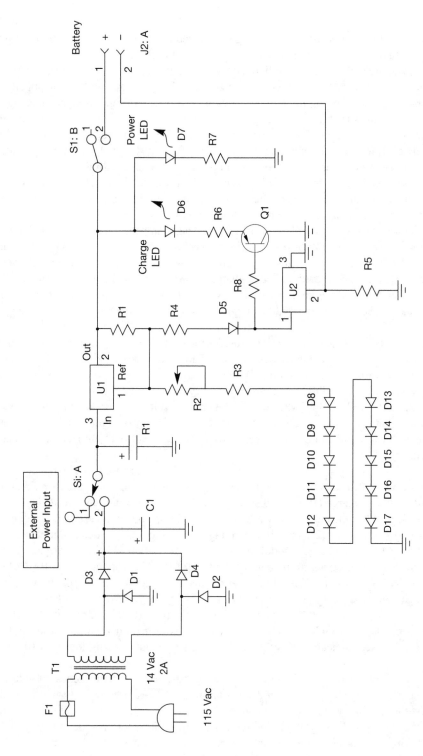

Figure 8-1 "Smart" 12 V lead–acid battery charger.

first needs the most charge, and as the battery charges up it requires less and less charge until the point where the battery is almost fully charged and just requires a "trickle charge."

This charging procedure works best with a flooded "wet" cell battery or one of the newer VRLA (valve regulated lead–acid), "gel" or "AGM" batteries. The battery being charged will automatically set the "smart" charger in one of two charging modes upon hookup. The circuit design takes into account the battery's current SOC and adjusts the terminal voltage at the output of J2 accordingly. The main charging circuit is very simple because, as we discussed before, the concept of lead–acid batteries has been around for centuries. The real secret to correctly charging a lead–acid battery system is to use a temperature-compensated voltage source that automatically varies its output in accordance with the battery's SOC. "Frying" a battery occurs when the charging unit fails to sense that the electrochemical rejuvenation (or charging) process has slowed to the point where the higher-voltage charging mode should end. Continual high-voltage charging will decrease the overall life of the battery.

Power for the "smart" charger is provided from T1, a 14 V ac, 2 ampere transformer. The input voltage is immediately presented to a full wave bridge rectifier consisting of diodes D1, D2, D3, and D4 and then filtered by C1 to reduce the voltage. If your application is to charge very-small-capacity batteries with a maximum charge current of only a few hundred milliamps, using a 14 V ac, 500 mA "wall wart" supply or a current-limited bench-top power supply set for 20 V dc will avoid excessive current draw that could damage a heavily discharged battery. Internal heating from excessive charge current will also degrade the overall battery life.

Integrated circuit U1 is a voltage regulator that provides the precision terminal voltage that is needed to charge the lead–acid cells. Unlike a standard voltage regulator that is designed for a fixed-level output, U1 lends itself well as a variable voltage source. With a maximum current source capability of about 1.3 amperes, U1 gives the user the flexibility to charge even very-large-capacity batteries. The other support components on the board help U1 know when to adjust its output voltage up or down to ensure the proper charging rate of the battery. These other components are grouped into two major sections, the SOC feedback loop and the ambient temperature compensation used during the "float" mode after the battery has been fully charged.

The SOC feedback loop consists mainly of U2 and R5 together to form a low-voltage comparator in conjunction with R1 and R4 to set the range of the charging voltage. Here's how the loop functions. Assume that the battery under charge (BUC) is discharged and drawing enough current to set the "smart" charger in charge mode. After the current drawn by the battery drops below a certain point, the need for high-voltage charging has ended. U2 monitors the voltage drop across R5 to determine when to switch U1's output at J2 from 14.4 V ("charge" mode) to 13.4 V ("float" mode). As the battery comes to a full charge, the charging current it draws drops below about 150 mA. The voltage across R5 (0.47 ohms) will then fall below 0.07 V thanks to Ohm's Law, $V = I \times R$. This trigger point causes the V+ pin (U2:1) to toggle from its charging mode high value of about 12.8 V to a charged float mode low value of about 0.7 V. When V+ (U2:1) toggles low, R4 is switched into the reference feedback circuit of VR1, causing its output voltage drop back to 13.4 V. The charged LED (D6) is turned on when the base–emitter junction of Q1 is forward biased, indicating that the battery is charged and is being topped-off by the Float mode operation.

As the battery is now charged, the ambient temperature compensation circuit comes into play. The effects of this circuit, formed by R2, R3, and diodes D8 through D17, are used only during the float mode operation to adjust the terminal voltage in accordance with the ambient temperature. If the temperature is not factored in, you run the risk of overcharging the battery when it is hot or undercharging the battery when it is cold. Taking advantage of the thermal characteristics of a PN diode ($\Delta2.2$ mV/°C), the diode matrix (D8 to D17) raises or lowers the reference terminal of U1 by 22 mV (10 × 2.2 mV/°C) for every 1°C change. This is just the right negative temperature compensation needed to properly charge lead acid batteries.

At the start of the charge cycle, you will notice that the heat sink used with U1 can get very warm. If you are charging a large-capacity battery. The fact that the temperature sensor matrix is on the same circuit board and in the same case will not negatively affect the compensation network because very little heat will be dissipated by the board components once the unit switches into float mode. The drop in charge current drawn by the battery is so low by the time float mode is entered that, the air cavity around the temperature sensor diodes will reacclimate to the surrounding ambient temperature.

Adjusting the "Smart" Battery Charger

After assembling the "smart" charger, you will have to alignment it. First, you will need to adjust R2 for tho proper float voltage with reference to the current air temperature around diodes D8 to D17. In order for the temperature compensation network to function properly and automatically adjust the float voltage as needed, you will need to know the air temperature in the room you are in. Note that this adjustment only needs to be done once. When R2 has been set, the unit will automatically track as needed from there on out.

To adjust R2, take a temperature reading for the current room temperature with a thermometer. Set your voltmeter on the 20 V dc scale and attach the probes (red = +, black = −) across the output of J2. Trim the pot (R2) to get a reading of 13.4 V dc (± 0.1 V dc) at a room temperature of 25°C (77°F) with no battery attached. If the room temperature is above or below 25°C, you will need to account for the difference by offsetting the alignment voltage by 22 mV (10 × 2.2mV/°C for each of the sensor diodes) for every 1°C of difference.

The parts list for the 12 V lead–acid battery charger follows.

R1	270 ohm, ¼ watt resistor
R2	1 kohm, potentiometer
R3, R6, R7	820 ohm, ¼ watt resistor
R4	18 kohm, ¼ watt resistor
R5	0.47 ohm, 1 watt resistor
R8	10 kohm, ¼ watt resistor
C1	1000 μF, 35 volt electrolytic capacitor
C2	10 μF, 35 volt electrolytic capacitor
D1, D2, D3, D4	1N4002 silicon diodes
D6, D7	LEDs
D5, D8, D9, D10, D11	1N4148 silicon diodes
D12, D13, D14, D17	1N4148 silicon diodes
Q1	2N3906 transistor, TO-92
U1	LM317 three-terminal regulator
U2	LM334 IC, TO-92
T1	Transformer, 14 V ac, 2 ampere
F1	Fuse, 2.5 ampere fast-blow
Miscellaneous	PC board, switch, wire, connectors, screws, nuts

Safety Considerations

Remember that the gases that form while charging lead–acid batteries are extremely explosive! Never charge a battery around an open flame or anything that can cause a spark that may ignite the venting gas! All lead–acid batteries produce hydrogen and oxygen gas during the electrochemical recharging process. The production of these gases is increased if overcharging occurs, commonly caused by a too high charge voltage. Sealed battery designs plan on the recombination of oxygen at the same rate it is produced, therefore eliminating the explosive mixture. Any hydrogen that is produced will diffuse through the plastic container and as long as the sealed battery is not in a sealed enclosure, the hydrogen will harmlessly disperse into the atmosphere. It is good practice to use adequate ventilation even with sealed batteries, due to the possibility of unforeseen problems.

The Solar Electric Battery Charger

Price and availability make solar panels an attractive way to maintain the charge on your batteries. Relatively small, low-power solar arrays provide a convenient way to charge a NiCad or sealed lead–acid battery for emergency and portable operation. You should always connect some type of charge controller between the battery and the solar array. This will prevent overcharging the battery and the possible resulting battery damage.

The solar electric battery charger is a photovoltaic (PV) controller that can be used at home, in your workshop, or in the field. With the solar charger you can keep your 12 volt lead–acid gel cells charged all the time. The solar battery charger is completely silent and handles up to 4 amperes of current from a solar panel and standby current to less than one milliamp.

The solar battery charger schematic is shown in Figure 8-2. Current from the solar panel is controlled by a power MOSFET that uses an International Rectifier IRF4905. This P-channel FET has a current rating of 64 amperes with an RDS_{on} of 0.02 W. It comes in a TO-220 case. Current sent from the solar panel is sent directly to the MOSFET source lead.

The P-channel MOSFET eliminates the need for a charge pump altogether. In order to turn on a P-channel MOSFET, all you have to do is pull the gate lead to ground. The P-channel MOSFET will dissipate 12 W of power. Current generated by solar panels is way too precious and expensive to have 12 W go up as heat in the charge controller.

With the P-channel MOSFET controlling the current, diode D4 (a 50SQ100 Schottky) prevents battery current from flowing into the solar panel at night. This diode also provides reverse polarity protection to the battery in the event you connect the solar panel backward. This protects the expensive P-channel MOSFET.

Zener diode D2 (a 1N4747) protects the gate from damage due to spikes on the solar panel's output. Resistor R12 pulls the gate up, ensuring that the power MOSFET is off when it is supposed to be. The solar charger never draws current from the battery. The solar panel provides all the power the solar charger needs, which means the solar charger "goes to sleep" at night. When the sun rises, the solar charger starts up again. As soon as the solar panel is producing enough current and voltage to start charging the battery, it will pass current into the battery. To reduce the amount of standby current, diode D3 passes current from the solar panel to U3, the voltage regulator. U3, an 78LO8 regulator, provides a steady +8 V to the

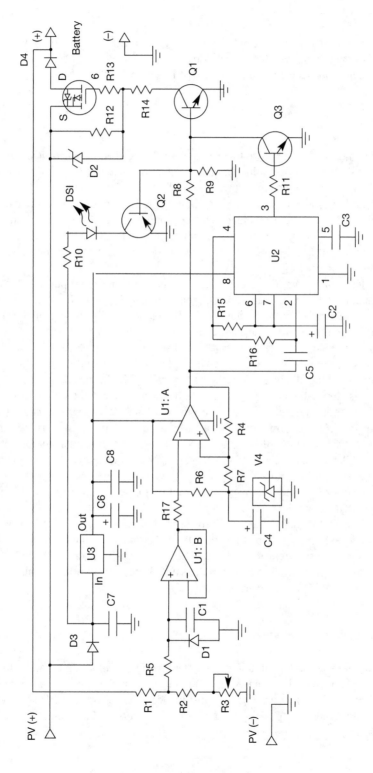

Figure 8-2 Solar electric battery charger. (Courtesy 2003 ARRL Handbook.)

solar charge controller. Bypass capacitors C6, C7, and C8 are used to keep everything working smoothly. As long as there is power being produced by the solar panel, the controller will be on. At sundown, the solar charger "goes to sleep" and sleep current is reduced to less than 1 mA.

Battery Sensing

The battery terminal voltage is divided down by resistors, R1, R2, and R3. Resistor R3, a 20 k trim potentiometer, sets the state-of-charge for the solar charger. A filter consisting of R5 and C1 helps keep the input free of noise picked up by the wires to and from the solar panel. Diode D1 protects the op-amp input in case the battery sense line is connected backward. An LM358 dual op-amp is used in the solar charger; section (U1B) buffers the divided battery voltage before passing it along to the voltage comparator, U1:A. Here, the battery sense voltage is compared to the reference voltage supplied by U4. Integrated circuit U4 is an LM3362-5.0 precision diode. To prevent U1:A from oscillating, a 10 megohm resistor is used to eliminate any hysteresis.

As long as the battery under charge and below the reference point, the output of U1:A will be high. This saturates transistors Q1 and Q2. Transistor Q2 conducts and lights LED DS1, the charging LED. Transistor Q1, also fully saturated, pulls the gate of the P-channel MOSFET to ground. This effectively turns the FET on and current flows from the solar charger into the battery via D4.

As the battery begins to take up the charge, its terminal voltage will increase. When the battery reaches the state-of-charge set point, the output of U1:A goes low. With Q1 and Q2 now off, the P-channel MOSFET is turned off, stopping all current into the battery. With Q2 off, the charging LED goes dark.

Since all hysteresis is eliminated in U1:A, as soon as the current stops, the output of U1:A pops back up high again, because the battery terminal voltage will fall back down as the charging current is removed. The output of U1:A is monitored by U2, an LM555 timer chip. As soon as the output of U1:A goes low, this low trips U2. The output of U2 goes high, fully saturating us transistor Q3. With Q3 turned on, it pulls the base of Q1 and Q2 low. Since both Q1 and Q2 are now deprived of base current, they remain off.

With the values shown for R15 and C2, charging current is stopped for about 4 seconds after the state-of-charge has been reached.

After the 4 second delay, Q1 and Q2 are allowed to have base drive from U1:A. This lights up the charging LED and allow Q4 to pass current once more to the battery. As soon as the battery hits the state-of-charge once more, the process is repeated. As the battery becomes fully charged, the V "on" time will shorten, whereas the "off" time will always remain the same 4 seconds. In effect, a pulse of current will be sent to the battery, which will shorten over time.

Final Adjustments

Locate a good digital voltmeter and a variable power supply with an output up to14.3 volts. Connect the solar charger's negative battery lead to the power supply negative lead. Connect the solar charger's positive and battery positive leads to the power supply positive lead. The charging LED should be on. If not, adjust trimmer R3 until it comes on. Check for +8 V at the Vcc pins of the LM358 and the LM555. You should also observe +5 volts from the LM336Z5.0 diode.

Quickly move the trimmer from one end if its travel to the other. At one point, the LED will go dark. This is the switch point. To verify that the "off pulse" is working, as the LED goes dark quickly reverse the direction of the trimmer. The LED should remain off for several seconds and then come back on. If everything seems to be working, it's time to set the state-of-charge trimmer.

Now, slowly adjust the trimmer until dark. You might want to try this more than once, as the closer you get the comparator to switch at exactly 14.3 V, the more accurate the solar charger will be. Set the power supply to slightly above the cut-off voltage that you want. For example, if you want 14.3 V, then set the supply to 14.5 V. Setting the supply higher takes this into account and usually you can get the trimmer set to exactly where you need it in one try. That's all you need to do. Disconnect the supply from the solar charger and you're ready for the solar panel.

The 14.3 V terminal voltage will be correct for just about all sealed and flooded-cell lead–acid batteries. You can change the state-of-charge set point if you want to recharge NiCads or captive-sealed lead–acid batteries. Keep the current from the solar panel within reason for the size of the battery you're going to be using. If you have a 7 Ah battery, then don't use a 75 W solar panel. In this situation, you will get much better results with a smaller solar panel. The tab of the power MOSFET is electrically hot, so it is advisable to install the solar charger in a small enclosure.

The parts list for the photovoltaic battery charger follows.

R1	100 kohms, 1%, ¼ watt resistor
R2	49.9 kohms, 1%, ¼ watt resistor
R3	20 kohm potentiometer
R4	10 megohm, 5%, ¼ watt resistor
R5, R7, R9	10 kohm, 5%, ¼ watt resistor
R13, R17	10 kohm, 5%, ¼ watt resistor
R6, R8, R10	1.8 kohm, 5%, ¼ watt resistor
R11	2.2 kohm, 5%, ¼ watt resistor
R12	27 kohm, 5%, ¼ watt resistor
R14	68 ohm, 5%, ¼ watt resistor
R15, R16	100 kohm, 5%, ¼ watt resistor
C1, C5, C7, C8	0.1 μF, 50 volt capacitor
C2, C4, C6	22 μF, 16 volt electrolytic capacitor
C3	0.01 μF, 5 volt capacitor
D1	1N914 silicon diode
D2	1N4747 20 V, 1 watt zener diode
D3	1N4002 silicon diode
D4	80SQ045 45 V, 8 ampere Schottky diode
DS1-	LED
Q1, Q2, Q3	2N4401 NPN transistor
Q4	IRF4905 P-channel MOSFET
U1	LM358AN op-amp
U2	LM555AN timer IC
U3	LM78L08, 8 V regulator
U4	LM336Z-5, 5 V zener diode in TO-92 case
Miscellaneous	wire, PC board, connectors, chassis box, etc.

An Emergency Power System

Emergency power capability is an important and useful complement to your overall electronics workbench and home. Emergency power will give you the capability to operate either 12 volt low-voltage lighting or 115 volt AC appliances in the event of a power failure. With emergency power, you will be able to operate your computer and most hand tools.

The emergency power system concept is a relatively straightforward project. Shown graphically in Figure 8-3, it is divided into four major components:

1. A "smart" 3-step, 10 ampere battery charger
2. A 12 volt "deep-cycle" battery
3. A dc power-distribution panel
4. A dc-to-ac inverter

In the event of a power failure, the emergency power system can be used to power many things around your home or electronics workbench. Depending upon how large you scale the system, you could power a computer, water pump, furnace fan, power tools, lighting, and test equipment. First, we will take a look at the battery, which is at the heart of the emergency power system.

The Battery

The battery is at the heart of an emergency power system. The problem with most storage batteries is that they produce dangerous hydrogen gas while charging and contain sulfuric acid, which can spill or leak

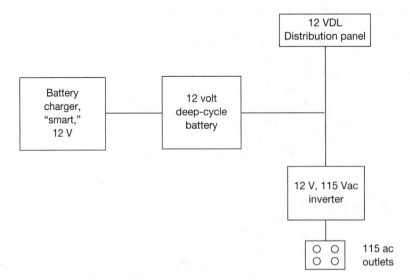

Figure 8-3   *Emergency power system.*

out. Safe placement of charging storage batteries is a major concern for an emergency power system. You normally would have to place the batteries outside or construct a forced venting system from the basement to send the gases outside. Otherwise, hydrogen gas would fill your basement and could be detonated by the furnace pilot light, causing an explosion.

Fortunately, in the past few years, lead–acid battery makers have quietly been advancing battery technology. There are now several technologies that provide for "recombinant" operation of storage batteries. In a recombinant battery, most of the hydrogen is not released, but recombines with oxygen within the battery to form water. Thus, you not only avoid the threat of explosion, but you never need to add water. These batteries are ideal for an emergency power system.

The recombinant technologies are found in batteries labeled AGM (absorbed glass mat), VRLA (valve-regulated lead–acid) or gel cell. These batteries hold the electrolyte against the plates in a way that avoids (but doesn't quite eliminate) the release of hydrogen during the charging cycle. A small amount of gas is released, but it is considered sufficiently small so that these batteries can be used with normal household ventilation. They are used to power indoor computer uninterruptible power supply systems and motorized wheelchairs, for example. They also do not freeze, spill, or leak acid.

Fortunately, these batteries are also of "deep-cycle" design. A deep-cycle battery, unlike the usual auto or marine starting battery, is designed so that it can be 75% discharged hundreds of times rather than just a few times, and still be recharged to provide full capacity.

When selecting a battery, look at the description carefully, since not all "sealed" or "no maintenance" batteries are recombinant. Some simply have no ports for water addition and provide a bit more water to start with, but they emit all the hydrogen of an open top battery. When the water level finally falls below the top of the plates they start to fail. They are not good for deep-cycle use -or for avoiding explosion. Look for AGM, VRLA, or gel cell batteries if you want to minimize hydrogen emission. Once you have settled on the battery "family, " there is the range of sizes and capacities to consider. The most important parameter for our application is "capacity" in ampere hours (Ah). Generally, the higher the Ah capacity, the higher the cost and weight. Consider purchasing an 80 to 100 Ah battery as it affords a balance between price and weight.

The Charger

The design of most chargers is such that they are current limited at their rated output. Thus, when drawing a load greater than the charger can supply, excess current will come from the battery, not the charger. The charger output should be fixed at its rating to protect the charger from excessive load if something goes awry.

In order to achieve the battery life described above, the charger needs to be able to support multiple stages of recharge, as well as different characteristics for different families of batteries (most gel cells should not be charged above 14.1 V, for example). The following is a description of a typical three-stage charger:

Stage 1—Bulk When the battery is at 75% capacity or lower, the charger pumps high amperage at a relatively low voltage.

Stage 2—Absorption. As the battery is charged to 75% capacity, the charger lowers the amperage and increases voltage (never exceeding the battery's designed voltage maximums) to gradually bring the batteries to full charge.

Stage 3—Maintenance (often called "float"). When batteries are fully charged, the charger drops the voltage to a maintenance level and gently maintains the battery at full charge.

In a linear charger, generally, when the battery is fully charged, the unit shuts off until the battery drops to 90% capacity and then turns on to bring it back to full charge. The result is that deep-cycle batteries have limited cycles built into them, thereby reducing the life of the batteries. Other types of batteries are charged at a higher voltage rate, which also reduces life. There are linear two- and three-mode "smart" chargers available, however, that do not use microprocessor control. These use analog comparators to sense voltage and current. There are many types of chargers out there for "deep-cycle" batteries. It is best to use a multistage charger; they are generally a bit more expensive but will allow longer battery life. As an example the, Guest Charger Model 2610 provides up to 10 A or two independent 5 A outputs to charge two 12 volt batteries simultaneously. The charge current is applied in the three stages defined above. A portable-style charger could also be used and moved to the garage, boat or RV, as needed.

Twelve Volt Power-Distribution Panel

A power-distribution panel allows you to connect a number of 12 volt devices to the 12 volt deep-cycle battery at any given time. The 12 volt distribution panel uses heavy-duty standard power connections at each power port. All power ports have large-diameter wire from the input on through to each output. Each power port has both negative and positive power terminals and they are colored and labeled. The distribution panel is a good way to power multiple devices. Many tools and equipment can utilize 12 volts dc directly and they can be connected through the dc power-distribution panel. These power panels generally run about $50 to $60 and are available through electronics and radio suppliers.

Power Inverters

Twelve volt dc to 115 volt ac inverters are quite easy to find and they add an extra dimension to your emergency power system. Power inverters generally are available in two major categories. There are sine-wave output and square-wave output types and a near-sine-wave variation. The sine-wave types are generally very expensive but are the closest to the power received from the ac output. The square-wave inverters are the cheapest type and the near-sine-wave types are priced in between. Inverters take a 12 volt dc battery voltage and convert the 12 volts dc to 115 volts ac, which can be used to power home appliances, tools, and so on. Motors generally do not like to operate on square-wave inverters and may cause both the inverter and the load to get hot. For motor operation, try to obtain a near-sine-wave or sine-wave inverter if at all possible. Inverters are rated in watts and usually the specification list both the peak power rating and the nominal or operating wattage. Make sure you size your system for the nominal rating, since the peak rating is only concerned with starting wattage from initial "spikes." When designing your own emergency power system, you will need to determine how much wattage you will draw from the system. The larger the wattage, the higher the price, and the axiom "you get what you pay for" applies here as well. The higher the load taken from the inverter and thus from the battery will determine how long you will be able to operate your appliance or equipment. Choose the best inverter you can afford and shop around. Often, discount stores sell name-brand, quality units but you have to know the "junk" from the quality. Choose name brands such as Tripplite.

Power Losses

We are conditioned to thinking that if there is 12 volts at one end of a pair of wires, there will also be the same 12 volts at the other end of the wires. Most technical types know that wire has resistance, and although the difference might be slight in light of low power demands, we are talking about real amperes when large amounts of power are drawn. The results of power loss caused by wire resistance can result in a significant voltage drop over a long distance between the 12 volt battery and the load. For example, if you had 6 feet of wire between the battery and the your power-distribution panel, and another 6 feet to the appliance or load, that's 12 feet of two wires in series or 24 feet of wire resistance to consider. Depending upon the wire sized used to connect the load, you might lose a significant amount of power in heating up the wire. As an example, if you have a 20 ampere load using 24 feet of 18 gauge wire, you would loose 3.125 volts (see Tables 8-1 and 8-2).

Table 8-1   Charger voltage calibration

Float mode voltage settings of potentiometer R2	
Temperature	J2 terminal voltage—12 volt battery
27°C (81°F)	13.356 V dc
25°C (77°F)	13.400 V dc
23°C (73°F)	13.444 V dc
21°C (70°F)	13.664 V dc

Rule: If the temperature is higher than 25°C (77°F), then reduce the alignment voltage at J2 by 22 mV for each 1°C difference. If the temperature is lower than 25°C (77°F), then increase the alignment voltage at J2 by 22 mV for each 1°C difference.

Table 8-2   Wire resistance versus voltage drop

Wire gauge	Resistance per 1000 ft	20 Ampere loss per 24 feet wire
8	0.0640 ohms	0.307 volts
10	1.018 ohms	0.489 volts
12	1.619 ohms	0.777 volts
14	2.575 ohms	1.23 volts
16	4.094 ohms	1.96 volts
18	6.510 ohms	3.12 volts
20	10.35 ohms	4.96 volts

Battery Safety

We like to think of 12 V systems as safe when compared to the 1000 V power supply. They certainly are from the point of view of an electrocution hazard, but storage batteries of this sort have significant energy and can do serious damage to people and objects. Our usual 12 V power supply will often "crowbar" to 0 V when shorted. The battery, however, will expend all of its energy in dramatic ways including the possibility of exploding.

Simple Battery Safety Rules

1. Wear safety glasses when working around storage batteries.
2. Do not have open flames near batteries, especially while under charge.
3. Remove all metal jewelry such as rings and bracelets when handling batteries.
4. Never use metal tools long enough to reach between the battery terminals or connections.
5. Protect the top of the battery (plastic battery box with lid, for example) so wires or equipment can't fall onto the terminals.
6. Use proper size ring terminals on all battery connections; use crimp-type connectors and solder them after crimping.
7. Install fuses as close to the battery terminals as possible.
8. Wash hands immediately following contact with the battery.

Considerations

You may want to consider other types of loads, depending on your environment. A key possibility is dc lighting. Note that high-efficiency focused lights typically draw 2 A, whereas a 50 W standard (12 V) bulb will draw more than 4 A. In a nonemergency power failure, I can imagine a request to run the refrigerator or furnace from time to time, and that may be a capability worth having if the load is reasonable. Use a recombinant 12 volt battery to avoid the risk of a hydrogen gas explosion. You may be able to obtain a surplus recombinant battery from a medical supply house.

CHAPTER 9

Electronic Components

In working with electronic circuits on your electronics workbench, you will need to become familiar with the electronic components such as resistors, capacitors, inductors, and semiconductors.

Resistors

The resistor is undoubtedly the most common and well known of all types of electrical components. Resistors are often used to drop voltage, limit current, attenuate signals, act as heaters, act as fuses, furnish electrical loads, and divide voltages. The voltage divider, for example, is used in a variety of networks to divide voltages in specified increments of the applied voltage for analog-to-digital converters and digital-to-analog converters. They are used as matched pairs with relative accuracy much greater than their absolute accuracy. Matching is used in building voltage dividers and Wheatstone and Kelvin bridges with extremely precise accuracy over a wide range of temperatures. This is done by matching the absolute value and the temperature coefficient of resistance (TCR). This accuracy is limited only by the ability to accurately measure these values and the stability of the resistors.

There are numerous varieties of resistors: carbon composition, carbon film, metal film, foil, precision wire-wound, power wire-wound, fuse resistors, filament-wound, and power film resistors. Each of these resistor types has a special purpose in electronic circuits. Resistors have numerous characteristics that determine their accuracy and their particular purpose, including tolerance at dc, temperature coefficient of resistance (TCR), frequency response, voltage coefficient, noise, stability with time and load, temperature rating, power rating, physical size, mounting characteristics, thermocouple effect, and reliability.

Types of Resistors, Mounting, and Physical Sizes

Carbon Composition Resistor. Carbon composition resistors were once the most common resistor on the market (Figure 9-1). They still have a very large market and prices are highly competitive. They are made from carbon rods cut to the appropriate length, then molded with leads attached. The mix of the carbon can be varied to change the resistivity for the desired values. High values are much more readily available. Very low values are more difficult to achieve. A 5% tolerance is available. This is usually achieved by measuring and selecting values. Normal tolerances without measurement and selection is in the area of 20%. The temperature coefficient of resistance is in the range of 1000 ppm/°C and is negative; that is, when the temperature goes up, the resistance goes down, and when the temperature goes down, the resistance goes up. This is due to the carbon particles being relaxed (with increase in temperature) and compressed (with the reduction in temperature).

These resistors also have a voltage coefficient. That is, the resistance will change with applied voltage; the greater the voltage, the greater the change. In addition to a power rating, they also have a voltage rating. The voltage rating of carbon composition resistors is determined by physical size as well as the value and wattage rating. (The wire-wound voltage rating is determined by the value and the wattage rating.) One more item to consider is that, due to their construction, they generate noise and this noise level varies with value and physical size. Their power capability in relation to physical size is greater than precision wire-wounds but less than power wire-wounds.

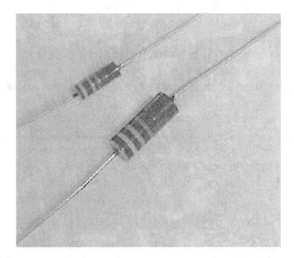

Figure 9-1   Carbon composition resistors.

Carbon Film Resistors. Carbon film resistors have many of the same characteristics as carbon composition resistors. The material is similar, therefore they have noise, a voltage coefficient, the TCR can be much lower because the formula can be varied to achieve this, the tolerance is much tighter due to the difference in manufacturing processes. The carbon film resistor is made by coating ceramic rods with a mixture of carbon materials. This material is applied to these rods in a variety of means. The ones most familiar to me are dipping, rolling, printing, or spraying the rods with the appropriate solution. The thickness of the coating can be determined by the viscosity of the solution. This as well as the material composition will determine the ohms per square. Some of you may not be familiar with this term. It simply means that if a material has a resistivity of 100 ohms/square, one square inch will have the same resistance as 1 square mm, or 1 square foot, or 1 square yard, or 1 square mile, all equaling 100 ohms, but the power handling capability is proportional to the size.

One batch of material can produce resistors in a wide range of values. These rods are cut to the length required for a specific size of resistor. They can then be spiral cut to a wide range of values. The original method of spiral cutting was done with grinding wheels on a machine similar to a lathe. Newer processes use lasers that are programmed to cut to specific values. The maximum ohmic value of this group is the highest in the discrete resistor group. Tolerance of 1% can be achieved without measuring and selecting. Tolerance of less than 1% can be achieved by measuring and selecting. You should use caution in getting tight tolerances of this type of resistor because the temperature coefficient, voltage coefficient, and stability may mean that it is only good for that tolerance at the time it was installed. The TCR of carbon film resistors is in the neighborhood of 100 to 200 ppm and is generally negative. Measuring and selecting can yield even tighter TCRs.

The frequency response of this type of resistor is among the best, far better than wire-wounds, and much better than carbon composition. The wire-wound resistors are inductive at lower frequencies and values and somewhat capacitive at higher frequencies regardless of value. Also, wire-wound resistors will have a resonant frequency. Carbon composition resistors will be predominately capacitive.

Metal Film Resistors. Metal film resistors are the best compromise of all resistors. They are not as accurate, have a higher temperature coefficient of resistance, and are not as stable as Precision wire-wounds. They are more accurate, do not have a voltage coefficient, and have a lower temperature coefficient than carbon film resistors. TCRs of 50 to 100 ppm can be achieved. They have a very low noise level when properly manufactured. In fact, some of the screening processes measure the noise level to determine if there are problems in a particular batch of resistors. Metal film resistors are manufactured by an evaporation/deposition process. The base metal is vaporized in a vacuum and deposited on a ceramic rod or wafer. Attempts have been made to vaporize low TCR materials and deposit them on substrates, but to date these attempts have not been successful. The very low TCR resistive materials are heat treated to achieve the resistivity and low TCR. This is not compatible with an evaporation process. The frequency characteristics of this type are excellent and better than carbon film resistors. The one area in which carbon film resistors are better than metal film resistors is the maximum values. Carbon film resistors can achieve higher maximum values than any other group.

Foil Resistors. Foil resistors are similar in characteristics to metal film resistors. Their main advantages are better stability and lower TCRs. They have excellent frequency response, low TCR, good stability, and are very accurate. They are manufactured by rolling the same wire materials as used in precision wire-wound resistors to make thin strips of foil. This foil is then bonded to a ceramic substrate and etched to produce the value required. They can be trimmed further by abrasive processes, chemical machining, or heat treating to achieve the desired tolerance. Their main disadvantage is that the maximum value is less than metal film resistors. The accuracy is about the same as metal film resistors, the TCR and stability approaches precision wire-wounds but are somewhat less because the rolling and packaging processes produce stresses in the foil. The resistive materials used in precision wire-wound resistors is very sensitive to stresses, which result in instability and higher TCRs. Any stresses on these materials will result in a change in the resistance value and TCR; the greater the stress, the larger the change. This type can be used as strain gauges, strain being measured as a change in the resistance. When used as a strain gauge, the foil is bonded to a flexible substrate that can be mounted on a part where the stress is to be measured.

Filament Resistors. Filament resistors are similar to bathtub or boat resistors except that they are not packaged in a ceramic shell (boat). The individual resistive element with the leads already crimped is coated with an insulating material, generally a high-temperature varnish. They are used in applications where tolerance, TCR, and stability are not important but the cost is the governing consideration. The cost of this type is slightly higher that of carbon composition and the electrical characteristics are better.

Power Film Resistors. Power film resistors are similar in manufacture to metal film or carbon film resistors. They are manufactured and rated as power resistors, with the power rating being the most important characteristic. Power film resistors are available in higher maximum values than the power wire-wound resistors and have a very good frequency response. They are generally used in applications requiring good frequency response and/or higher maximum values. Generally, for power applications the tolerance is wider. The temperature rating is changed so that under full load, the resistor will not exceed the maximum design temperature. The physical sizes are larger and, in some cases, the core may be made from a more heat conductive material and other means employed to help radiate heat.

Precision Wire-Wound Resistors. The precision wire-wound resistor is a highly accurate resistor (within 0.005%) with a very low TCR. A TCR of as little as 3ppm/°C can be achieved. However these components are too expensive for general use and are normally used in highly accurate dc applications. The frequency response of this type is not good. When used in an RF application, all precision wire-wound resistors will have a low Q resonant frequency and the power handling capability is very small. They are generally used as reference resistors for voltage regulators and decoding networks. Their accuracy is maintained at 25°C and will change with temperature. The maximum value available is dependent upon physical size and is much lower than that of most other types of resistors. Their power rating is approximately 1/10 that of a carbon composition resistor of similar physical size. They are rated for operation at +85°C or +125°C, with maximum operating temperature not to exceed +145°C. This means that full rated power can be applied at +85 or +125°C with no degradation in performance. It may be operated above +125 or +85°C if the load is reduced. The derating is linear; the rated load at +125 or +85°C and no load at +145°C. Life is generally rated for 10,000 hours at rated temperature and rated load. The allowable change in resistance under these conditions is 0.10%. Extended life can be achieved if operated at lower temperatures and reduced power levels. End-of-life requirements are generally defined by the manufacturer or in some case by user specification. Some degradation in performance can be expected. In some cases, particularly if the tolerance is very low and the TC is low, the rated power is reduced to improve resistor stability throughout life. Precision resistors, regardless of type, are designed for maximum accuracy, not to carry power. The materials used in these resistors are highly stable heat-treated materials that do change under extended heat and mechanical stress. The manufacturing processes are designed to remove any stresses induced during manufacture. There is little detectable noise in this type of resistor. The stability and reliability of these resistors is very good and their accuracy can be enhanced by matching the absolute value and the temperature coefficient over their operating range to achieve very accurate voltage division.

Power Wire-Wound Resistors. Power wire-wound resistors are used when it is necessary to handle a lot of power. They will handle more power per unit volume than any other resistor. Some of these resistors are free wound, similar to heater elements. These require some form of cooling in order to handle any appreciable amount of power. Some are cooled by fans and others are immersed in various types of liquids, ranging from mineral oil to high-density silicone. Most are wound on some type of winding form. These winding forms vary. Some examples are ceramic tubes, ceramic rods, heavily anodized aluminum, and fiberglass mandrels. To achieve the maximum power rating in the smallest package size, the core on which the windings are made must be of a material with high heat conductivity. It may be steatite, alumina, beryllium oxide, or in some cases hard anodized aluminum. Theoretically, the anodized aluminum core has a better heat conductivity than any other insulated material, with beryllium oxide being very close. There are specific problems with the anodized aluminum cores such as nicks in the coating, abrasion during capping, and controlling the anodized thickness. There are various shapes available, including oval, flat, and cylindrical; most shapes are designed to optimize heat dissipation. The more heat that can be radiated from the resistor, the more power that can safely be applied. A group of these resistors called "chassis mounted resistors" are generally cylindrical power resistors wound on a ceramic core molded and pressed into an aluminum heat sink, usually with heat-radiating fins (Figure 9-2). These are designed to be mounted to metal plates or a chassis to further conduct heat. This results in a rating approximately five times or more its normal rating. These resistors come in a variety of accuracies and TCRs. They can be custom made as a cross between a precision resistor and a power resistor, capable of handling more power than the standard precision wire-wound resistor but not as accurate. Practically speaking, tolerances of 1% and temperature coefficients of 20 ppm can be achieved on all except the parts that are coated with vitreous enamel having low values. The curing process for vitreous enamel (a type of glass) requires extremely high heat and pressure to the winding. This particular group normally will have tolerances of 10% with a TCR of 100 ppm/°C. Power resistors

Figure 9-2 Chassis-mounted resistor.

come in a variety of ratings. Most are rated at +25°C and derated linearly to either +275°C or +350°C. Again, if the ambient temperature of operation is +275°C, no power can be applied, and at +125°C half the rated power can be applied. These power ratings are based on mounting the resistor in free air with the leads terminated at the recommended point. On axial-lead components, this is ⅜ of an inch from the body. If they can be mounted closer, the resistor will run cooler, or you can apply slightly more power, and if mounted further out, you must reduce the power. *Caution:* if mounted directly over and in contact with a printed circuit board, the heat from the resistor can char the board if full power is applied.

Fuse Resistors. Fuse resistors serve a dual purpose: they are used as resistors and fuses. They are designed so that they will open with a large surge current. The fusing current is calculated based on the amount of energy required to melt the resistive material (the melt temperature plus the amount of energy required to vaporize the resistive material). These resistors will normally run hotter than a normal precision or power resistor so that a momentary surge will bring the resistive element up to the fusing temperature. Some designs create a hot spot inside the resistor to assist in this fusing. Calculations are made and samples are produced to verify the calculations. The major unknown is the heat transfer of the materials, which can be quite significant for pulses of long duration and is very difficult to calculate. Mounting of these devices is critical because it will affect the fusing current. These are quite often made to mount in fuse clips for more accurate fusing characteristics.

Identification and Configuration

Resistors are identified by their color bands (see Table 9-1). The first color band is the first significant number value, the second band is the second significant number value, the third band is the multiplier value, and the forth band is the tolerance of the resistor in percent. For example, if the first color is yellow, then the first number is 4. If the second color is violet, then the second number is then 7. If the third color band is orange, then the multiplier value is three zeros or (000). If the forth color band is gold, then the tolerance is 5%. Taken in order, $4 + 7 + (000) =$ a resistor value of 47,000 with a 5% tolerance.

Resistors can be configured in series or parallel configurations, as shown in Figure 9-3, to increase or decrease resistance values. Placing two or more resistors in series will add the values together as $R_{total} = R1 + R2 + R3$. Placing resistors in parallel will decrease the total overall value of resistance. If two 10 ohm resistors are placed in parallel, the total resistance will be half or 5 ohms. If you place different val-

Table 9-1 Resistor/capacitor color codes

Color*	1st Significant #	2nd Significant #	Multiplier	Tolerance
Black	0	0	1	
Brown	1	1	10	1[†]
Red	2	2	100	2[†]
Orange	3	3	1000	3[†]
Yellow	4	4	10,000	4[†]
Green	5	5	100,000	5[†]
Blue	6	6	1,000,000	6[†]
Violet	7	7	10,000,000	7[†]
Gray	8	8	100,000,000	8[†]
White	9	9	1,000,000,000	9[†]
Gold		—	0.1	5
Silver		—	0.01	10

*Fourth color band: no color band = 20% tolerance; silver band = 10 % tolerance; gold band = 5% tolerance.
[†]Applies to capacitors only.

ues of resistance in parallel, you will have to use the equation for parallel resistances:

$$R_{total} = \frac{1}{R1} + \frac{1}{R2} + \frac{1}{R3} + \cdots \frac{1}{R_n}$$

Resistor Parameters

Resistor Tolerance. Resistor tolerance is expressed as the deviation from nominal value in percent and is measured at 25°C only, with no appreciable load applied. It will change depending on other conditions

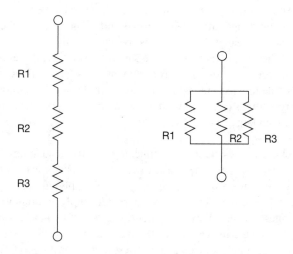

Figure 9-3 Series (left) and parallel (right) configurations.

when in use. For example, a 100 ohm resistor with a tolerance of 10% can range in value from 90 ohms to 110 ohms and this will change as power is applied and the temperature varies.

Temperature Coefficient of Resistance. The temperature coefficient of resistance (TCR) is expressed as the change in resistance in ppm (0.0001%) with each degree of change in temperature Celsius (°C). This change is not linear—with the TCR is lowest at +25°C and increases as the temperature increases (or decreases). It can be expressed as either a bell-shaped curve or an S-shaped curve. It is treated as being linear unless very accurate measurements are needed, then a temperature correction chart is used. Normally, a resistor with a TCR of 100 ppm will change 0.1% over a 10 degree change and 1% over a 100 degree change. The expression ppm, one part in a million, is similar to percent, 1 part in 100 (or percentile)

Frequency Response. Frequency response is the change in resistance with changes in frequency and is more difficult to measure. Where exact values are needed, these changes can be plotted, but not very accurately, and normally in dB change. These measurements can be made with a Boonton RX meter, which is designed for measuring low-Q circuits.

Noise. Noise levels are measured with very specialized equipment. It is extremely difficult to measure accurately and does not effect the value of the resistor but can have a devastating effect on low signals, digital amplifiers, high-gain amplifiers, and other applications sensitive to noise. The best approach is to use resistor types with low or no noise in applications that are sensitive to noise.

Voltage Coefficient. The voltage coefficient is the change in resistance with applied voltage and is associated with carbon composition resistors and carbon film resistors. It is a function of value and the composition of the carbon mixture used in the manufacture of these resistors. This is entirely different from and in addition to the effects of self-heating when power is applied.

Thermocouple Effect. The thermocouple effect is due to the thermal electromotive force (emf) generated by the change in the temperature at the junction of two dissimilar metals. This emf is due to the materials used in the leads or, in the case of wire-wound resistors, the resistive element also. It can be minimized by keeping both leads at the same temperature. The thermal emf is the result of the difference between the temperature of the two leads. One lead will cause a positive emf and the opposite lead will generate a negative emf (or visa versa). When both leads are at the same temperature, the emfs generated will cancel each other and the same is true where the resistive element joins the leads. Resistors with nickel leads used in certain welded module applications will generate the highest thermal emf. The resistive element (the wire) of wire-wound resistors is designed with a low thermal emf, but some of the wire used for high TCR resistors will have a much larger thermal emf.

Stability. Stability is the change in resistance with time at a specific load, humidity level, stress, and ambient temperature. The lower the load and the closer to +25°C the resistor is maintained, the better the stability. Humidity will cause the insulation of the resistor to swell, applying pressure (stress) to the resistive element causing the change. Changes in temperature alternately apply and relieve stresses on the resistive element, thus causing changes in resistance. The wider the temperature changes and the more rapid these changes are, the greater the change in resistance. If severe enough, it can literally destroy the resistor. Rapidly and continuously subjecting a device to its lowest and highest operating temperatures (called a thermocycle test) is considered a destructive test.

Reliability. Reliability is the degree of probability that a resistor (or any other device) will perform its desired function. There are two ways of defining reliability. One is mean time between failures (MTBF) and the other is failure rate per 1000 hours of operation. Both of these means of evaluating reliability must be determined with a specific group of tests and a definition of what is the end of life for a device, such as a maximum change in resistance or a catastrophic failure (short or open). Various statistical studies are used to arrive at these failure rates and large samples are tested at the maximum rated temperature with rated load for up to 10,000 hours (24 hrs per day for approximately 13 months).

Temperature Rating. The temperature rating is the maximum allowable temperature at which the resistor may be used. There are generally two temperatures in the rating. For example, a resistor may be rated at full load up to +85°C and derated to no load at +145°C. This means that with certain allowable changes in resistance over its life, the resistor may be operated at +85°C at its rated power. It also may be operated at temperatures in excess of +85°C if the load is reduced, but in no case should the temperature exceed the design temperature of +145°C with a combination of ambient temperature and self-heating due to the applied load. A word of caution: some rated loads are at +25°C and must be derated if the ambient temperature exceeds +25°C.

Power Rating. Power ratings are based on physical size, allowable change in resistance over life, thermal conductivity of insulating and resistive materials, and ambient operating conditions. Again, note that all resistors are not rated alike. To be safe, use the largest physical size and never use it at its maximum temperature and power ratings unless you are prepared to accept the maximum allowable changes in resistance. Another thing to note: the majority of change under those conditions will occur during the first 100 hours of operation. It is important that all of the above characteristics be considered when selecting a particular style and tolerance for each application.

Resistors Sizes. Resistors are available in almost any size from 0.065 inches diameter by 0.125 inches long to 12 inches in diameter to several feet high (for very high voltage resistors). They come in almost any shape that is imaginable. The most common form is cylindrical with leads coming out of either end. They can be manufactured in custom shapes to fit the available space when quantities justify.

Types of Resistor Mountings. Resistors can be made with almost any type of mounting. If the need arises, special mountings can be designed to fit the customer's needs. Some of the more common means of mounting are listed below. The term "leads" is used in the general sense as a means of connecting the resistor. They may be lugs, wire leads, pins, or any means of connecting the resistor to the circuit.

> **Axial Leads.** Axial lead mounting is what most of us are familiar with using. It consists of a cylindrical, rectangular, or any other body shape with leads extending from either end, parallel to the resistor's major axis.
> **Radial Leads.** Radial lead mounting is similar to axial lead mounting except that the leads come out of the body perpendicular to its major axis.
> **Surface Mount.** Resistors are available in a surface mounting configuration. They are generally chip resistors that are mounted by solder reflow techniques. They consist of a flat resistive element of a ceramic substrate (or a cylindrical ceramic core) with a solder pad on each end. Sizes range from 0.163 inches in diameter to 0.555 inch long cylinders to 0.020 high by 0.031 wide by 0.062 long chips.

Fuse Clip Mounting. The fuse clip type is made such that it will mount directly into a fuse clip. Fuse resistors are sometimes made like this.

Single Inline Packaging (SIP). The single inline package is normally associated with resistor networks consisting of several resistors in the same package. It is a flat rectangular package with the several leads coming out of one surface, generally the narrow, long surface.

Dual Inline Package (DIP). The dual inline package is again normally associated with resistor networks. The main difference is that the leads extend from both narrow, long surfaces and are formed to either flush mount on a PC board or through-hole mount on a PC board.

Flat Packs. The flat pack is roughly the same as the dual inline package except that the leads come straight out and are not formed for surface mounting or through-hole mounting. This is just a variation of DIP mounting.

PC Mounting. PC mounting consists of both leads of the resistor coming out the same surface so that it is easier to mount a resistor (or any other device) vertically. The resistor may be rectangular or cylindrical.

Capacitors

A capacitor is a device that stores an electrical charge or energy on its plates. These positive and a negative plates are placed very close together with an insulator in between to prevent them from touching each other. Figure 9-4 presents a wide variety of capacitor types, shown schematically in Figure 9-5. Usually, a capacitor has at least two more plates, depending on the capacitance or dielectric type (see Figure 9-6). The larger the plate area and the smaller the area between the plates, the larger the capacitance, which also depends on the type of insulating material between the plates. The interplate distance is the smallest when air is the insulation. Replacing the air space with another insulation material will increase the capacitance many times over. The capacitance ratio using an insulation material is called di-

Figure 9-4 *Various capacitors.*

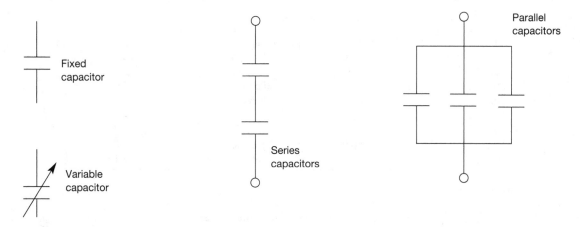

Figure 9-5   Capacitor types.

electric constant, whereas the insulation material itself is just called the dielectric. Looking at Table 9-2, note, for example, that if a polystyrene dielectric is used instead of air, the capacitance will be increased 2.60 times.

Capacitors are absolutely necessary for almost every electronic circuit but they can create enormous problems when troubleshooting circuits. The basic unit of capacitance is the farad. The farad value is quite large and not very practical to work with, so capacitance is usually measured in microfarads, abbreviated μF, or picofarads (pF). The farad is used in converting formulas and other calculations. A 1 μF is on millionth of a farad (10^{-6} F) and 1 pF is one-millionth of a microfarad (10^{-12} F).

Capacitors come in a variety of sizes, shapes, and models, and can be custom manufactured to your own specifications. They also come in a variety of materials, such as aluminum foil, polypropylene, polyester (Mylar), polystyrene, polycarbonate, kraft paper, mica, Teflon, epoxy, oil-filled, electrolyte, and tantalum. The value of a capacitor can vary from a fraction of a picofarad to more than a million microfarads. Voltage levels can range from a couple to a substantial couple of hundred thousand volts. The latest research and development with capacitors has involved the material niobium.

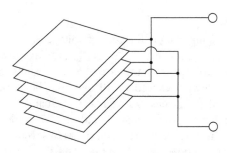

Figure 9-6   Capacitor plate construction.

Table 9-2 Dielectric constant of materials

Air	1.00	Paper	3.00
Alismag 196	5.70	Plexiglass	2.80
Bakelite	4.90	Polyethylene	2.30
Cellulose	3.70	Polystyrene	2.60
Fiber	6.00	Porcelain	5.57
Formica	4.75	Pyrex	4.80
Glass	7.75	Quartz	3.80
Mica	5.40	Stealite	5.80
Mycalex	7.40	Teflon	2.10

When discussing capacitors, you will often hear the term "charge," which is the amount of stored electricity on the plates or, actually, the electric field between theses plates, and is proportional to the applied voltage and the capacitor's capacitance. The formula used to calculate the amount of capacitance is

$$Q = C \times V$$

where
Q = charge in coulombs
C = capacitance in farads
V = voltage in volts

There is also something else involved when there is charge present—something stored called energy. The formula used to calculate the amount of energy is:

$$W = V^2 \times C/2$$

where:
W = energy in joules
V = voltage in volts
C = capacitance in farads

Charging a capacitor is simple: simply apply the correct voltage on the terminals of the capacitor and then wait until the current stops flowing. It doesn't take very long at all. However, be careful not to exceed the capacitor's working breakdown voltage or, in the case of an electrolytic capacitor, it will explode. The breakdown voltage is the voltage that when exceeded will cause the dielectric (insulator) inside the capacitor to break down and conduct. If that happens, the results can be catastrophic. When charging polarized capacitors, you must watch the orientation of the positive and negative terminals. A good quality capacitor can hold a charge for a long time, from seconds to several hours to several days, depending on size. A capacitor, in combination with other components, can be used as a filter that blocks dc or ac, current, frequency, and so on.

Capacitor Codes

Reading capacitor codes is not as difficult as it may first appear. Most of the smaller capacitors have two or three numbers printed on them, some with one or two letters added to that value; see Tables 9-1 and 9-3, which will assist in determining capacitor values.

If a capacitor is marked with the number 105, it means 10 + 5 zeros = 10 + 00000 = 1,000,000 pF = 1000 nF = 1 μF, and that's exactly the way you write it too. The value is in Picofarads. The letters added to the value is the tolerance and in some cases a second letter is the temperature coefficient, but is mostly only used in military applications (basically industrial stuff). So, for example, if you have a ceramic capacitor with 474J printed on it it means 47 + 4 zeros = 470,000 = 470,000 pF, J = 5% tolerance. 470,000 pF = 470 nF = 0.47 μF. Pretty simple, isn't it? The only major thing to get used to is to recognize if the value is in μF, nF, or pF.

Other capacitors may just have 0.1 or 0.01 printed on them. If so, this means a value in μF. Thus, 0.1 means just 0.1 μF. If you want this value in nanofarads, just move the comma three places to the right, which makes it 100 nF. The average hobbyist uses only a couple of capacitor types like the common electrolytic and ceramic capacitors and, depending on the application, a more - type like metal-film or polypropylene.

When working with capacitors, you will often see them represented as a number that may not seem familiar to you. In Table 9-4, for example, you see a value of 0.001 nF listed. The 0.001 nF capacitor value can also be represented as 1 pF. Many people are not familiar with the nanofarad values but rather the picofarad values instead. This table will help you to convert between the three most common capacitance representations.

Types of Capacitors

Ceramic capacitors are constructed with dielectric materials such as titanium acid barium. Internally, these capacitors are not constructed as a coil, so they are well suited for use in high-frequency applications. Typically used to bypass high-frequency signals to ground, they are shaped like a disk and are available in very small capacitance values and very small sizes and are very cheap and reliable. They are sub-

Table 9-3 Capacitor code values

Third digit	Multiplier	Letter	Tolerance
0	1	D	0.5pF
1	10	F	1%
2	100	G	2%
3	1000	H	3%
4	10,000	J	5%
5	100,000	K	10%
6, 7	Not used	M	20%
8	0.01	P	+100, –0%
9	0.1	Z	+80, –20%

Table 9-4 Capacitor conversion values

Microfarads (μF)		Nanofarads (nF)		Picofarads (pF)
0.000001 μF	=	0.001 nF	=	1 pF
0.00001 μF	=	0.01 nF	=	10 pF
0.0001 μF	=	0.1 nF	=	100 pF
0.001 μF	=	1 nF	=	1000 pF
0.01 μF	=	10 nF	=	10,000 pF
0.1 μF	=	100 nF	=	100,000 pF
1 μF	=	1000 nF	=	1,000,000 pF
10 μF	=	10,000 nF	=	10,000,000 pF
100 μF	=	100,000 nF	=	100,000,000 pF

ject to drifting, depending on ambient temperature. NPO types are the temperature-stable types. They are identified by a black stripe on top. Ceramic and electrolytics are the most widely available and used capacitors.

The electrolytic capacitor is constructed using an electrolyte, basically a conductive salt in a solvent. Aluminum electrodes are used by using a thin oxidation membrane. Electrolytic capacitors are most always polarized, that is, they have a plus terminal and minus terminal. Applications for electrolytic capacitors include Ripple filters and timing circuits. Electrolytic capacitors are inexpensive and readily available but not very accurate. They are subject leakage, drifting, and generally not suitable for use in HF radio circuits. They are available in very small or very large values measured in microfarads. When you use this type capacitor in one of your projects, the rule of thumb is to choose one that is twice the supply voltage. Care must be taken when installing these capacitors since polarity is very important. If the leads are reversed on these capacitors, they will explode.

The polyester film capacitor uses a thin polyester film as a dielectric. They do not have as high a tolerance as polypropylene capacitors, but cheap, temperature stable, readily available, and widely used. Tolerance is approximately 5% to 10%, but can be quite large depending on capacity or rated voltage and so they may not be suitable for all applications.

Polypropylene capacitors are mainly used when a higher tolerance is needed than polyester capacitors can offer. The polypropylene film is the dielectric. There is very little change in capacitance when these capacitors are used in applications around a frequency of 100 KHz. Tolerance is about 1%. Very small values are available.

Polystyrene capacitors use polystyrene as a the dielectric material. They are constructed like a coil inside, so they are not suitable for high-frequency applications, but are good when used in filter circuits or timing applications of a couple of hundred kilohertz or less. Electrodes may be reddish in color when copper leaf is used for the electrodes or silver when aluminum foil is used.

The silver–mica capacitor uses mica as its dielectric. Used in resonance circuits, frequency filters, and military RF applications, they are highly stable, have good temperature coefficients, and excellent endurance because of their frequency characteristics. However, no large values or high-voltage types are available. They can be expensive but worth the extra expense.

The metalized polyester film capacitor utilizes a dielectric made of a metal oxide. These capacitors are good quality, low drift, and temperature stable. Because the electrodes are thin, they can be made extremely small.

Tantalum capacitors are electrolytic capacitors with tantalum pentoxide electrodes. They are superior to other electrolytic capacitors, with excellent temperature and frequency characteristics. When tantalum powder is baked in order to solidify it, a crack forms inside. An electric charge can be stored in this crack. Like other electrolytics, tantalums are polarized, so watch for the "+" and "−" indicators. Mostly used in analog signal systems because of their lack of current-spike noise, their small size fits anywhere, they are reliable, and most common values are readily available. However, they are expensive, easily damaged by spikes, and although large values exists, they may be hard to obtain. The largest one in my own collection is a 220 μF/35 V, beige colored capacitor.

Epoxy capacitors are manufactured using epoxy-based polymers as the dielectric. They are widely available, stable, and cheap. They can be quite large depending on capacity or rated voltage and so they may not be suitable for all applications.

Supercapacitors or electric double-layer capacitors are quite a marvel to behold. Capacitance is 0.47 farad (470,000 μF). Despite the large capacitance value, its physical dimensions are relatively small. It has a diameter of 21 mm (almost an inch) and a height of 11 mm ($\frac{1}{2}$ inch). Like other electrolytic capacitors, the supercapacitor is also polarized, so exercise caution with regard to the breakdown voltage. Care must be taken when using this capacitor. It has such large capacitance that, without precautions, it could destroy part of a power supply such as the bridge rectifier or volt regulators because of the huge inrush current at charge. For a brief moment, this capacitor acts like a short circuit when it is charged. Protection circuitry is a must for this type.

In multilayer ceramic capacitors, the dielectric material is made up of many layers. They are small in size, with very good temperature stability and excellent frequency stability. Used in applications to filter or bypass the high frequency to ground, multilayer capacitors suffer from high-Q internal (parallel) resonances, generally in the VHF range. The CK05 style 0.1 μF/50V capacitors, for example, resonate around 30 MHz. The effect of this resonance is effectively no apparent capacitance near the resonant frequency. As with all ceramic capacitors, be careful bending the legs or spreading them apart too close to the disc body or they may be damaged.

Adjustable capacitors are also called trimmer capacitors or variable capacitors. They use ceramic or plastic as the dielectric. Most of them are color coded to easily recognize their tunable size: yellow (5 pF), blue (7 pF), white (10 pF), green (30 pF), and brown (60 pf). The ceramic type have values printed on them.

Tuning or "air-core" capacitors use the surrounding air as a dielectric. Mostly used in radio and radar equipment, some older variable capacitors of this type have incredibly large dimensions. Applications using this type usually have several (air) capacitors combined (ganged), so when the adjustment shaft is turned, the capacitance of all of them changes simultaneously.

Capacitors in Series

Multiple capacitors connected in series will have a total capacitance lower than the lowest-value capacitor in that circuit. Note that this is just opposite to the way of calculating resistors. For two capacitors in a series combination, use the simple formula below:

$$C_{total} = C1 \times C2/C1 + C2$$

If you have two identical capacitors in series then use $C_{total} = \frac{1}{2} C$.

Capacitors in Parallel

Capacitors connected in parallel, which is the most desirable method, have their capacitance added together, which is just the opposite of parallel resistors. It is an excellent way of increasing the total storage capacity of an electric charge, since $C_{total} = C1 + C2 + C3$.

Keep in mind that only the total capacitance changes, not the supplied voltage. Every single capacitor will see the same voltage, no matter what. Be careful not to exceed the specified voltage on the capacitors when combining those with different voltage ratings or they may explode. For example, say you have three capacitors with voltages of 16 V, 25 V, and 50 V. The voltage must not exceed the lowest voltage, in this case the 16 V. As a matter of fact and as a rule of thumb, always choose a capacitor that is twice the supplied input voltage. For example, if the input voltage is 12 V, you would select a 24 V type (in real life, 25 V).

Inductors

The inductor is one of the most important components in the electronics field, it utilizes ac waveforms to reduce current; the more current, the more induced resistance there will be in the circuit.

An inductor is a passive electronic component that stores energy in the form of a magnetic field. In its simplest form, an inductor consists of a wire loop or coil. The inductance is directly proportional to the number of turns in the coil. Inductance also depends on the radius of the coil and on the type of material around which the coil is wound. For a given coil radius and number of turns, air cores result in the least inductance. Materials such as wood, glass, and plastic are known as dielectric materials and are essentially the same as air for the purposes of inductor winding. Ferromagnetic substances such as iron, laminated iron, and powdered iron increase the inductance obtainable with a coil having a given number of turns. In some cases, this increase is on the order of thousands of times. The shape of the core is also significant. Toroidal (donut-shaped) cores provide more inductance, for a given core material and number of turns, than solenoidal (rod-shaped) cores.

How Inductors Work

When current flows through a conductor, a magnetic field surrounds the conductive wire (see Figure 9-7). The more current traveling through the wire, the greater the amount of flux lines that will be present. These lines of flux can generate voltage on surrounding conductors. This voltage can be used to alter waveforms in circuits where ripple current is present. Ripple current is dc current that does not stay at one level, but instead varies like the ripples on the surface of a vibrated container of water. Since flux lines are little more then static charges that can never intersect with flux lines of the same charge, as you increase current, these lines of force start to extend further and further outward. If you have a field surrounding a conductor (wire) and it comes in contact with another conductor (wire), then the flux lines will push electrons in a predictable direction, essentially creating an emf. However, if there is only one coil that is the primary wire, then the emf is induced on itself, generating voltage in the opposite

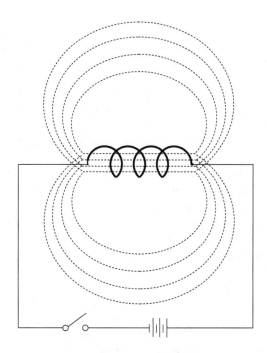

Figure 9-7   Magnetic field.

direction. Known as counterelectromagnetic force (cemf) or countervoltage, this phenomenon is what makes the inductor possible.

As you increase current, the magnetic field starts expanding outward, moving through nearby coils. When this happens, voltage is generated in the opposite direction, making it harder and harder to increase current flow. As the current reaches its peak, the inductor's opposition to current flow also becomes high, since in order to generate voltage there must be movement in either the magnetic fields or the coil wires. If you are using an inductor in an audio application with an alternating voltage, you should pay attention to what happens when current declines as well when it increases. This is just like what happens when voltage is introduced, but instead of generating voltage in one direction, it generates it in the opposite direction, making the waveform look more oval than peaks and valleys. When current has reached its peak, the inductor stops working until there is a disruption in current flow, as in the case of ripple current or in the case of audio applications or power supplies.

In audio applications, you would refer to these waves as frequencies or audio waves. In the case of speaker systems, you would want to discriminate high frequencies from low frequencies. To do this, you would use the inductor to filter out high frequencies, which is good when you are connecting a bass speaker. If you were making a power supply, you would not use the inductor to flatten down the ac current. What you would do is use a related device known as a transformer, which steps down current or voltage. Next, you would add a device known as a diode to make the current travel in only one direction without making the waveforms flat. To flatten down the waveform, you would use a device related to the inductor, known as the capacitor, which is basically an inefficient battery that only holds a charge for fractions of a second but is perfect in an alternating dc circuit. This would be a good power source for analog radios, but in the case of computer devices, you would want to further flatten the current with the inductor.

Inductors, chokes, and coils are passive devices. Inductors store energy in the form of a magnetic field. In their simplest form, inductors consist of a wire loop or coil. The inductance is directly proportional to the number of turns in the coil. Inductance also depends on the radius of the coil and on the type of material around which the coil is wound. For a given coil radius and number of turns, air cores result in the least inductance. Dielectric materials such as wood, glass, and plastic are essentially the same as air for the purposes of inductor winding. Ferromagnetic substances such as iron, laminated iron, and powdered iron increase the inductance obtainable with a coil having a given number of turns. In some cases, this increase is on the order of thousands of times. The shape of the core is also significant. Toroidal (donut-shaped) cores provide more inductance, for a given core material and number of turns, than solenoidal (rod-shaped) cores. Toroids look exactly like doughnuts and come in various diameters, thicknesses, permeabilities, and types, depending upon the frequency range of interest. They have a high inductance for the physical space occupied.

Important physical specifications to consider when searching for inductors, chokes, and coils include mounting options, core materials, lead types, and inductance types. Mounting options include through hole and surface mount. Coils can be wound on various core materials, the most popular being iron (or iron alloys, laminations, or powder) and ferrite, a black, nonconductive, brittle magnetic material. The inductance of a given coil increases with the "permeability" of the core material. The core may be in the shape of a rod, a toroid (doughnut), or other shapes. Other core materials include air, ceramic, and phenolic. Lead types can be axial, radial, flying, no leads (SMT), tab, gull wing, and J-leads. The inductance can be fixed or variable.

Important electrical specifications to consider when searching for inductors, chokes, and coils include inductance range, inductance tolerance, maximum dc resistance, and operating current range. Common applications for inductors, chokes, coils and ferrite beads include common-mode, general-purpose, high-current, high-frequency, power, and RF chokes. Common-mode choke coils are useful in a wide range of applications to prevent electromagnetic interference (EMI) and radio frequency interference (RFI) from power supply lines and to prevent various types of electronic equipment from malfunctioning. Operating temperatures are also important to consider.

Types of Inductors

There are numerous styles of inductors, all with specific applications. Figure 9-8 shows a range of inductors from micro-inductors to mini-transformers. The inductors shown in Figure 9-9 illustrates three different coil types. The coil marked A is a toroid, which is often used in power supplies or current-sensing circuits. The coil marked B is an air-core coil commonly used in power supplies, and the coil marked C is a molded inductor that looks much like a resistor. This type of coil is often used in oscillators and sensing circuits. Figure 9-10 shows variable-core inductors for precision radio tuning or frequency discrimination. Figure 9-11 shows a range of power inductor coils with ferrite cores for boosting inductance values in power supply designs.

Figure 9-8   Various inductors.

Units of Inductance

The standard unit of inductance is the henry (H). This is a large unit. More common units are the microhenry (μH; 1 μH = 10^{-6} H) and the millihenry (mH; 1 mH = 10^{-3} H). Occasionally, the nanohenry (nH) is used (1 nH = 10^{-9} H). Table 9-5 lists the color codes used in determining coil values.

Inductors are used with capacitors in various wireless communications applications. An inductor connected in series or parallel with a capacitor can provide discrimination against unwanted signals. Large in-

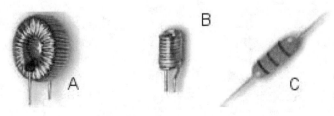

Figure 9-9   Inductor coil types.

Figure 9-10 Variable-core inductors.

ductors are used in the power supplies of electronic equipment of all types, including computers and their peripherals. In these systems, the inductors help to smooth out the rectified utility ac, providing pure, battery-like dc.

It is difficult to fabricate inductors on integrated circuit (IC) chips, due to their size. Resistors can sometimes be substituted for inductors in most microcircuit applications and, in some instances, inductance can be simulated by simple electronic circuits using transistors, resistors, and capacitors fabricated on IC chips.

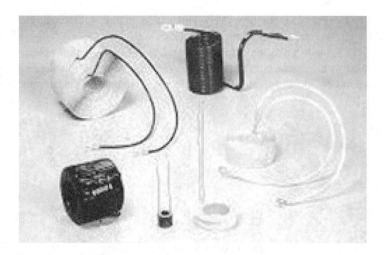

Figure 9-11 Power inductor coils.

Table 9-5   Inductor color code

Color	Digit	Multiplier	Tolerance
Black	0	1	
Brown	1	10	
Red	2	100	
Orange	3	1000	
Yellow	4		
Green	5		
Blue	6		
Purple	7		
Gray	8		
White	9		
None			20%
Silver			10%
Gold			5%

The capacity of an inductor is controlled by four factors:

1. The number of coils. More coils means more inductance.
2. The material that the coils are wrapped around (the core)
3. The cross-sectional area of the coil. More area means more inductance.
4. The length of the coil. A short coil means narrower (or overlapping) coils, which means more inductance.

Putting iron in the core of an inductor gives it much more inductance than air or any nonmagnetic core would.

The equation for calculating the number of henries in an inductor is:

$$H = (4 \times \pi \times \# \text{Turns} \times \text{Coil Area} \times \mu)/(\text{Coil Length} \times 10,000,000)$$

The area and length of the coil are in meters. μ is the permeability of the core. Air has a permeability of 1, whereas steel might have a permeability of 2000.

Another major use of inductors is to team them up with capacitors to create oscillators.

Series and Parallel Configurations

In a dc circuit, there are three basic setups for inductors: parallel, series, and series parallel. The reason for having three different types is because inductors can be found in so many different values and some are cheaper then others. In a series circuit, all the inductors and resistors are in the same circuit, placed one after the other like a train and its cargo.

In an ac circuit, the inductor acts more like a resistor then any other component because it resists any kind of current change, whether it is positive or negative current flow. So when power starts to charge

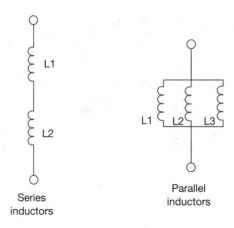

Figure 9-12 *Basic inductor circuits.*

to a peak voltage of, say, 10 V and before the inductor can fully charge, the current changes direction. But the nature of the inductor is that the flux lines create a countervoltage until the flux lines can get established in a directional flow. When current changes direction within the inductor, the flux lines are still rotating in one direction and it takes energy to make these flux lines change to the opposite direction. Because ac changes direction many tines a second (60 times from U.S. wall sockets), the inductor doesn't fully establish its full direction and, in affect, is always generating a small amount of current in the opposite direction. This makes waveforms just a little flatter but this is perfect in applications like power supplies and audio applications where you want to distort alternating dc "ripple" power or voice signals. If you were to make an inductor perfectly designed with no resistance and have a source of infinite inductance, then the outcome would be zero volts. This is because in ac power, voltage fluctuates from a positive voltage to an equal amount of negative voltage and the center point or average voltage is zero. Figure 9-12 illustrates series and parallel connections between inductors. When placing inductors in series, the total inductance will add up, the same as for resistances: L_{total} = L1 + L2 + L3. Placing inductors in parallel will reduce the overall inductance; use the formula

$$L_{total} = \frac{1}{L1} + \frac{1}{L2} + \frac{1}{L3} + \cdots \frac{1}{L_n}$$

Winding air-wound coils using thin-insulation magnet wire can be accomplished by using the values given in Table 9-6. The coil is tightly wound over a cylinder-shaped form. It must be stiffened with varnish or glyptol to keep its final shape. The coil must be removed from the coil form before use. Finally, the insulation must be scraped off both ends of the finished coil before it is soldered into the circuit.

Transformers

A transformer is an energy transfer device. It has an input side (primary) and an output side (secondary). Electrical energy applied to the primary is converted into a magnetic field, which in turn induces a current in the secondary, which carries energy to the load connected to the secondary. The energy applied to the

Table 9-6 Inductor design chart

Wire gauge (AWG)	Wire size (inch)	Resistance (ohms/foot)	Length per lb/foot
8	0.1285	0.6405	20.01
9	0.1144	0.8077	25.23
10	0.1019	1.018	31.82
11	0.0907	1.284	40.12
12	0.0808	1.619	50.59
13	0.0720	2.042	63.80
14	0.0641	2.575	80.44
15	0.0571	3.247	101.4
16	0.0508	4.094	127.9
17	0.0453	5.163	161.3
18	0.0403	6.510	203.4
19	0.0359	8.210	256.5
20	0.0320	10.35	323.4
21	0.0285	13.05	407.8
22	0.0253	16.46	514.2
23	0.0226	20.76	648.4
24	0.0201	26.17	817.7

primary must be in the form of a changing voltage, which creates a constantly changing current in the primary, since only a changing magnetic field will produce a current in the secondary. The Figure 9-13 depicts a transformer both pictorially and schematically.

A transformer consists of at least two sets of windings wound on a single magnetic core. There are two main purposes for using transformers. The first is to convert the energy on the primary side to a different voltage level on the secondary side. This is accomplished by using differing turns counts on primary and secondary windings; the voltage ratio is the same as the turns ratio. The second purpose is to isolate the energy source from the destination, either for personal safety or to allow a voltage offset between the source and load.

Transformers are generally divided into two main types. Power transformers are used to convert voltages and provide operating power for electrical devices, whereas signal transformers are used to transfer some type of useful information from one form or location to another.

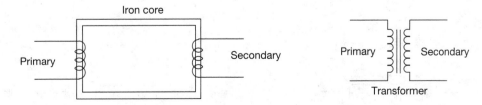

Figure 9-13 Transformer. Left, pictorial representation. Right, schematic.

How Semiconductors Work

Semiconductors have had a monumental impact on our society. They are at the heart of diodes and transistor microprocessors. Anything that is computerized or uses radio waves depends on semiconductors. Today, most semiconductor chips and transistors are created with silicon, the main element in sand and quartz. Carbon, silicon, and germanium (germanium, like silicon, is also a semiconductor) have a unique property in their electron structure—each has four electrons in its outer orbital. This allows them to form nice crystals. The four electrons form perfect covalent bonds with four neighboring atoms, creating a lattice. In carbon, the crystalline form is diamond shaped. In silicon, the crystalline form is a silvery, metallic-looking substance.

In a silicon lattice, all silicon atoms bond perfectly to four neighbors, leaving no free electrons to conduct electric current. This makes a silicon crystal an insulator rather than a conductor. Metals tend to be good conductors of electricity because they usually have "free electrons" that can move easily between atoms, and electricity involves the flow of electrons. Although silicon crystals look metallic, they are not, in fact, metals. All of the outer electrons in a silicon crystal are involved in perfect covalent bonds, so they can't move around. A pure silicon crystal is nearly an insulator—very little electricity will flow through it.

A semiconductor such as silicon has properties somewhere between those of a conductor and an insulator. The ability of a semiconductor to conduct electricity can be changed dramatically by adding small amounts of a different element to the semiconductor crystal. This process is called doping. Early experiments showed that an electric current through a semiconductor was carried by the flow of positive charges as well as negative charges (electrons).

Doping Silicon

You can change the behavior of silicon and turn it into a conductor or semiconductor by doping it. In the process of doping, you mix a small amount of an impurity into the silicon crystal. Basically there are two types of impurities used in the doping process (see Figure 9-14).

N-Type. In N-type doping, phosphorus or arsenic is added to the silicon in small quantities. Phosphorus and arsenic each have five outer electrons, so they are out of place when they get into the silicon lattice. The fifth electron has nothing to bond to, so it is free to move around. It takes only a very small quantity of the impurity to create enough free electrons to allow an electric current to flow through the silicon. N-type silicon is a good conductor. Electrons have a negative charge, hence the name N-type.

P-Type. In P-type doping, boron or gallium is the dopant. Boron and gallium each have only three outer electrons. When mixed into the silicon lattice, they form "holes" in the lattice in which a silicon elec-

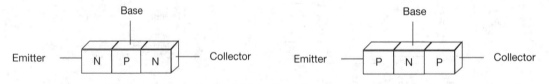

Figure 9-14   *Semiconductor transistors. Left, N-type. Right, P-type.*

tron has nothing to bond to. The absence of an electron creates the effect of a positive charge, hence the name P-type. Holes can conduct current. A hole happily accepts an electron from a neighbor, moving the hole over a space. P-type silicon is a good conductor.

A minute amount of either N-type or P-type doping turns a silicon crystal from a good insulator into a viable (but not great) conductor, hence the name "semiconductor." N-type and P-type silicon are not that amazing by themselves, but when you put them together, you get some very interesting behavior at the junction.

Creating a Diode

A diode is the simplest possible semiconductor device. A diode allows current to flow in one direction but not the other. When you put N-type and P-type silicon together as shown in Figure 9-15, you get a very interesting phenomenon that gives a diode its unique properties. Even though N-type silicon by itself is a conductor, and P-type silicon by itself is also a conductor, the combination of the two shown in the figure does not conduct any electricity. The negative electrons in the N-type silicon get attracted to the positive terminal of the battery. The positive holes in the P-type silicon get attracted to the negative terminal of the battery. No current flows across the junction because the holes and the electrons are each moving in the wrong direction. If you flip the battery around, the diode conducts electricity just fine. The free electrons in the N-type silicon are repelled by the negative terminal of the battery. The holes in the P-type silicon are repelled by the positive terminal. At the junction between the N-type and P-type silicon, holes and free electrons meet. The electrons fill the holes. Those holes and free electrons cease to exist, and new holes and electrons spring up to take their place. The effect is that current flows through the junction.

When reverse-biased, an ideal diode would block all current. A real diode lets perhaps 10 microamperes through—not a lot, but still not perfect. If you apply enough reverse voltage, the junction breaks down and lets current through. Usually, the breakdown voltage is a lot more voltage than the circuit will ever see, so it is irrelevant. When forward-biased, there is a small amount of voltage necessary to get the

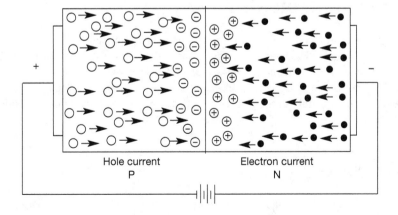

Hole current
P

Electron current
N

Figure 9-15   P-N diode junction.

diode going. In silicon, this voltage is about 0.7 volts. This voltage is needed to start the hole–electron combination process at the junction.

The junction diode is useful in a wide variety of applications including the rectification of ac signals (producing dc from ac), the detection of radio signals, the conversion of solar power to electricity, and the generation and detection of light. It also finds use in a variety of electronic circuits as a switch, as a voltage reference, or even as a tunable capacitor. The P–N junction is also the basic building block of a host of other electronic devices, of which the most well-known is the junction transistor. For this reason, a study of the properties and behavior of the P–N junction is important.

Transistors and Integrated Circuits

A transistor is created by using three layers rather than the two layers used in a diode. You can create either an NPN or a PNP sandwich. A transistor can act as a switch or an amplifier. A transistor looks like two diodes back to back. You would imagine that no current could flow through a transistor because back-to-back diodes would block current both ways, and this is true. However, when you apply a small current to the center layer of the sandwich, a much larger current can flow through the sandwich as a whole. This gives a transistor its switching behavior. A small current can turn a larger current on and off, much like how a relay is used to leverage a low current to control a larger current.

A silicon chip is a piece of silicon that can hold thousands of transistors. With transistors acting as switches, you can create Boolean gates, and with Boolean gates you can create microprocessor chips. The natural progression from silicon to doped silicon to transistors and finally to integrated circuit chips is what have made microprocessors and other electronic devices so inexpensive and ubiquitous in today's society. The fundamental principles are surprisingly simple. Integrated circuits can be fabricated into many different functional building blocks used in modern electronic circuits, such as amplifiers, filters, regulators, and comparators. The miracle is the constant refinement of those principles to the point where, today, tens of millions of transistors can be inexpensively formed on a single chip.

Protecting Components

You need to take steps to protect the electronic and mechanical components you need in circuit construction. Some components can be damaged by rough handling. Dropping a ¼ W resistor causes no harm, but dropping a vacuum tube or other delicate subassemblies usually causes damage. Some components are easily damaged by heat. Some of the chemicals used to clean electronic components (such as flux removers, degreasers, or control-lubrication sprays) can damage plastic. Check them for safety before you use them.

Electrostatic Discharge

Some components, especially high-impedance components such as FETs and CMOS gates, can be damaged by electrostatic discharge (ESD). Protect these parts from static charges. Most people are familiar with the static charge that builds up when one walks across a carpet, then touches a metal object; the resultant spark can be quite lively. Walking across a carpet on a dry day can generate 35 kV! A worker sitting at a bench can generate voltages up to 6 kV, depending on conditions, such as when relative humidity is less than 20%.

You don't need this much voltage to damage a sensitive electronic component; damage can occur with as little as 30 V. The damage is not always catastrophic. A MOSFET can become noisy or lose gain; an IC can suffer damage that causes early failure. To prevent this kind of damage, you need to take some precautions.

The energy from a spark can travel inside a piece of equipment to effect internal components. Protection of sensitive electronic components involves the prevention of static buildup together with the removal of any existing charges by dissipating any energy that does build up. Several techniques can be used to minimize static buildup. First, remove any carpet in your work areas. You can replace it with special antistatic carpet, but this is expensive. It is less expensive to treat the carpet with antistatic spray, which is available from Chemtronics, GC Thorsen, and other lines carried by electronic wholesalers.

Even the choice of clothing you wear can affect the amount of ESD. Polyester has a much greater ESD potential than cotton. Many builders who have their workbench on a concrete floor use a rubber mat to minimize the risk of electric shocks from the ac line. Unfortunately, the rubber mat increases the risk of ESD. An antistatic rubber mat can serve both purposes.

Many components are shipped in antistatic packaging. Leave components in their conductive packaging. Other components, notably MOSFETS, are shipped with a small metal ring that temporarily shorts all of the leads together. Leave this ring in place until the device is fully installed in the circuit. These precautions help reduce the buildup of electrostatic charges. Two other techniques are to offer a slow discharge path for the charges or to maintain and handle the components at the same ground potential. One of the best techniques is to connect the operator and the devices being handled to earth ground or a common reference point. It is not a good idea to directly ground an operator working on electronic equipment, though; the risk of shock is too great. If the operator is grounded through a high-value resistor, ESD protection is still offered but there is no risk of shock. The operator is usually grounded through a conductive wrist strap, such as the one made by 3M . This wrist band is equipped with a snap-on ground lead. A 1 Mohm resistor is built into the snap, which is connected to a charge-dissipating mat that is connected to ground. The mat should be an insulator that has been impregnated with a resistance material. Suitable mats and wrist straps are made by 3M, GC Electronic, and others; they are available from most electronics supply houses.

The work area should also be grounded to a mat. Use a soldering iron with a grounded tip to solder sensitive components. Most irons that have three-wire power cords are properly grounded. When soldering static-sensitive devices, use two or three jumpers to ground you, the work, and the iron. If the iron does not have a ground wire in the power cord clip, run a jumper from the metal part of the iron near the handle to the metal box that houses the temperature control. Another jumper connects the box to the work. Finally, a jumper goes from the box to an elastic wrist band for static grounding.

Use antistatic bags to transport susceptible components or equipment. Keep your workbench free of objects such as paper, plastic, and other static-generating items. Use conductive containers with a dissipative surface coating for equipment storage. All of the antistatic products described above are available from Newark Electronics and other suppliers.

Testing Electronic Components

In this chapter, you will learn how to test systems and components using a multimeter and an oscilloscope. The multimeter and oscilloscope can help you to determine whether a component is good or defective. If you have a large "junk box" of parts and if you wish to determine if those parts could be used in a new circuit you are building or repairing, then you will want to read this chapter.

Why Testing Is Important

Testing tells you whether something works or if it doesn't. Testing is also instructive. If you measure the input of an amplifier and see 0.1 volts and then measure the output and see10 volts, you learn that the amplifier has a gain of 100. If you see a glitch when a circuit fails, you learn that the glitch is the problem. Testing lets you explore a circuit. Most electronic enthusiasts and engineers enjoy probing a new piece of gear to see what it does.

Testing also collects data, which is needed to determine if something is operating properly. Understanding how test instruments work allows you to understand instrumentation. Instrumentation is fundamental to signal analysis, and signal analysis is used on nearly every electronic device. So, it becomes apparent why testing and understanding testing fundamentals are so important at both a practical and a theoretical level.

Multimeter

A good volt–ohmmeter or multimeter will quickly become your best friend when testing electronic components. A decent multimeter will allow you to test resistors, capacitors, inductors, semiconductors, and open and shorted circuits. You can usually get by with a good digital multimeter and an analog meter. The analog meter is very useful for showing slow variations or averaging out unusual waveshapes. The digital mulitmeter is perfect for precision measurements that no analog meter can match. A digital multimeter with a 3.5 to 4.5 digit display and diode and capacitance measuring capabilities is well worth a few more dollars.

Sometimes, paying a little extra for a special function in a meter can save paying for a separate instrument. Look closely at what is available. There is a huge variety out there. Always look at the DC input resistance of a meter, sometimes called "input impedance" and not to be confused with resistance ranges. It should be as high as possible, preferably at least 10 megohms. This will ensure that the meter does not affect the operation of the circuit under test.

Oscilloscope

The oscilloscope is a very effective measuring instrument that provides significantly more information about a signal than an ordinary analog or digital voltmeter. The oscilloscope essentially "paints" a picture of the waveform's characteristics. It can measure and display voltage relative to time, as well as frequency and signal waveforms.

Oscilloscopes are broken down into two major classifications: analog and digital. Digital scopes have many more features such as data logging, delay, and storage features. Newer digital oscilloscopes are much more accurate, smaller, and lighter, and will often run on batteries and utilize LCD displays.

There are many oscilloscope models to choose from, depending on the type of measurement task at hand. A general-purpose scope is usually adequate for most basic measurements. Most any oscilloscope can be used to perform the basic component tests shown in this chapter. A dual-channel oscilloscope with internal and external triggering as well as X1 and X10 probes would be all you would need for testing electronics components.

Logical Thinking Is Your Best Test Instrument

Good test equipment makes testing easier. Making testing better requires thought and understanding. You can improve the accuracy and resolution of many measurements with simple statistical procedures. It is tedious, awkward, and time-consuming, but 10-fold to 100-fold improvements in accuracy and resolution can be accomplished this way. This means that you have the potential to turn the fast 8-bit A/D in your microprocessor into a very slow 12-bit A/D. Under some circumstances, this may be a perfect answer to serious problems.

You need to understand both the instrument and the test to get the most out of your gear. Do you know how dual time bases work? Or how a universal frequency counter differs from a basic one? Or that a tracking generator makes filter testing a snap? These points are learned through experience and a willingness to learn. Often, reading the instruction manual will show you many different ways to use that piece of test gear that maybe you didn't know about. Watch and learn from others, and don't be afraid to ask questions. Testing is fundamental to many aspects of engineering and to the business of engineering. Understanding how testing is accomplished, how test instruments work, and the limits of test equipment will enable you to make the most out of your time, whether you are a hobbyist or design engineer. Learning about the numerous techniques of testing is always worthwhile for the beginner as well as the seasoned veteran.

Making Resistor Measurements

Resistance in a circuit is the opposition to current. All materials used in an electric circuit will offer some kind of resistance. Components with low resistance are called conductors and are used as paths for current. Components with extremely high resistance provide no path for current and are called insulators. Parts manufactured specifically to be placed in a circuit to provide resistance are called resistors and are used most frequently in electrical circuits to limit current or to create voltage drops, whereas insulators are used to resist or prevent current in certain paths.

Resistors come in different values, shapes and sizes and are classified in two ways: by resistance measured in ohms and by power rating measured in watts. The power ratio of resistors for electronic circuits range from $\frac{1}{8}$ watt to hundreds of watts. The ohmic values range from hundredths of ohms (0.01) to hundreds of megohms (100×10^6). Manufacturers have adopted a standard color-code system for indicating the resistance or ohmic value of low-power resistors (normally below 2 watts). Higher-power resistors usually have the resistance value imprinted on their bodies. An assortment of different types of resistors with different wattage ratings is shown in Figure 10-1.

Low-wattage resistors (usually also physically small) are usually of carbon composition or deposited carbon. Carbon composition resistors are made of finely ground carbon mixed with a binder molded into a cylindrical shape and have pigtail leads. The compressed mixture can be made to have a resistance from one ohm to tens of megohms. This type resistor usually has wattage ratings of 2 watts or less, and the resistance value is indicated by color-coded bands encircling the body of the resistor. The deposited carbon resistor consists of carbon vapor deposited on a glass or ceramic form. Spiral paths are etched into the carbon until the desired resistance is obtained. Such resistors are usually higher-precision types (accurate to 1% or less).

Wire-wound resistors are made of resistance wire wound on a ceramic core. Fusible resistors are a special type of wire-wound resistor made to burn out and open the circuit to protect other components if current becomes greater than the fuse resistor is designed to handle.

A variable resistor is generally either carbon or wire-wound with a movable wiper arm that will contact the resistance element at any point between the end extremes. When the arm is moved, usually with a shaft, the resistance between the wiper and either end varies. These variable resistors are commonly called potentiometers or rheostats. If three terminals are used in the circuit, it is referred to as a potentiometer. If only two terminals are used, it is referred to as a rheostat.

Color Codes

Resistor values are measured in ohms with an ohmmeter. A common way of indicating resistance values for composition resistors is to print color bands on the body of the resistor. A standard color code for the bands has been adopted by resistor manufacturers (see Figure 10-2). The numerical values they represent

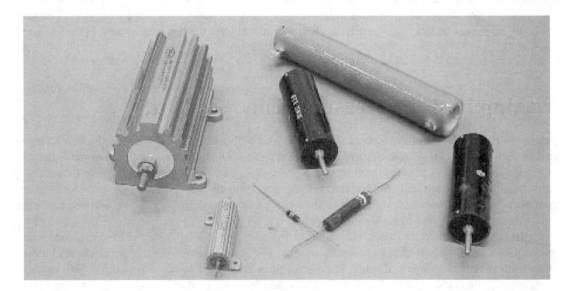

Figure 10-1   Assorted resistors.

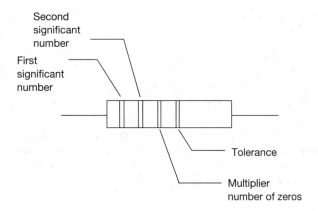

Figure 10-2   Resistor color code.

are given in Table 10-1, which shows how to evaluate the bands. The first band will be nearer one end of the resistor. Read from this band toward the other end. The first band is the first significant number; the second band, the second significant number; and the third band, the multiplier (number of zeros that follow the fist two numbers) to indicate the resistor value. If the third band is gold or silver, this indicates a multiplier of 0.1 or 0. 01, respectively, rather than additional zeros.

As the resistors are mass-manufactured, they all will not be of exactly the same value. The fourth band indicates the tolerance of the resistor's value from the indicated (color code) value. A gold fourth band indicates a tolerance of ±5%, a silver band ±10%, and no fourth band ±20%. The resistor value will be within a tolerance band of ±5%, ±10%, or ±20% of the color-coded value.

Table 10-1   Resistor color codes

Color	First and second significant #	Multiplier	Tolerance
Black	0	1	
Brown	1	10	
Red	2	100	
Orange	3	1000	
Yellow	4	10,000	
Green	5	100,000	
Blue	6	1,000,000	
Violet	7	10,000,000	
Gray	8	100,000,000	
White	9	—	
Gold		0.1	5%
Silver		0.01	10%
No band		—	20%

Example: Resistor color code (yellow—blue—orange—silver) = 4—6—000—10%. Resistor = 46,000 ohms with 10% tolerance.

Measuring a Fixed Resistor

One of the easiest measurements to make with a VOM is to measure the value of a fixed resistor in a circuit. Use the VOM as an ohmmeter and connect the test leads of the ohmmeter across the resistor as shown in Figure 10-3. Note that the figure illustrates both a fixed resistor and a variable resistor or potentiometer. Resistors, as you may remember, are represented by two leads and a stepped sawtooth. You can also perform an in-circuit resistor test. First, disconnect the equipment from the power source and disconnect one end of the resistor from any circuit or additional component so that only the resistance of the single resistor is measured. For a fixed-resistor measurement, place the meter probe leads at each end of the resistor leads (points A and B in the figure). Be careful not to have your fingers cross the resistor leads since this can affect the measurement of resistance.

Measuring a Variable Resistor

In order to test a potentiometer with an ohmmeter, the total resistance is first measured from end to end, that is, from point A to point B in Figure 10-3. This reading could vary within a tolerance of ±20% of the stated value. Next, the resistance should be tested from the wiper arm (point C) to one end of the potentiometer at point A. The potentiometer should be rotated through its full range. A test can then be made from the wiper arm (point C) to point B. The resistance should vary smoothly from near zero to the full

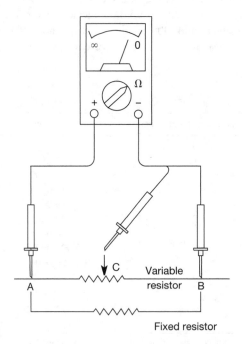

Figure 10-3 Measuring a resistor or potentiometer.

value of the resistance. Since the reading should vary smoothly and continuously, this is one application in which a VOM has the edge over the DVM. Any sudden jump to either a higher or lower value, or any erratic reading, would indicate a defective spot on the resistance element. If the wiper arm is not making firm contact with the resistance element, simply tapping the case of the potentiometer may produce erratic resistance readings. Any of these erratic indications mean a defective or dirty potentiometer. A spray cleaner may salvage the unit; however, if after cleaning it does not produce a smooth resistance change over its entire range, it should be discarded.

Rheostats may be tested by a similar process, with the exception that the only measurement is from the wiper to one end since the rheostat only has two terminals. In general, rheostats will be open if they are defective because they usually are high-wattage units and must handle high current.

Thermally Intermittent Resistors

All electronic components, including fixed and variable resistors, can be thermally intermittent. Measure the resistance while subjecting the suspected component to extreme temperature change to detect this type of defect. Radio Shack stores stock an aerosol spray component cooler to spray on a resistor to cool it. The tip of the soldering iron can be used to heat it. Be careful not to apply excessive heat to resistors. A sudden or erratic change in resistance as temperature is changed indicates that the resistor is thermally intermittent and defective.

"Shifty" Resistors. There is a class of resistors whose resistances change as the operating conditions change. Common ones are thermistors, varistors, and photoconductors. Thermistors and photoconductors can be measured with an ohmmeter. The resistance of a varistor is calculated from voltage and current measurements.

> Thermistors. A thermistor is a resistor whose resistance varies with temperature. It exhibits large negative temperature characteristics; that is, the resistance decreases as the temperature rises and increases as the temperature falls.
> Varistors. A varistor is a resistor whose resistance is voltage dependent. Its resistance decreases as voltage across it is increased.
> Photoconductors. A photocell's (photoconductor) resistance varies when light shines on it. When the cell is not illuminated, its "dark" resistance may be greater than 100 kilohms. When illuminated, the cell resistance may increase to a few hundred ohms. These values can be measured with an ohmmeter.

Making Capacitor Measurements

Capacitance is the property whereby two conductors separated by a nonconductor (dielectric material) have the ability to store energy in the form of an electric charge and oppose any change in that charge. A number of different types of capacitors are shown in Figure 10-4. The operation of a capacitor depends on the electrostatic field set up between the two oppositely charged parallel plates.

Figure 10-4   *Various capacitors.*

The unit of capacity is the farad, named in honor of Michael Faraday. It is the amount of capacitance that will cause a capacitor to attain a charge of one coulomb when one volt is applied. Expressed as a mathematical equation,

$$C = Q/V$$

where C will be one farad when Q is one coulomb and V is one volt. The farad is very large for practical applications; therefore, smaller values are used. A microfarad is 10^{-6} farads, a nanofarad is 10^{-9} farads, and a picofarad is 10^{-12} farads. Microfarads and picofarads are very common in electronic circuits.

The physical factors that determine the amount of capacitance a capacitor offers to a circuit are:

1. The type of dielectric material (K)
2. The area in square meters of the plates (A)
3. The number of plates (n)
4. The spacing of the plates in meters, which also is the thickness of the dielectric (t)

Note that before a capacitor is to be measured with an ohmmeter, remove it from the circuit and short across its leads or plates to make sure it has no residual charge. Such residual charge could damage an ohmmeter.

Relative Amount of Capacitance

Although an ohmmeter is not an actual capacitance checker or meter, it will give a relative indication of its condition, and two capacitors can be compared as to their relative capacitance by using a VOM. The amount of needle deflection of an ohmmeter can be used to indicate a relative amount of capacitance. By connecting the ohmmeter to the capacitor as shown in Figure 10-5, the ohmmeter battery charges the capacitor to its voltage. The meter will deflect initially and then fall back to infinity as the capacitor charges. In Figure 10-6, af-

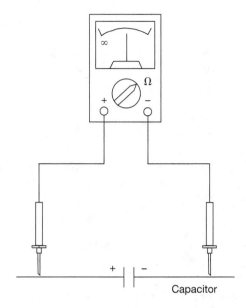

Figure 10-5   Measuring a capacitor. Initial test.

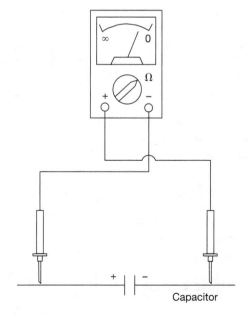

Figure 10-6   Measuring a capacitor. Charging test.

ter the initial charge of the capacitor, the ohmmeter leads are reversed and the capacitor voltage is now in series with the voltage inside the ohmmeter. The charge on the capacitor is aiding the ohmmeter battery. The needle now deflects a larger deflection proportionally to the amount of capacitance, and then decays as the charge is redistributed on the capacitor.

Leaky and Shorted Capacitors

Paper, mica, and ceramic capacitors fail in two ways. The dielectric breaks down and the capacitor plates short together, or the capacitor becomes "leaky." When "leaky," the dielectric still supports a voltage but the dielectric resistance becomes much lower than normal. Both of these conditions can be detected with a VOM or DVM. There are two checks that can be made. The first is simply a resistance measurement using the VOM or DVM as an ohmmeter across the terminals of the capacitor. If the capacitor is shorted, the ohmmeter will read zero or a very low value of resistance. If the capacitor has become "leaky," then the resistance measurement will be much less than the normal—nearly infinite for a good capacitor. Leaky capacitors need to be replaced before they turn into shorted capacitors.

In some capacitors, the dielectric does not become "leaky" until a voltage is applied. That is, it breaks down under load. This defect cannot be detected with an ohmmeter but can be found by using the VOM or DVM as a voltmeter. A dc voltage is placed across the series combination of the voltmeter and the paper, mica, or ceramic capacitor (not electrolytic capacitors unless proper polarity is maintained). A good capacitor will show only a momentary deflection on the voltmeter, then the reading will decay to zero volts as the capacitor charges to the supply voltage. A defective capacitor will have a low insulation resistance, R_{ins} (it may be at a particular voltage), and will maintain a voltage reading on the meter. The lower the insulation resistance of the capacitor, the higher the voltmeter will read. When insulation resistance is checked by this method, it is in series with the meter. Because R_{ins} is normally high, it limits the current; therefore, a change in the VOM voltmeter range does not significantly affect the total resistance of the circuit, and the percentage of meter-scale deflection remains fairly constant with different voltmeter ranges. The power supply voltage V_s should be set for the rated working voltage of the capacitor for this test.

If the insulation resistance is such that it produces a scale reading on a VOM or DVM, the R_{ins} may be calculated by using the following equation:

$$R_{ins} = R_{input}(V_s - V_m)/V_m$$

where V_s = supply voltage, V_m = VOM or DVM measured voltage, R_{input} = VOM or DVM input resistance, and R_{ins} = capacitor insulation resistance.

Measuring An Electrolytic Capacitor

Special care must be taken when measuring electrolytic capacitors because they are polarized. As a result, when using a VOM as an ohmmeter to test an electrolytic capacitor, the ohmmeter test lead polarity must be correct to give the proper indication. In most cases, this occurs when the positive ohmmeter test lead is connected to the positive electrolytic capacitor terminal. In any case, the test lead arrangement that gives the highest resistance reading is the one to use.

Leakage Current of Electrolytic Capacitors

Measuring the leakage current of an electrolytic capacitor is the best way to judge whether the capacitor is still useful. The circuit shown in Figure 10-7 is used for measuring leakage current of electrolytic capacitors using an ammeter. V_s, the capacitor rated voltage, is applied across the capacitor and the leakage current is indicated by the series ammeter. The maximum permissible leakage current of a new electrolytic capacitor is related to the voltage rating (WVDC) and capacitance of the capacitor according to the following equation:

$$I = kC + 0.030$$

where I = leakage current in mA, C = rated value of capacitor in microfarads, and k = a constant as given in Table 10-2.

Many factors affect the amount of leakage, such as the age of the capacitor, how long it has been uncharged in a circuit, how near to its rated voltage it has been working at, or how long a new one has been on the shelf. If the capacitor exceeds the permissible leakage values, it should be discarded. Experience will help make this test more conclusive.

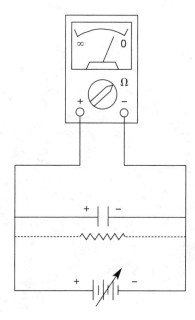

Figure 10-7 Measuring leakage current of a capacitor with an ammeter.

Table 10-2 Electrolytic capacitor k constant values

Constant of k	dc working voltage
0.010	3 to 100
0.020	101 to 250
0.025	251 to 350
0.40	351 to 500

Making Inductance Measurements

The property of an electric circuit or component that opposes changes in circuit current is called inductance. The ability of the circuit or component to oppose changes in current is due to its ability to store and release energy that it has stored in a magnetic field. Every circuit has some inherent inductance but devices that purposely introduce inductance to a circuit are called inductors. Let's look at some basics of inductance and how to test inductors with a VOM or DVM.

Inductor Basics

Inductance and Impedance. An inductor may have any number of physical forms and shapes. An assortment of inductor types is shown in Figure 10-8. An inductor is basically nothing more than a coil of wire. Inductors are sometimes referred to by such names as choke, impedance coil, reactor, or combinations such as choke coil or inductive reactor The amount of inductance of a coil is measured by a unit called the

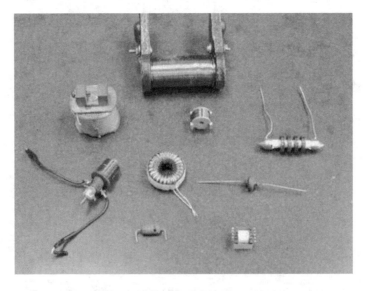

Figure 10-8 Assorted inductors.

henry. Smaller units, as for capacitors, are practical; the millihenry (1×10^{-3} henries) and micro-henry (1×10^{-6} henries) are very common units in electronic circuits.

The amount of inductance in an inductor depends on the magnetic flux produced and the current in the coil. Mathematically, this may be expressed by the following equation:

$$L = N\phi/I$$

where L is the inductance in henries
I is the current through the coil in amperes
N is the number of turns of wire
ϕ is the magnetic flux linking the turns

Basically, all inductors are made by winding a length of conductor around a core made either of magnetic material or of insulating material. When a magnetic core is not used, the inductor is said to have an air core. The physical characteristics or geometry of both the core and the windings around the core affect the amount of inductance produced—more turns, better magnetic core material, larger core cross-section area, and shorter coil length all increase the inductance.

In a dc circuit, the only changes in current occur when the circuit is closed to start current and when it is opened to stop current. However, in an ac circuit, the current is continuity changing each time the voltage alternates. Since inductance in a circuit opposes a change in current, and since ac is continually changing, there is an opposition offered by the inductor to the ac current that is called reactance. The amount of inductive reactance is given by the following equation:

$$X_L = 2\pi fL$$

where X_L is the inductive reactance in ohms, f is the frequency of the ac in hertz, and L is the inductance in henries.

The quantity $2\pi f$ represents the rate of change of current in radians per second. It is called angular velocity.

In ac circuits that contain only inductance, the inductive reactance is the only thing that limits the current. The current is determined by Ohm's Law with X_L replacing R, as follows:

$$I = V/X_L$$

If an ac circuit contains both resistance and inductance (reactance), then the total opposition to current flow is termed impedance and is designated by the letter Z. When a voltage V is applied to a circuit that has an impedance Z, the current I is

$$I + V/Z \quad \text{(where } Z = R2 + XL2)$$

Continuity

A series RL circuit may be formed by one or more resistors connected in series with one or more coils. Or, since the wire used in any coil has some resistance, a series RL circuit may consist of just a coil or coils by themselves. The resistance of the coils, which effectively is in series with the inductance, supplies the circuit resistance. Using a VOM as an ohmmeter, a simple continuity test will quickly locate an open inductor. The resistance can be measured easily for any inductance that is not open—just place the meter used as an ohmmeter across the coil terminals and measure the resistance just as a resistor was measured in Figure 10-4. Normal resistance values depend on the wire size and number of turns (length) of the wire that makes up the inductor. Some coils with fine wire and a large number of turns will have hundreds of ohms resistance; large coils with large wire and a small number of turns will have tens of ohms. If no resistance is measured at ail, the inductance is open.

Other Failures

Inductors become defective because insulation breaks down and turns short together or the coil shorts to the core. Simple continuity checks with an ohmmeter between one end of the coil and the core detect the coil shorted to the core, but a few shorted turns on an inductor are very difficult to detect. If one-half the coil shorts out, resistance checks should detect it, but for a few turns, very accurate measurements must be made in order to detect that the coil is defective.

Transformer Basics

A transformer is a device for coupling ac power from a source to a load. A conventional transformer consists of two or more windings on a core that are isolated from each other. Energy is coupled from one winding to another by a changing magnetic field. An ac voltage applied across the primary results in primary current. The changing current sets up an expanding and collapsing magnetic field that cuts across the turns of the secondary winding. This changing magnetic field induces an ac voltage in the secondary that produces a current in any load connected across the secondary. The core around which the primary and secondary are wound may be iron for low frequencies, as in the case of power and audio transformers. Primary and secondary windings on an air core may be employed for transformers that couple energy in higher frequency circuits.

Ideal Transformer

If the transformer were ideal, there would be no power loss from primary to secondary and 100% of the source power would be delivered to the load. Since voltage times current equals power, the power relationship is given by

$$V_p \times I_p = V_s \times I_s$$

where V_p = primary voltage in volts, I_p = primary current in amperes, V_s = secondary voltage in volts, and I_s = secondary current in amperes.

In an ideal transformer, the ratio of primary voltage V_p to V_s, the voltage induced in the secondary, is the same as the ratio of the number of turns in the primary N_p to the number of turns in the secondary N_s. The following equation expresses the relationship:

$$V_p/V_s = N_p/N_s$$

The turns ratio of a transformer is B_s/B_p, the ratio of the secondary turns to the primary turns.

Step-Up and Step-Down Transformers

If the number of turns in the primary and the secondary are equal, then the voltages appearing across the primary and secondary are equal. This type of transformer with a one-to-one turns ratio is called an isolation transformer. If a lower voltage appears across the secondary than across the primary, it is called a step-down transformer and the turns ratio would be less than 1. However, if a higher voltage appears across the secondary than across the primary it is called a step-up transformer and the turns ratio would be greater than 1. According to the primary and secondary power relationship equation given previously, the secondary current will be stepped down if the secondary voltage is stepped up; and if the secondary voltage is stepped down, the secondary current is stepped up.

Resistance Testing of Transformer Windings

Resistance continuity testing with an ohmmeter can be done on most small transformers to determine the continuity of each winding (see Figure 10-9). Connect your ohmmeter or VOM across the primary winding between points A and B. You should get a resistance reading. Next connect your meter across the secondary windings between points D and E. Again, you should get a resistance reading. Comparison of the measured resistance with the published data from the manufacturer should determine if a suspected transformer is defective. Power transformers and audio output transformers usually have their windings color coded so that the respective winding can be measured with an ohmmeter to determine if there is continuity and to measure the winding resistance. If the winding measures infinite resistance, the winding is open. The break may occur at the beginning or end of the winding where the connections are made to the terminal leads. This type of break can possibly be repaired by resoldering the leads to the winding. If the discontinuity is deeper in the transformer, the transformer will have to be replaced.

If the winding resistance is very high compared to its rated value, there may be a cold solder joint at the terminal connections. If the condition cannot be corrected, the transformer will have to be replaced..

Shorts—Primary and Secondary

A short from a winding to the core or to another winding may be found by measuring the resistance on a high-ohms scale from the core to the winding or from winding to winding (see Figure 10-9). Place the ohmmeter leads on the winding lead at point A and the other meter lead on the core at point C. Now, lift

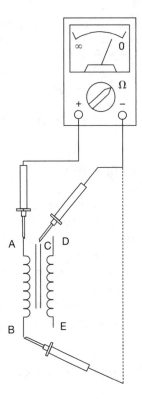

Figure 10-9 Testing a transformer.

the meter lead from point A, move the meter probe to point B, make another measurement from point B to the core, and look for a short. Next, check the secondary winding for a short by connecting one meter probe to point D and the second probe lead to point C or the core, and take a reading. Finally, lift the probe lead from point D and move it to point E and take a second reading between point E and the core, looking for a short. Any continuity reading at all would indicate leakage to the winding from the core or between windings and indicate a defective transformer. A few shorted turns are difficult to detect but if a large percentage of the transformer is shorted out, resistance measurements will detect it.

Measuring Semiconductor Devices

Diodes

The diode is a two-terminal, nonlinear device that presents a relatively low resistance to current in one direction and a relatively high resistance in the other. A "perfect" diode would act like a switch—either on (conducting) or off (not conducting) depending on the voltage polarities applied to the terminals. The cathode of a diode is usually identified by some means of marking. On small-signal glass or plastic diodes, a colored band or dot is often used. For rectifiers, sometimes a + is used to indicate the cathode, or metal can devices have a large flange on the cathode.

Figure 10-10 illustrates a forward-bias ohmmeter test of a diode. Note that the plus (+) lead of the meter is connected to the anode of the diode. The meter should read a low resistance of about 500 to 700 ohms. Figure 10-11 shows a reverse-bias test of a diode. Note that the meter plus (+) probe is now connected to the cathode of the diode and the minus (–) meter probe lead is connected to the anode of the diode. Your ohmmeter should now read a high resistance. A low resistance in both directions indicates a shorted diode; a high resistance in both directions indicates an open diode, a condition in which high current has destroyed internal connections or high voltage has broken down the junctions.

Transistors

The transistor is a three-terminal device that has virtually replaced the vacuum tube. There are two basic types of transistors: bipolar and field-effect.

Bipolar Junction Transistors (BJTs). A transistor is a device made of two P-N junctions. The transistor is basically an off device and must be turned on by applying forward bias to the base–emitter junction. Think of a transistor as two diodes connected back to back. Therefore, each junction, like a diode, should show low forward resistance and high reverse resistance. These resistances can be measured with an ohmmeter, as we

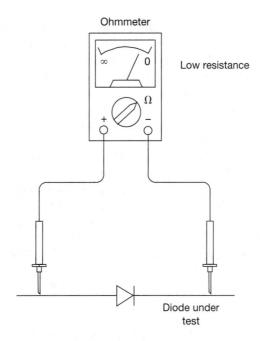

Figure 10-10   Forward-bias ohmmeter test of a diode.

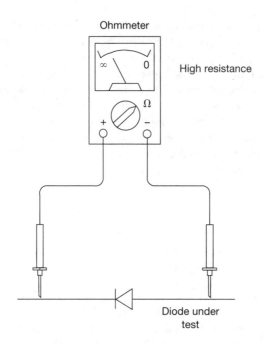

Figure 10-11 *Reverse-bias ohmmeter test of a diode.*

did with diodes. The diagram shown in Figure 10-12 depicts ohmmeter testing of a PNP transistor. The polarities of the voltages applied are shown to indicate forward or reverse bias on the NPN and PNP transistors. The same ohmmeter range can be used for each pair of measurements (base to collector, base to emitter, emitter to collector). For most transistors, any ohmmeter range is acceptable. However, in some meters, the $R \times 1$ range may provide excessive current for a small transistor. Also, the highest resistance range may have excessive voltage at the terminals of some ohmmeters. Either of these conditions may damage the transistor being tested. As a result, it is best to start with the mid-ranges for the resistance measurements.

Testing each transistor requires six individual tests. First, start out by making the two base-to-collector tests. Place the plus (+) ohmmeter lead on the base and the minus (−) lead on the collector lead; you should get a high resistance reading. Now, reverse the leads. Place the plus (+) meter lead on the collector and the minus (−) lead on the base lead; you should get a low resistance reading on your meter. Now, repeat the two tests for the base-to-emitter terminals. Finally, you will need to make two tests from between the collector-to-emitter leads.

Note that the above tests are for an PNP transistor. If you wish to test an NPN transistor, look at Figure 10-13 and you will notice that the polarity is opposite that of the PNP transistors.

Defects

If the reverse resistance reading is low but not shorted, the transistor is leaky. If both forward and reverse readings are very high, the transistor is open. If the forward and reverse readings are the same or nearly

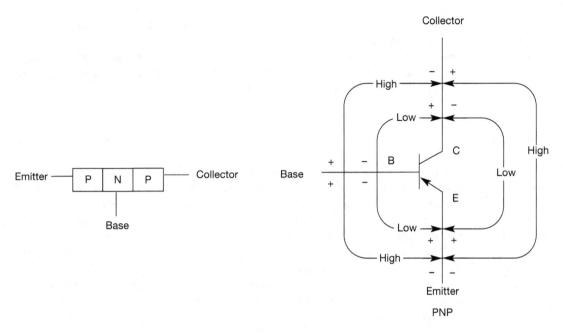

Figure 10-12 Testing a PNP transistor.

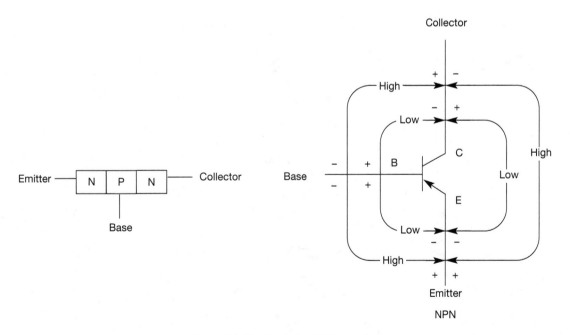

Figure 10-13 Testing a NPN transistor.

equal, the transistor is defective. A typical resistance in the forward direction is 100 to 500 ohms. However, a low-power transistor might show only a few ohms resistance in the forward direction, especially at the base–emitter junction. Reverse resistances are typically 20K to several hundred thousand ohms. Typically, a transistor will show a ratio of at least 100 or so between the reverse resistance and forward resistance. Of course, the greater the ratio, the better the device is for an application.

Testing a UJT

A unijunction transistor (UJT) is solid-state, three-terminal semiconductor used in timing, pulse, oscillator-sensing, and thyristor-triggering circuits. The UJT has two base leads and an emitter lead as shown in Figure 10-14. Testing a UJT is actually quite simple. Take your ohmmeter or VOM, set in the ohmmeter position, and connect one probe lead to base 1 (B1) and the other meter probe lead to base 2 (B2), and read the resistance. Now, reverse the leads and measure. You will note that both readings will indicate a high resistance. Now, place the minus (–) lead of the meter on the emitter and the plus (+) lead on B1 and take a reading, then lift the probe from B1 and place it on B2 and take a reading. Both readings should indicate high resistance readings. Finally, place the plus (+) meter lead on the emitter and the minus (–) lead on the B1 lead of the UJT and take a reading, then lift the plus lead from B1 and place it on B2 and take a reading. Both measurements should indicate a low resistance. If you get similar results on these tests, your UJT is good.

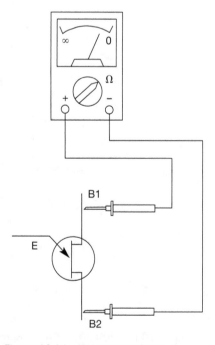

Figure 10-14   Testing a UJT circuit.

Field-Effect Transistors

A field-effect transistor (FET) is a voltage operated device that requires virtually no input current. This gives them an extremely high input resistance. There are two major categories of field-effect transistors: junction FETs and insulated gate FETs, more commonly known as MOS (metal–oxide–semiconductor) field-effect transistors. In the bipolar junction transistor, the FET is available in two polarities: P-channel and N-channel.

The schematic symbol for an N-channel field-effect transistor is shown in Figure 10-15 and a P-channel FET is depicted in Figure 10-16. Notice that the terminals are identical for N-channel and P-channel FETs but the arrow on the gate terminal is reversed. This also indicates that the current direction from source to drain depends on the polarity type of the FET. Since there is no set designation for the source and drain terminals of the FET, the reference manual or the equipment schematic should be consulted to identify the terminals on the FET being tested.

Testing FETs is somewhat more complicated than testing a bipolar junction transistor. First, determine from the markings whether the device is a JFET or a MOS type; otherwise, the terminal measurements will have to indicate the type. Do not attempt to remove the FET from the circuit or handle it unless you are certain that it is a JFET or a MOSFET with protected inputs. If you touch the leads of these devices, static electricity can damage an unprotected device very quickly. Make certain all static electricity is grounded out before handling FETs.

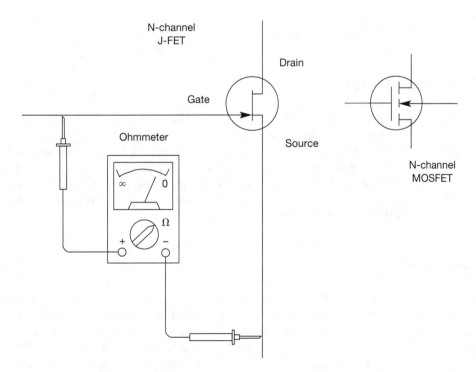

Figure 10-15 Forward-bias test of N-channel FET.

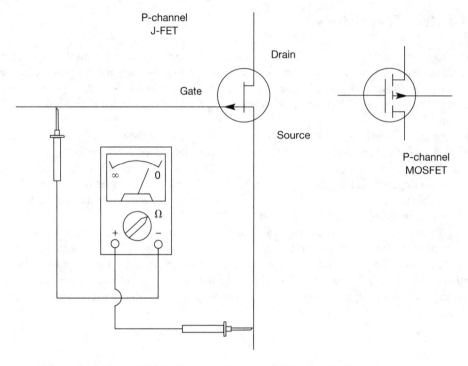

Figure 10-16   Forward-bias test of P-channel FET.

JFET Measurements

To test the forward resistance of the JFET gate-to-source junction, use a low-voltage ohmmeter on the $R \times$ 100 scale. For an N-channel JFET, connect the positive lead to the gate (see figure 10-15) and the negative lead to either the source or drain. Reverse the leads for a P-channel JFET. The resistance should be less than 1 kohms. To test the reverse resistance of the N-channel JFET junction, reverse the meter leads and connect the negative lead of the ohmmeter to the gate and the positive lead to the source or drain. The device should show almost infinite resistance. Lower readings indicate either leakage or a short. Reverse the leads for a P-channel JFET as seen in Figure 10-16.

Testing a JFET

The following simple out-of-circuit test will demonstrate if a junction FET is operational but will not indicate if the device is marginal. Operational means it is not shorted or open, which is by far the most common occurrence when a FET becomes defective. After the JFET has been removed from the circuit, connect the ohmmeter between the drain and source terminals. Touch the gate lead with a finger and observe the ohmmeter polarity connections to the source and drain terminals and the channel type (P or N). Reverse the leads of the ohmmeter to the terminals and again touch the gate terminal. The ohmmeter should indicate a small change in the resistance opposite to that previously observed if the FET is operational. The change in resistance will be very slight and some operational (good) FETs will not appear to change.

Testing a MOSFET

When testing a MOSFET, the device must be handled with caution and the hands and instruments must be discharged to ground before measurements are made. If a MOSFET is to be checked for gate leakage or breakdown, a low-voltage ohmmeter on its highest resistance scale should be used. The MOSFET has an extremely high input resistance and should measure "infinity" from the gate to any other terminal. Lower readings indicate a breakdown in the gate insulation. The measurements from source to drain should indicate some finite resistance. This is the distinguishing characteristic of a MOSFET; it has no forward and reverse junction resistance because the metal gate is insulated from the source and drain by silicon oxide. There should be a very high resistance with both polarities of voltage applied.

A simple in-circuit test for a MOSFET is shown in Figure 10-17. The MOSFET gate is connected in series with a 100 kohm resistor through a switch to a 12 volt battery. A 1 megohm resistor is also placed across the gate and source leads. The drain lead of the MOSFET is connected in series with a small 12

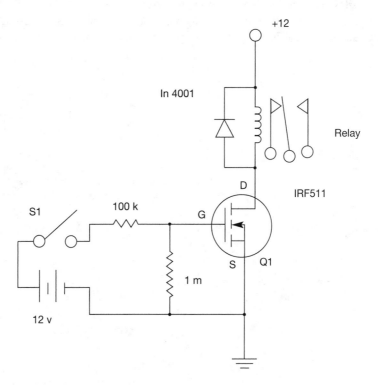

Figure 10-17 In-circuit test for MOSFET.

volt relay to the 12 volt power source. Note the protection diode across the relay coil. When the switch at S1 is pressed, the relay should pull in if the MOSFET is good.

SCR Testing

An SCR is a gated diode that is used for the control of dc circuits. If a positive voltage is applied to the anode relative to the cathode, the diode will not conduct in the forward direction until triggered by current in the gate. Once triggered on, the diode is turned off by the voltage between the anode and cathode going to zero. Testing with an ohmmeter is not recommended for high-current SCRs and should only be used as a relative indication in low-current SCRs. The current supplied by the ohmmeter may not be enough to "fire" or "hold" the SCR and, therefore, may not always indicate the true junction condition of the device.

A simple test of low-power SCRs can provide an approximate evaluation of their gate-firing capabilities by connecting an ohmmeter as shown in Figure 10-18. The negative lead is connected to the cathode and the positive lead to the anode. Use the $R \times 1$ scale on the ohmmeter. Short the gate to the anode with S1. This should turn the SCR on; a reading of 10–50 ohms is normal. When S1 is opened and the gate-to-anode short is removed, the low-resistance reading should remain until the ohmmeter lead is removed from the anode or the cathode. Reconnecting the ohmmeter leads to the anode and cathode should show a high resistance until S1 is closed again to short the gate to the anode.

The circuit shown in Figure 10-19 illustrates a "real world" test setup circuit used for testing an SCR under load. The gate of the SCR is connected through the normally open switch at S1 and in series with a

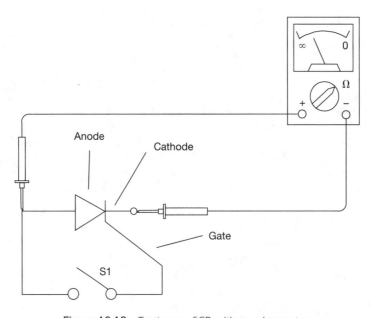

Figure 10-18   Testing an SCR with an ohmmeter.

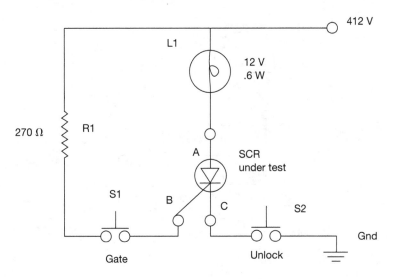

Figure 10-19   SCR test setup.

resistor to a 12 volt dc power source. A 12 volt, 0.6 watt incandescent lamp is connected in series with the anode of the SCR to the 12 volt power source. The cathode of the SCR is connected to one end of a normally closed pushbutton switch connected to ground. To fire the SCR, push S1 and the lamp should light up. When you want to turn off the lamp, push switch S2 and the SCR is unlatched.

Testing a Triac

In order to test a triac you will first need to set your VOM or ohmmeter to the lowest resistance range. Attach the minus (–) meter probe to the triac's T1 lead (see Figure 10-20). Next, attach the plus (+) meter probe to the triac's T2 lead. Take a jumper lead and momentarily short T2 and the gate lead. This should turn the Triac on and you should get a low-resistance reading of about 20 to 50 ohms. Next, disconnect one of the meter probe leads from the Triac momentarily and you should get a meter reading of infinity. Finally, reverse the meter leads and repeat the test once again. If you get the same results as the first test, then the triac is probably OK.

Testing a Battery

A battery is a single cell or group of cells that generate electricity from an internal chemical reaction. Its purpose is to provide a source of steady dc voltage of fixed polarity. The battery, like every power source, has an internal resistance that affects its output voltage. For a good cell, the internal resistance is very low, with typical values less than an ohm. As the cell deteriorates, its internal resistance increases, preventing the cell from producing its normal terminal voltage when there is load current. A dry cell loses its ability

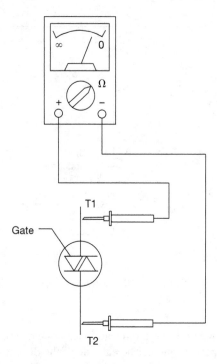

Figure 10-20 Testing a triac.

to produce an output voltage even when it is out of use or stored on a shelf. There are several reasons for this, but mainly it is due to self-discharge within the cell and loss of moisture in the electrolyte. Therefore, batteries should be used as soon after manufacture as possible. The shelf life is shorter for smaller cells and for used cells.

Our first series of measurements use the multimeter. The first test determines if a battery has the proper voltage output using a multi-meter and load resistor. First you will need to turn on your multimeter and select a voltage range. Here is where common sense must be observed so that you will not burn out your multimeter. Pay attention to what it is you are trying to measure before actually taking a measurement. If you are measuring a 1.5 volt battery, then you would select the 10 volt range; if you have a lower range such as a 2 volt range, you could select that range when measuring 1.5 volts. Always start with the highest scale and move down when measuring voltages to ensure that you do not damage your meter.

Testing a Battery under Load

A very "weak" (high internal resistance) battery can have almost normal terminal voltage with an open circuit or no load current. Thus, a battery should be checked under its normal load condition, that is, in the equipment that it powers with the power switch on. Out of the equipment, the only meaningful test is with

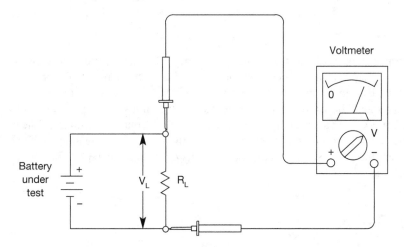

Figure 10-21   Testing a battery under load.

a load resistor across the battery as in Figure 10-21. The value of the load resistor depends on the battery being tested. For a standard D cell, R_L = 10 ohms; for a C cell, R_L = 20 ohms; for an AA cell, R_L = 100 ohms; and for a 9 volt battery, R_L = 330 ohms. The terminal voltage should not drop to less than 80% of its rated value under load. The internal resistance may be calculated by the equation

$$R_I = V_{NL} - V_L / I_L$$

Measure V_{NL} first without a load, then measure V_L with a load and calculate the internal resistance.

Current Drain

The current drain on a battery should be measured in its actual operating condition; that is, with its normal load. It can be measured by connecting the current meter in series between the battery and the equipment it is powering. The battery clip may be disconnected from the battery on one side and the current meter connected to complete the circuit. The correct polarity must be observed. Start on a high current range in case of an excessively high current due to a malfunction.

Testing with an Oscilloscope

The oscilloscope is one of the most versatile pieces of test equipment found on the test bench and it can be used to perform many different tests of individual components and systems. It can look at voltages and waveforms and test for system gain, noise, distortion, and so on.

Battery Test

Figure 10-22 illustrates an oscilloscope test of a 12 volt battery. Turn on your oscilloscope and center the horizontal trace line so that it sits on the center graticule. Set the vertical gain/attenuator to 2 volts/div and set the horizontal time base to 0.1 ms/div. Next, set the trigger source to internal (auto). Coupling should be set to dc and slope should be set to (+). You can now attach the minus (–) scope probe lead to the minus (–) terminal of the battery. Next attached the scope's X1 probe to the plus (+) terminal of the battery. Once the battery is connected to the oscilloscope, you should see the scope trace line jump and by counting the vertical divisions, that the scope should read about 12 volts dc. If your scope has an internal LED voltmeter, it should also read about 12 volts or more.

Transformer Testing

Our next oscilloscope test will be to test the output of a transformer's secondary winding (Figure 10-23). Turn on your oscilloscope and set the vertical amplifier/attenuator to 10 volts/div. Next, set the coupling control to ac, the slope to (+), and the trigger source to "line." The horizontal time base should be set to 2 ms/div. Adjust the trace line to the center graticule. Finally, connect the primary of a small 12 volt transformer to an ac wall outlet, then connect the secondary of the transformer to the oscilloscope's probe and ground leads. Since the transformer is producing an ac signal output, there are no particular polarity concerns when connecting the scope's probes. On the scope's display, you should see a sine wave (right-hand side of Figure 10-23). The sine wave should have a peak-to-peak reading of about 35.6 volts.

$$V_p = (12.6 \text{ V}) \times 1.414 = 17.8 \text{ V peak}$$

$$V_{p\text{-}p} = V_p \times 2 = 17.81 \text{ } V \times 2 = 35.6 \text{ V}$$

$$T = 1/f = 1/60 = 16.67 \text{ ms}$$

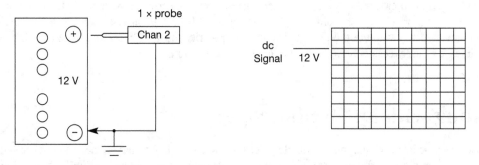

Figure 10-22   Testing a 12 volt battery with an oscilloscope.

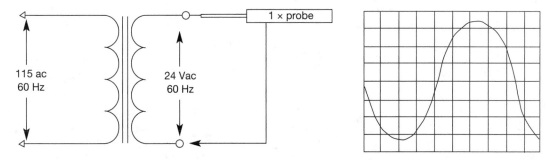

Figure 10-23 *Testing a transformer with an oscilloscope.*

In the above equations, the 12 volt transformer is multiplied by the rms value of 1.414 and we get 17.8 volts. Multiply this value by two we get the peak-to-peak value of the signal of approximately 35.6 volts ac with a time of 16.67 ms.

Figure 10-24 depicts an oscilloscope test of a half-wave rectifier connected to a small transformer output. First, set up your oscilloscope for the measurement by adjusting the vertical gain/attenuator to 5 volts/div. Next, set the horizontal time base control to 2 ms/div and the trigger source to line (auto). Set the coupling to dc and the slope to (+). Rotate the vertical position to center the trace on the center graticule and you are now ready to make a measurement. Connect the scope's ground lead to one end of the transformer's secondary as shown. For this test you will need two oscilloscope test probes. Take the first scope probe (probe 1) and place the probe on the input to the rectifier. Connect the ground lead of the second scope probe (probe 2) to the transformer secondary terminal along with the first probe ground. Now connect the second scope probe (probe 2) to the output of the silicone diode. You should have a 60 ohm load resistor across the secondary of the transformer.

You should now adjust your scope to display both channels and you should see two sine wave traces (bottom of Figure 10-24). Note that in a half-wave rectifier setup, the output frequency is the same as the input frequency:

$$V_{in} = 2.25 \text{ div} \times (2 \text{ volts/div}) = 4.5 \text{ V peak or } 9 \text{ V peak to peak}$$

$$V_{out} = 1.9 \text{ div} \times (2 \text{ volts/div}) = 3.8 \text{ V peak } (0.7 \text{ V drop across the diode})$$

$$T_{input} = 16.67 \text{ ms } (60 \text{ Hz})$$

$$T_{output} = 16.7 \text{ ms } (60 \text{ Hz}) - \text{output frequency} = \text{input frequency with half-wave rectifier circuit}$$

Measuring Voltage Gain

Our next test measures the gain of an amplifier. Look at Figure 10-25. Turn on your oscilloscope and make the following adjustments. First, set the vertical gain/attenuator to 1 volt/div. Next, set the horizon-

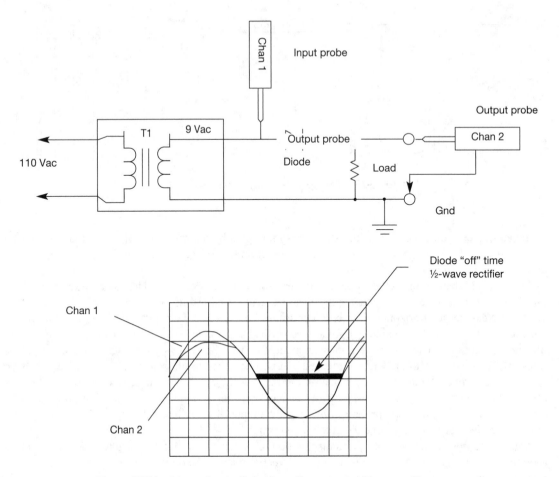

Figure 10-24 Measuring half-wave rectifier output with an oscilloscope.

tal sweep rate to 0.1 ms/div and set the trigger source to internal (auto), the slope to (+), and the coupling to ac. Set the display for two channels (ALT). Finally, set the trace lines to the center graticule. Once you see both traces, you can begin the test.

Locate a signal source such as a waveform from a signal generator. Connect the signal source to the input of the audio amplifier and place the first oscilloscope probe (probe 1) on the signal output lead from the signal source as shown. Next, connect the second scope probe (probe 2) to the output lead of the audio amplifier. Be sure to have a load on the audio amplifier, with an 8 to 16 ohm resistor or speaker. Before applying power to the signal generator and audio amplifier, make sure that the gain control of both devices is set to minimum. With the test setup ready, turn up the signal source output just a bit and also adjust the audio amplifier volume control to one-quarter or less of the range. Now take a look at your oscilloscope and you should see two waveform traces. Channel 1 should display a smaller signal than the output of the amplifier on channel 2. Mathematically, this can be expressed as follows:

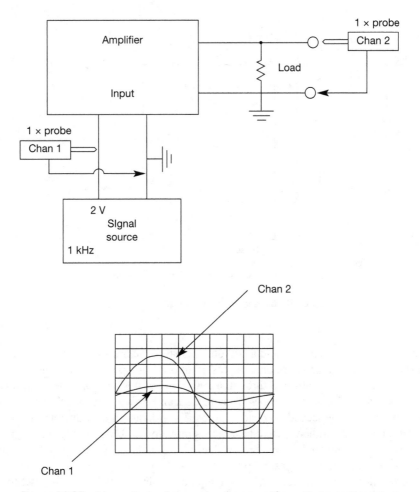

Figure 10-25 *Measuring voltage gain of an amplifier with an oscilloscope.*

$$V_{output} = 5 \text{ div} \times (1 \text{ volt/div}) = 5 \text{ volts}$$

$$V_{input} = 2 \text{ div} \times (1 \text{ volt/div}) = 2 \text{ volts}$$

$$\text{Average output gain} = V_{output}/V_{input} = 5/2 = 2.5 \text{ gain}$$

Build Your Own Diode and Transistor Tester

The transistor and diode curve tracer circuit shown in Figure 10-26 allows you to use your oscilloscope to obtain a visual indication of the condition of these components. The heart of the transistor/diode visual

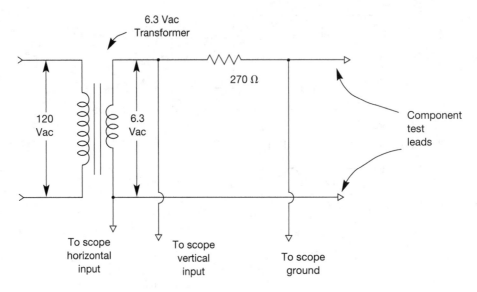

Figure 10-26   Diode/transistor curve tracer circuit.

tester is a 120 to 6.3 V ac transformer and a 270 ohm resistor. The transformer's secondary windings are connected to both the vertical and horizontal inputs of an oscilloscope. The secondary winding, which is coupled to the vertical input, is also fed through a 270 ohm resistor and forms the "ground" lead to the scope; it also serves as one of the test probes. The other secondary winding, which is fed to the horizontal scope input, also serves as the second test lead or probe. The primary of the transformer is plugged into an 115 V ac outlet. The transistor/diode visual tester is a simple to build, easy to use, fast "go, no-go" tester. This tester is great for testing all those extra parts that you might have in your "junk" box and can be easily assembled in a small aluminum chassis box.

Set up your oscilloscope with the vertical gain/attenuator set to 1 volt/div, slope set to plus (+), and coupling set to dc. To test a diode, connect the test probes to both ends of a diode and you should see some sort of waveform displayed on your scope. If you see vertical and horizontal lines intersecting at a right angle you will know that your diode is working and is good. Figure 10-27 shows the waveforms for open–shorted and leaky–good semiconductors. A straight line depicts a shorted semiconductor, and a horizontal line indicates an open semiconductor. An intersecting wavy vertical/horizontal line with an imperfect junction indicates a leaky semiconductor that should be replaced.

The visual semiconductor tester can also be used to test transistors as well as diodes. In order to test transistors, you will have to test pairs of junctions as we did with VOM or ohmmeter testing. The visual semiconductor tester will save you lots of time when testing those questionable diodes or transistors.

Simple Extended-Range Measurements

There is a simple, low-cost method to measure resistors, capacitors, and inductors that are beyond the range of your multimeter. Say that you want to measure a 50 megohm resistor and your multimeter range

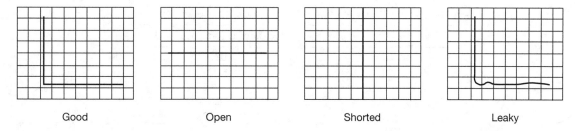

| Good | Open | Shorted | Leaky |

Figure 10-27 *Oscilloscope waveforms from component tester.*

is limited to only 2 megohms. If you take a 2 megohm resistor and place it in parallel with an unknown resistor, the resulting resistance would be less than the 2 megohm resistor, and therefore, you can use your multimeter with the 2 megohm range to measure a resistance value greater than your meter range. It would then be simple to make a graph of the resulting resistance or calculate the unknown resistance using Ohm's Law. The formula for parallel resistance is

$$R_{total} = (R1 \times R2)/(R1 + R2)$$

Using a 2 megohm resistance, you can calculate that your DVM would read as shown in Table 10-3 (with parallel unknown resistance). Using Ohm's Law and one resistor, the measurement capacity of your DVM can be increased from 2 megohm to over 100 megohm.

You can also apply this principle to measure capacitors beyond the range of your meter. Using a formula similar to the one for series capacitors, you could take a 20 μF capacitor and measure it with your limited-range DVM. This time however, you would connect the capacitors in series rather than parallel as you did with the resistors. Be sure to observe the correct polarity when using two electrolytic capacitors connected in series. The formula for connecting capacitors in series is

$$C_{resulting} = (C1 \times C2)/(C1 + C2)$$

Using a 20 μF electrolytic capacitor, you can calculate that the DVM would read as shown in Table 10-4 with the unknown series capacitor. For both resistors and capacitors, it is possible to multiply the useful range of an inexpensive meter and use it to measure components that would not otherwise be measurable

Table 10-3 *Actual versus measured resistance
(parallel resistances)*

Unknown resistance	Measured resistance
2,500,000	1,111,111
5,000,000	1,420,000
10,000,000	1,666,666
20,000,000	1,818,000
100,000,000	1,960,000

Table 10-4   Actual versus measured capacitance (series capacitors)

Actual capacitance in μF	Measured capacitance in μF
20	10
50	14.29
100	16.66
200	18.8
500	19.23
1000	19.6

using only one resistor or capacitor. Neither the resistor nor capacitor needs to be at exactly the upper limit of the DVM. Simply choose a value as close as possible to the upper limit, measure the value, and use the measured value in the formula. For convenience, you can use a spreadsheet to calculate numerous values and draw a graph that you can then carry with your DVM. The same procedure can be used to measure inductor values beyond the range of your multimeter.

CHAPTER 11

Electronic Troubleshooting Techniques

When faced with an inoperative circuit or equipment failure, this chapter will assist you in asking some major questions such as: "Should I fix the equipment or send it back to the dealer for repair?" "What kind of test equipment do I need?" "What do I need know to be able to fix the equipment myself?" The answers to these questions will depend on the type of test equipment you have available, the availability of a schematic or service manual, and your own experience. It is usually worth a few minutes of your time to check over the questionable circuit or equipment. The problem may turn out to be simple and the solution readily apparent, thereby avoiding the cost and effort of shipping the equipment back to the manufacturer. Look for a mentor or a friend with some experience to guide through your first troubleshooting encounters to help you gain some confidence.

Troubleshooting is an art, part detective work and part common sense. You must develop or have the ability to read a schematic diagram and to visualize a signal flowing through the circuit. The confidence and experience you gain by repairing your own equipment can prove to be extremely valuable in your electronics career or hobby. The more practice you gain, the easier the troubleshooting process will become over time. It is also most gratifying to save some money by repairing equipment yourself. Looking at a complex circuit for the first time can be quite intimidating. In your mind, you have to mentally divide the circuit into smaller parts or blocks. In this way, you can sort of "trick" your mind into thinking that you are working on a smaller or simpler circuit. Try to divide up the circuit into smaller blocks. Identify oscillators, power supplies, and amplifiers and mentally separate these as small, independent circuits. Look at each section of the circuit as a simple circuit connected to another simple circuit and this will help you to divide the circuit into smaller, manageable pieces.

In order to fix electronic equipment, you need to understand the system and circuits you are troubleshooting. A working knowledge of electronic theory, circuitry, and components is an important part of the process. If necessary, review the electronic and circuit theory explained in the other chapters of this book. When you are troubleshooting, you are looking for the unexpected. Knowing how circuits are supposed to work will help you to look for things that look out of place or suspect. Figure 11-1 shows a pro-

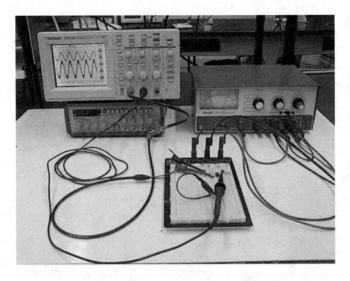

Figure 11-1   Circuit under test.

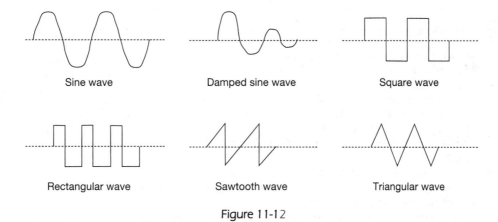

Sine wave   Damped sine wave   Square wave

Rectangular wave   Sawtooth wave   Triangular wave

Figure 11-12

totype circuit under test, using an oscilloscope. Figure 11-2 depicts some typical waveforms you will encounter using your oscilloscope to troubleshoot electronics equipment.

Test Equipment

Many of the steps involved in troubleshooting require the use of electronic test equipment. Some people think they need expensive test instruments to repair their own equipment. In fact, you probably already own the most important instruments and others may be purchased inexpensively, rented, borrowed, or built at home. The test equipment available to you may limit the kind of repairs you can do, but you will be surprised at the kinds of repair work you can do with simple test equipment and a little thinking. Most equipment troubleshooting can be accomplished using the "big four": a multimeter, an oscilloscope, a function generator, and a frequency counter. Often, just a multimeter and an oscilloscope will be all you need to do basic troubleshooting. Chapters 2 through 6 discuss the "essential" pieces of test equipment as well as other useful equipment that you might come across in designing, building, and utilizing your electronics workbench.

Safety

Before beginning any troubleshooting or circuit probing, we should first review some important safety concerns. Safety should be the first and foremost thing discussed in troubleshooting, since death is permanent! Some of the voltages found in electronic equipment can often be fatal. Only 50 mA flowing through the body is painful. A current of 100 to 500 mA is usually fatal. Under certain conditions, as little as 12 volts can kill.

Make sure you are 100% familiar with all the safety rules and the dangerous conditions that might exist in the equipment you are servicing. Remember, if the equipment is not working properly, dangerous conditions may exist where you don't expect them. Treat every component as potentially "live."

Table 11.1   *Safety rules*

1. Keep your left hand in your pocket or at your side.
2. Replace fuses only with those of the proper rating.
3. Be careful with tools that may cause short circuits.
4. Wear a hard hat when someone is working above you.
5. Wear rubber-sole shoes or use a rubber mat when standing on a concrete floor.
6. Wear shoes with nonslip soles that will support you when climbing.
7. Always use a safety belt when climbing.
8. Wear safety glasses for protection against sparks or metal fragments.
9. Do not work alone.
10. Test leads should be well insulated.
11. Do not subject electrolytic capacitors to excessive voltages, ac voltages, or reverse voltages.
12. Use only grounded plugs and receptacles.
13. Switch off the power, disconnect power sources, and discharge capacitors before doing circuit work.
14. Use GFI protected circuits when working outside.
15. Install a GFI circuit breaker for your electronics workbench.

Some older equipment uses "ac/dc" circuitry. In these circuits, there is no isolation from a power transformer—one side of the chassis is connected directly to the ac line. This is an electric shock waiting to happen. Remember the "left-hand rule": when you are troubleshooting circuits, keep your left hand in your pocket and use your right hand to do the probing. Keeping your left hand in your pocket or at your side prevents a current path from traveling through your left hand and arm to your heart. See Table 11-1 for a list of safety rules.

Documentation

Once you have determined that a piece of equipment is indeed broken, you need to do some preparation before you diagnose and attempt to repair it. First, locate a schematic diagram and/or service manual. It is possible to troubleshoot without a manual, but a schematic is almost indispensable. The original equipment manufacturer is the best source of a manual or schematic. However, many old manufacturers have gone out of business. Several sources of equipment manuals are listed in the Appendix. If all else fails, you can sometimes reverse engineer a simple circuit, tracing wiring paths and identifying components to draw your own schematic. However, this takes practice as well as patience. If you have access to the data books for the active devices used in the circuit, the pin-out diagrams and applications notes will sometimes be enough to help you understand and troubleshoot the circuit.

Using All Your Senses

Your senses, although not strictly defined as "test equipment," will tell you much about the equipment you are trying to repair. Consider your senses as an effective spectrum analyzer. We all have the ability to use these natural "test instruments."

Eyes

Use them constantly. Scan circuit boards, connectors, and components. Look for evidence of heat and arcing, burned components, broken connections or wires, poor solder joints, or other obvious visual problems.

Ears

Severe audio distortion can be detected by ear. The "snaps" and "pops"of arcing or the sizzling of a burning component may help you track down circuit faults. An experienced troubleshooter can diagnose some circuit problems by the sound they make. For example, the sound due a bad audio-output IC sounds slightly different than that due to a defective speaker.

Nose

Your nose can tell you a lot. With experience, the smells of ozone, an overheating transformer, or a burned carbon-composition resistor each become unique and distinctive.

Fingers

Carefully use your fingers to measure low heat levels in components. Small signal transistors can be fairly warm to the touch; anything hotter can indicate a circuit problem. One very useful aid to troubleshooting is the "freeze" spray. If you spray a circuit area with the spray, the components will quickly take on a white powdery look. If a component is very hot, the white will quickly fade away. Be aware, however, that some high-power resistors and transistor devices can get downright hot during normal operation. Usually, these components are isolated from the main circuit or are housed in a heat sink.

Brain

More troubleshooting problems have been solved with a VOM and brain than with the most expensive spectrum analyzer. You must use your brain to analyze data collected by all the other human senses.

Defining the Problem

To begin troubleshooting, you must define the problem accurately. Ask yourself these three important questions:

1. What functions of the equipment do not work as they should; what does not work at all?
2. Has the trouble occurred in the past? It is a good idea to keep a record or log of troubles and maintenance in the owner's manual. Calibration notes or test point information are always very helpful

for later use. Write the answers to the questions. The information will help with your work, and may help service personnel if their advice or professional services are required.

3. What kind of performance can you realistically expect?

Simplify the Problem

If the broken equipment is part of a system, you need to find out exactly which part of the system is bad. For example, a bad channel on your home stereo system could be due to anything from a bad cable to a bad amplifier or a bad speaker component. Simplify the system as much as possible. To diagnose the stereo system, start troubleshooting by checking just the amplifier with a set of known good headphones. Take little steps, one at a time, dividing circuits and problems into small blocks or tests. Simplifying the problem will often isolate the bad component more quickly, and you will not be overwhelmed by a large or complex circuit problem.

Look for Obvious Problems

Try the simple things first. If you are able to solve the problem by replacing a fuse or reconnecting a loose cable, you will avoid a lot of effort. Many experienced technicians have spent hours troubleshooting a piece of equipment only to learn the hard way that the power was off or that they were not using the equipment properly. Make sure the equipment is plugged in, that the ac outlet does indeed have power, that the equipment is switched on, and that all of the fuses are good. If the equipment uses batteries or an external power supply, make sure these are working.

Check to make sure that all wires, cables, and accessories are working and plugged into the right connectors or jacks. In a system, it is often difficult to be sure which component or subsystem is bad.

Connector faults are generally more commonplace than component troubles. Consider poor connections as prime suspects in your troubleshooting detective work. Do a thorough inspection of each of the connectors and connections. Are all transistors and integrated circuits firmly seated in their sockets? Are all interconnection cables sound and securely connected? Many of these problems are obvious to the eye, so look around carefully and you could avoid spending unnecessary money and time.

Make sure that your equipment is operating as designed. Many electronic "problems" are caused by a switch that is set in the wrong position or a unit that is being asked to do something it was not designed to do. Before you open up your equipment for major surgery, make sure you are using it correctly and that you understand how the equipment functions. Be sure to read the manual before tearing into equipment.

Open the Equipment

Once all of the initial preparation work has been done, it is now time to take a look at the circuit and dig in. You usually will have to start by taking the equipment apart. This is the part that can trap an unwary technician. Most experienced service technicians can tell you a story about how they took apart a piece of equipment and were later unable to easily put it back together. Don't let this happen to you.

Take lots of notes as you take it apart and list each component you remove. Draw a picture of the subsystems or cabling layouts and/or write down the order in which you did things, the color codes, part placements, hardware notes, and anything else you think you might need to be able to reassemble the equipment weeks later when the back-ordered part comes in. Put all of the screws in one place. A plastic jar with a lid works, as does a sandwich baggie. If you drop it, the plastic jar is not apt to break and the lid will keep all the parts from flying around the work area. It may pay to have a separate labeled container for each subsystem.

Looking Inside the Equipment

Many service problems are visible, if you look carefully for them. Many times, an electronics technician will spend hours tracking down a failure, only to find a bad solder joint or a burned component that would have been spotted in a careful inspection of the printed-circuit board. Start troubleshooting by carefully inspecting the equipment. Scan all the components carefully, and look for cracked, broken, or black sooty resistors and capacitors. Look for dented or "bloated" capacitors; this often indicates a problem. Also look for burned spots on the printed-circuit board or PC board "lands" that have been "lifted" above the board. Carefully observe all the solder joints. Look for cracks in the solder joints and check the smoothness, composition, and color of the solder joint. These are all tell-tale signs of problems. In rare cases, defects occur during the wave soldering process and sometimes joints will have bad color or not look smooth but "pitty." This is a sign of likely problems.

It is time-consuming, but as a good detective, you really need to look at every connector, every wire, every solder joint, and every component. A connector may have loosened, resulting in an open circuit. Sometimes, the connector pins will come loose and cause an open circuit condition. You may spot broken wires or see a bad solder joint. Flexing the printed-circuit board or tugging on components a bit while looking at their solder joints will often locate a defective solder job. Look for scorched components. Make sure all of the screws securing the printed-circuit board are tight and making good electrical contact. See if you can find evidence of previous repair jobs; these may not have been done properly.

Make sure that each integrated circuit is firmly seated in its socket. Look for pins folded underneath the IC rather than making contact with the socket. If you are troubleshooting a newly constructed circuit, make sure each part is of the correct value or type number and that it is orientated or installed correctly.

If your careful inspection doesn't reveal anything, it is time to apply power to the unit under test and continue the process. Observe all safety precautions while troubleshooting equipment. There are voltages inside some equipment that can kill you. If you are not qualified to work safely with the voltages and conditions inside the equipment, do not proceed.

Troubleshooting Approaches

There are two fundamental approaches to troubleshooting equipment: the systematic approach and the intuitive approach. The systematic approach uses a defined process or set of steps to analyze and isolate the problem. The intuitive approach relies on troubleshooting experience to guide you in selecting which circuits to test and which tests to perform. The systematic approach is usually chosen by beginning troubleshooters.

Whether you choose the systematic or instinctive or intuitive approach to troubleshooting, you will find that using a block diagram approach will help you to "step" through the circuit easily. The block diagram is a road map. It shows the signal paths for each circuit function. These paths may run together and sometimes cross. Those blocks that are not in the paths of faulty functions can be eliminated as suspects. Sometimes the symptoms point to a single block, and no further search is necessary.

In some cases where more than one block is suspect, several approaches may be used. Each requires testing a block or stage. Signal injection, signal tracing, instinct, or a combination of all techniques may be used to diagnose and test electronic equipment.

The intuitive approach works well for those with years of troubleshooting experience. Electronics enthusiasts who are new to this troubleshooting game can often use some extra guidance. A systematic approach is a disciplined procedure that allows us to tackle problems in unfamiliar equipment with a reasonable hope of success. There are two common systematic approaches to troubleshooting at the block level: signal tracing and signal injection. The two techniques are very similar. Differences in test equipment and the circuit under test determine which method is best in a given situation. They are often combined.

Signal Tracing

When tracing a signal, start at the beginning of a circuit or system and follow the signal through to the end. When you find the signal at the input to a specific stage, but not at the output, you have located the defective stage. You can then measure voltages or see waveforms and perform other tests on that stage to locate the specific failure. This is much faster than testing every component in the unit to determine which is bad.

Signal tracing is suitable for most types of troubleshooting of receivers and analog amplifiers. It is the best way to check transmitters, receivers, and amplifiers because all of the necessary signals are present by design.

It is sometimes possible to use over-the air signals in signal tracing; in a radio receiver, for example. However, if a good signal generator is available, it is best to use it as the signal source, and a modulated signal source is best.

Equipment

A voltmeter with an RF probe is the most common instrument used for signal tracing, but low-level signals cannot be measured accurately with this instrument. Signals that do not exceed the junction drop of the diode in the probe will not register at all, but the presence, or absence, of larger signals can be observed.

A dedicated signal tracer can also be used. It is essentially a type of audio amplifier. An experienced technician can usually judge the level and distortion of the signal by ear. You cannot use a dedicated signal tracer to follow a signal that is not amplitude modulated (single sideband is a form of amplitude modulation). Signal tracing is not suitable for tracing CW signals, FM signals, or oscillators. To trace these, you will have to use a voltmeter and RF probe and/or an oscilloscope.

An oscilloscope is the most versatile signal tracer. It offers high input impedance, variable sensitivity, and a constant display of the traced waveform. If the oscilloscope has sufficient bandwidth, RF signals can be observed directly. Alternatively, a demodulator probe can be used to show demodulated RF signals on a low-bandwidth oscilloscope. Dual-trace scopes can simultaneously display the waveforms, including their phase relationship, present at the input and output of a circuit.

Testing Procedures

First, make sure that the circuit under test and test instruments are isolated from the ac line by transformers. Set the signal source to an appropriate level and frequency for the unit you are testing. For a receiver, a signal of about 10 μV should work fine. For other circuits, use the schematic, an analysis of circuit function, and your own good judgment to set the signal level.

Switch on power to the test circuit and connect the signal source output to the test-circuit input. Place the tracer probe at the circuit input and ensure that you can hear the test signal. Using an oscilloscope, observe the characteristics of the signal you are monitoring while progressing through each stage (see Figure 11-3). Start with the first stage and progress though each stage, one step at a time, using the test points in each stage, if available. Compare the detected signal to the source signal during tracing; the waveforms should look clean and the same from the input through to the output stage.

Move the tracer probe to the output of the next stage and observe the signal. The signal level should increase in amplifier in stages and may decrease slightly in other stages. The signal will not be present at the output of a "dead" stage.

Low-impedance test points may not provide sufficient signal to drive a high-impedance signal tracer, so tracer sensitivity is important. Also, in some circuits the output level appears low where there is impedance change from input to output of a stage.

Note that in a receiver there are two signals that are combined at the mixer—the input signal and the local oscillator signal—which should be present in the mixer stage. Loss of either one of these signals will result in no output from the mixer stage. Switch the signal source on and off repeatedly to make sure that the tracer reading varies and does not disappear.

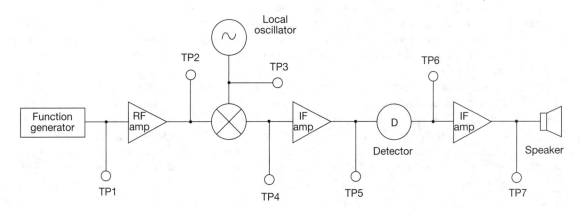

Figure 11-3 *Signal tracing a receiver.*

Signal Injection

Like signal tracing, signal injection is particularly suited to some situations. It is a good choice for troubleshooting amplifiers and receivers, since the receivers already have a detector as part of the design. It is suitable for either high- or low-impedance circuits and can be used with vacuum tubes, transistors, or ICs.

Equipment

If you are testing equipment that does not include a suitable detector as part of the circuit, some form of signal detector is required. Any of the instruments used for signal tracing are adequate. Most of the time, your signal injector will be a signal generator. There are other injectors available, some of which are square-wave audio oscillators rich in RF harmonics. These are usually built into a pen-sized case with a test probe at the end. These "pocket" injectors do have their limits because you cannot vary their output level or determine their frequency. They are still useful, though, because most circuit failures are caused by a stage that is completely dead.

Consider the signal level at the test point then choosing an instrument. The signal source used for injection must be able to supply appropriate frequencies and levels for each stage to be tested. For example, a typical superheterodyne receiver requires AF, IF, and RF signals that vary from 6 V at audio frequencies (AF), to 0.2 μV at radio frequencies (RF).

Testing Procedure

If an external detector is required, set it to the proper level and connect it to the test circuit. Set the signal source for AF and inject a signal directly into the signal detector to test operation of the injector and detector. Move the signal source to the input of the preceding stage and observe the signal. Continue moving the signal source to the inputs of successive stages.

When you inject the signal source to the inputs of the defective stage, there will be no output. In order to prevent the stage from overloading, you will need to reduce the level of the injected signal as testing progresses through the circuit to the final stages. Use suitable frequencies for each tested stage.

Make a rough check of each stage gain by injecting a signal at the input and output of an amplifier stage. You can then compare how much louder the signal is when injected at the input. This test may mislead you if there is a radical difference in impedance from stage input to output. Understand the circuit operation before testing.

Mixer stages present a special problem because they have two inputs rather than one. A lack of output signal from a mixer can be caused by either a faulty mixer or faulty local oscillator (LO). Check oscillator operation with a oscilloscope or absorption wavemeter, or by listening on another receiver. If none of these instruments are available, inject the frequency of the LO at the LO output. If a dead oscillator is the only problem, this should restore operation. If the oscillator is operating but off frequency, a multitude of spurious responses will appear. A simple signal injector that produces many frequencies simultaneously is not suitable for this test. Use a well-shielded signal generator set to the appropriate level at the LO frequency.

Dividing and Conquering

Under certain conditions, the block search may be speeded up by testing at the middle of each successively smaller circuit section. Each test limits the fault to one-half of the remaining circuit. As an example, suppose a receiver has 14 stages and the fault is in stage 12. This approach requires only four tests to locate the faulty stage, a substantial saving of time and effort.

This "divide and conquer" method, unfortunately, cannot be used in equipment that splits the signal path between the input and the output. Test readings taken inside feedback loops are quite misleading unless you understand the circuit operation and the type of waveform you are expecting at each point in the test circuit. It is best to consider all stages within a feedback loop as a single block during the block search. Both signal tracing and signal injecting procedures may be speeded by taking some diagnostic shortcuts. Rather than precheck each stage sequentially, check a point halfway through the system.

Suppose you have a high-frequency (HF) receiver that is not working correctly. You observe that there is no response from the speaker. First, substitute a working speaker to rule out the speaker. If there is still no sound, then proceed to check the power supply. If no clues indicate any particular stage, signal tracing or injection can be used at this point to help provide the answer. Get out the signal generator and power it up. Set the generator for a low-level RF signal, switch the signal off, and connect the output to the receiver. Switch the signal on again and place a high-impedance signal-tracer probe at the antenna connection. Instantly, the tracer will emit a strong audio note and it will appear that the test equipment is functioning. Move the probe to the input of the receiver's detector. As the tracer probe touches the circuit, the familiar note will sound. Next, set the tracer for audio and place the probe halfway through the audio chain. It will be silent. Move the probe halfway back to the detector and the note will appear once again but no signal will be present at the output of the stage. You now know that the defect is somewhere between the two points tested. In this case, the third audio stage is faulty.

The Intuitive Approach

In the "intuitive" approach to troubleshooting, you rely on your judgment and experience to decide where to start testing and what and how to test. When you immediately check power supply voltages or the ac fuse in a unit that is completely nonfunctional, that is an example of the intuitive approach. If you are faced with a receiver that has distorted audio and immediately start testing the speaker and audio output stage, or if you immediately start checking the filter and bypass capacitors in an audio stage that is oscillating or "motor-boating," you are intuitively troubleshooting the device. The check for connector problems mentioned at the beginning of this section is a good starting point. Experience has shown that connector faults are so common that they should be checked even before a systematic approach begins.

When intuition is based on experience, an intuitive search may be the fastest procedure. If your intuition is correct, repair time and effort may be reduced substantially. As experience and confidence grow, the merits of the intuitive approach grow with them. However, inexperienced technicians who choose this approach are at the mercy of chance.

Testing within Intermediate Stages

Once you have followed all of the troubleshooting procedures and have isolated your problem to a single defective stage or circuit, a few simple measurements and tests will usually pinpoint one or more specific components that need adjustment or replacement.

First, check the parts in the circuit against the schematic diagram to be sure that they are reasonably close to the design values, especially in a newly built circuit. Even in a commercial piece of equipment, someone may have incorrectly changed them during attempted repairs. A wrong-value part is quite likely in new construction such as a homemade project.

Testing for Voltage Levels

Check the circuit voltages. If the voltage levels are printed on the schematic, this makes it simple. If not, analyze the circuit and make some calculations to see what the circuit voltages should be. Remember, however, that the printed or calculated voltages are nominal; measured voltages may vary from the calculations.

When making measurements, remember the following points

Make measurements at device or connector leads, not at circuit-board traces or socket lugs. This tests the entire circuit including connector and connecting wires and subassemblies all at once.

Use small test probes to prevent accidental shorts between close components leads; this is especially critical between IC pins.

Never connect or disconnect power to solid-state circuits with the power switch on.

Consider the effect of the meter on voltages you are trying to measure. A 20 kΩ/V meter may change the voltage reading of the circuit. A vacuum tube voltmeter or an FET input voltmeter is a good choice for this type of circuit troubleshooting, since they have a very high input impedance that is not likely to effect the circuit under test.

Voltages may give you a good clue to what is wrong with the circuit. If they don't, check the active devices in the circuit, such as transistors and ICs. If a bad device is found, remove it and substitute a good device. After connections, most circuit failures are caused directly or indirectly by a bad active device. The experienced troubleshooter usually tests these first.

Analyze the other components and determine the best way to test each one. There is additional information about electronic components in Chapter 9. There are two voltage levels in most circuits: V+ and ground. Most component failures (opens and shorts) will shift dc voltages near one of these levels. Typical failures that show up as incorrect dc voltages include open coupling transformers; shorted capacitors; open, shorted, or overheated resistors; and open or shorted semiconductors.

Oscillation Problems

Oscillations occur whenever there is too much positive feedback in a circuit that has gain and also in digital circuits. It can occur at any frequency, from a low-frequency audio buzz often called "motorboating", well up into the RF region.

Unwanted oscillations are usually the result of changes in the active device (increased junction or interelectrode capacitance), failure of an oscillation-suppressing component (open decoupling or bypass capacitors or neutralizing components) or new feedback paths caused by improper lead dressing or dirt on the chassis or components. It can also be caused by improper design, especially in home-designed circuits. A shift in bias or drive levels may aggravate oscillation problems. Oscillations that occur in audio stages do not change as the radio is tuned because the operating frequency, and therefore the component impedances, do not change. However, RF and IF oscillations usually vary in amplitude as operating frequency is changed. Oscillation stops when the positive feedback is removed. Locating and replacing the defective (or missing) bypass capacitor can often solve this type of problem. The defective oscillating stage can be found more reliably with a signal tracer or oscilloscope, as opposed to meter testing.

Amplitude Distortion Problems

Amplitude distortion is the product of nonlinear operation. The resultant waveform contains not only the input signal, but new signals at other frequencies as well. All of the frequencies combine to produce the distorted waveform. Distortion in a transmitter gives rise to splatter, harmonics, and interference.

Figure 11-4 shows a test setup for creating a reference square wave signal. Figure 11-5 depicts some typical examples of distortion found through measurement with an oscilloscope. Clipping, which is also called flat-topping, is the result of excessive drive signals. The corners on the waveform show that harmonics are present. Note that a square wave signal contains the fundamental and all odd harmonics. In radio circuits, these odd harmonics are often heard well away from the desired operating frequency.

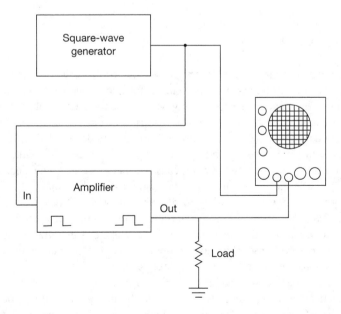

Figure 11-4   Test setup for creating a reference square wave signal.

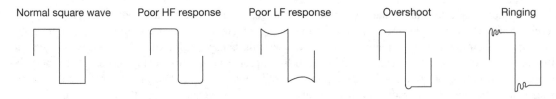

Figure 11-5   Distorted waveforms.

Harmonic distortion produces radiation at frequencies far removed from the fundamental; it is a major cause of electromagnetic interference (EMI). Harmonics are generated in nearly every amplifier. When they occur in a transmitter, they are usually caused by insufficient transmitter filtering, caused either by poor design or filter component failure.

Incorrect bias brings about unequal amplification of the positive and negative wave sections. The resultant waveform is rich in harmonics.

Frequency and Distortion Problems

If a "broadband" amplifier, such as an audio amplifier, does not amplify all frequencies equally, the result is frequency distortion. In many cases, this distortion is deliberate, as in a transmit microphone amplifier that has been designed to pass only frequencies from 200 to 2000 Hz. In most cases, the ability to detect and measure distortion is limited by the test equipment available to the electronics technician or hobbyist.

Noise Problems

Noise is produced whenever current flows through a conductor that is warmer than absolute zero. A slight hiss is present in all electronic circuits and is normal. Unfortunately, noise is compounded and amplified by succeeding circuit stages. Repair is necessary only when noise threatens to obscure normally clear signals.

Semiconductors can produce two kinds of hiss. The first is normal—an even, white noise that is much quieter than the desired signal. The second is noise from a faulty device. This is usually erratic, with pops and crashes that are sometimes louder than the desired signal. In an analog circuit, the end result of noise is usually sound. In a control or digital circuit, noise causes erratic operation, unexpected switching, and so on.

Noise problems usually increase with temperature, so localized heat may help you find the source. The use of "freeze" spray may help locate faulty localized components. Noise from any component may be sensitive to mechanical vibration. Tapping various components with an insulated screwdriver may quickly isolate a bad part. Noise can also be traced with an oscilloscope or signal tracer. Nearly any component or connection can be a source of noise. Defective components are the most common cause of crackling noises. Defective connections are a common cause of loud, popping noises.

Check connections at cables, sockets, and switches. Look for dirty variable-capacitor wipers and potentiometers. Mica trimmer capacitors often sound like lightning when arcing occurs. Test them by installing a series 0.0 1 μF capacitor. If the noise disappears, replace the trimmer. Potentiometers are particularly prone to noise problems when used in dc circuits. Clean them with spray cleaner and rotate the shaft several times.

Rotary switches may be tested by jumpering the contacts with a clip lead. Loose contacts may sometimes be repaired, either by cleaning, carefully rebending the switch contacts, or gluing loose switch parts to the switch deck. Operate variable components through their range while observing the noise level at the circuit output.

Distortion Measurement

A distortion meter is used to measure distortion of audio frequency (AF) signals. A spectrum analyzer is the best piece of test gear to measure distortion of RF signals. If a distortion meter is not available, an estimation of AF distortion can sometimes be made with a function generator capable of producing both sine and square waves and a general-purpose oscilloscope.

To estimate the amount of frequency distortion in an audio amplifier, set the generator for a square wave and look at it on the oscilloscope. Be sure to use a low-capacitance probe for these tests. The wave should show square corners and a flat top. Next, inject a square wave at the amplifier input and again look at the input wave on the oscilloscope. Any new distortion is a result of the test circuit loading the generator output. Note that if the waveshape is severely distorted, the test is not valid. Now, move the test probe to the test circuit output and look at the waveforms, shown in Figure 11-6, which will help in evaluating square wave distortion.

The above applies only to audio amplifiers without frequency tailoring. In RF or radio frequency gear, the transmitter may have a very narrow audio passband; therefore, inserting a square wave into the microphone input may result in an output that is difficult to interpret. The frequency of the square wave will have a significant effect.

Anything that changes the proper bias of an amplifier can cause distortion. This includes failures in the bias components, leaky transistors, or vacuum tubes with interelectrode shorts. These conditions may mimic automatic gain control (AGC) trouble. Improper bias often results from an overheated or open resistor. Heat can cause resistor values to permanently increase. Leaky or shorted capacitors and RF feedback can also produce distortion by disturbing bias levels. Distortion is also caused by circuit imbalance in Class AB or B amplifiers.

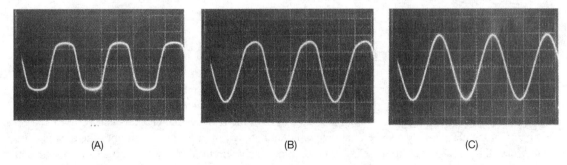

(A) (B) (C)

Figure 11-6   Distorted waveforms.

Oscillations in an IF amplifier may produce distortion. They cause constant, full AGC action, or generate spurious signals that mix with the desired signal. IF oscillations are usually evident on the signal strength (S) meter, which will show a strong signal even with the antenna disconnected.

Alignment

Alignment is not often the cause of an electronics problem. As an example, if an AM receiver suddenly begins producing weak and distorted audio, an inexperienced person frequently suspects that poor PC alignment is the problem. If the manufacturer's instructions and the proper test equipment are not available, our inexperienced person now begins "adjusting" the transformer cores. Before long, the receiver is hopelessly misaligned. Now our misguided person must send the radio to a shop for an alignment that was not needed before "repairs" were attempted. Alignment does not shift suddenly. A normal signal-tracing procedure would have shown that the signal was good up to the audio output IC, but badly distorted after that. The defective IC that caused the problem would have been easily found and quickly replaced.

Arcing Problems

Arcing is definitely a serious sign of trouble that is occurring or has occurred in a particular circuit. It may also create a potential fire hazard. Arc sites are usually easy to find because an arc that generates visible light or noticeable sound also pits and discolors conductors and components. Arcing is caused by component failure, dampness, dirt, or lead orientation. If the dampness is temporary, dry the area thoroughly and resume operation. Dirt may be cleaned from the chassis with a residue-free cleaner. Arrange leads so that high-voltage conductors are isolated from other parts of the circuit. Keep them away from sharp corners and screw points.

Arcing occurs in capacitors when the working voltages are exceeded. Air-dielectric variable capacitors can sustain occasional arcs without damage, but arcing indicates operation beyond circuit limits. Transmatches or antenna tuners that are working beyond their ability or power ratings may suffer from arcing and cause permanent circuit damage. A failure or high standing wave ratio (SWR) in an antenna circuit may also cause transmitter or antenna tuner arcing.

Contamination Problems

Contamination is another common service problem. Cold soda spilled into a "hot" piece of electronics is an extreme example that actually occurs often. Conductive contaminants range from water to metal filings. Most can be removed by a thorough cleaning. Any of the residue-free cleaners can be used, but remember that the cleaner may also be conductive. Do not apply power to the circuit until the area is completely dry; this may take up to three days. Often placing the circuit directly in sunlight or under a hair dryer will speed up the drying process.

Keep cleaners away from variable-capacitor plates, transformers, and parts that may be harmed by the chemicals. The most common conductive contaminant is solder, either from a printed circuit board "solder bridge" or a loose piece of solder that often surfaces at the most inconvenient time. Look for odd-looking solder joints—those that look black, nonshiny, rough, or pitted. Poor manufacturing processes such as using contaminated solder, not cleaning surfaces prior to wave soldering, or incorrect heating temperatures can cause poor solder joints.

Solder "Bridges"

In a typical PC-board solder bridge, the solder that is used to solder one component forms a short circuit to another PC-board trace or component. Unfortunately, they are common in both new construction and repair work. Look carefully for them after you have completed any soldering, especially on a PC board. It is even possible that a solder bridge may exist in equipment you have owned for a long time, unnoticed until it suddenly decided to become a short circuit. Often, items such as loose solder blobs, loose hardware, or small pieces of component leads can show up in the most awkward and troublesome places and cause short circuits or intermittent problems.

Replacing Defective Components

If you have located a defective component within a circuit stage, you will need to replace it. When replacing socket-mounted components, be sure to align the replacement part correctly. Make sure that the pins of the device are properly inserted into the socket.

Some special tools can make it easier to remove soldered parts. A chisel-shaped soldering tip helps pry leads from printed circuit boards or terminals. A desoldering iron or desoldering bulb operates by suction to remove excess solder, making it easier to remove the component. Spring-loaded desoldering pumps are more convenient than bulbs. A desoldering wick draws solder away from a joint when pressed against the joint with a hot soldering iron.

In all cases, remember that soldering tools and melted solder can be hot and dangerous, and can and often cause serious burns and scars. Wear protective goggles and clothing when you are soldering. If you do burn your fingers, run the burn area immediately under cold water and seek first aid or medical attention. Always seek medical attention if you burn your eyes; even a small burn can develop into serious trouble. Seek medical attention at once.

Typical Symptoms and Faults

Power Supplies

Many equipment failures are caused by power supply circuit failures. Fortunately, most power supply problems are easy to find and repair. Use a voltmeter to measure the power supply output. Loss of output voltage is usually caused by an open circuit. Note that a short circuit most often draws excessive amounts of current that causes fuses to "open," thus becoming an open circuit.

Most fuse failures are caused by a shorted diode in the power supply or a shorted power device (RF or AF) in the failed equipment. More rarely, one of the filter capacitors can short. If the fuse has opened, turn off the power, replace the fuse and measure the load-circuit dc resistance. The measured resistance should be consistent with the power supply ratings. A short or open load circuit indicates a problem.

If the measured resistance is too low, check the load circuit with an ohmmeter to locate the trouble. Sometimes, nominal circuit resistances are listed in some older equipment manuals; this can be extremely helpful. If the load circuit resistance is normal, suspect a defective IC regulator or problem in the rest of the unit. Electrolytic capacitors fail with long disuse (two years or more).

Integrated circuit regulators can oscillate and sometimes cause failures. The small-value capacitors on the input, output, or adjustment pins of the regulators are designed to prevent oscillations within the power supply circuit. Check or replace these capacitors, especially after a regulator has failed.

AC ripple (hum) is usually caused by low-value filter capacitors in the power supply that have begun to fail over time. Less likely, hum can also be caused by excessive load, a regulation problem, or RF feedback in the power supply. Look for defective filter capacitors, which are usually measured as open or with a low value, often caused by a defective regulator or shorted filter choke. In older equipment, defective filter capacitors will often have visible leaking electrolyte. Look for corrosion residue at the capacitor leads. In new construction projects, make sure RF energy is not getting into the power supply. Defective electrolytic capacitors are the most common problem in old power supplies and older radios that have not operated in many years.

A simple filter-capacitor test can be performed quickly by temporarily connecting a replacement capacitor of about the same value and working voltage across the suspect capacitor. If the hum goes away, replace the bad component permanently.

Once the faulty component is found, inspect the surrounding circuit and consider what may have caused the problem. Sometimes, one bad component can cause another to fail. For example, a shorted filter capacitor increases current flow and burns out a rectifier diode. While the defective diode is easy to find, the capacitor may show no visible damage.

Switching Power Supplies

Switching power supplies are quite different from conventional linear power supplies. In a "switcher" or switching power supply, a switching transistor is used to change dc voltage levels. They usually have AF oscillators and complex feedback paths. Any component failure in the rectifiers, switch, feedback path, or load usually results in a completely dead supply. Every part is suspect in a switcher. Although active device failure is still the number one suspect, it pays to carefully test all the components if a diagnosis cannot be made with traditional techniques.

Some equipment, notably TVs and monitors, derive some of their power supply voltages from the proper operation of other parts of the circuit. In the case of a TV or monitor, voltages are often derived by adding secondary low-voltage windings to the flyback transformer and rectifying the resultant ac voltage (usually about 15 kHz). These voltages will be missing if there is any problem with the circuit they are derived from.

Amplifiers

Amplifiers are the most common circuits in electronics, so it is recommended that you understand how they function. This will greatly aid you in repairing any equipment with amplifiers, such as guitar amplifiers, receivers, and so on. The output of an ideal amplifier would in theory match the input signal in every respect except magnitude of the signal. Ideally, no distortion or noise would be added but, unfortunately, real amplifiers always add a small amount of noise and distortion. Figure 11-6 shows an amplified sine wave signal. The sine wave shown in (A) is a "clipped" sine-wave signal, the one in (B) shows nonlinear amplification and (C) depicts a pure sine wave signal.

Amplifier Gain

Gain is the measure of amplification. It is usually expressed in decibels (dB) over a specified frequency range, known as the bandwidth or passband of the amplifier. When an amplifier is used to provide a stable load for the preceding stage, or as an impedance transformer, there may be little or no voltage gain.

Amplifier failure usually results in a loss of gain or excessive distortion at the amplifier output. In either case, check external connections first. Is there power to the stage? Has the fuse opened? Check the speaker and leads in audio output stages, the microphone and push-to-talk (PTT) line in transmitter audio sections. Excess voltage, excess current, or thermal runaway can cause sudden failure of semiconductors. The failure may appear as either a short or open circuit of one or more P–N junctions.

Thermal runaway occurs most often in bipolar transistor circuits. If degenerative feedback, in which the emitter resistor reduces base–emitter voltage as conduction increases, is insufficient enough, thermal runaway will occur and allow excessive current flow and device failure. Check transistors by substitution, if possible.

Faulty coupling components can reduce amplifier output. Look for component failures that would increase series, or decrease shunt, impedance in the coupling network. Coupling faults can be located by signal tracing or parts substitution. Other passive component defects reduce amplifier output by shifting bias or causing active-device failure. These failures are evident when the dc operating voltages are measured.

In a receiver, a fault in the AGC loop may force a transistor into cutoff or saturation. Open the AGC line to the device and substitute a variable voltage for the AGC signal. If amplifier action varies with voltage, suspect the AGC-circuit components; otherwise, suspect the amplifier. In an operating amplifier, check carefully for oscillation noise. Oscillations are most likely to start with maximum gain and the amplifier input shorted. Any noise that is induced by 60 Hz sources can be heard or seen with a oscilloscope synchronized to the ac line.

Unwanted amplifier RF oscillation can be cured with changes of lead orientation or circuit components. Separate the input leads from output leads, use coaxial cable to carry RF between stages, and neutralize If interelement or junction capacitance. Ferrite beads on the control element of the active device often stop unwanted oscillations.

Low-frequency oscillations or "motor-boating" often indicates poor stage isolation or inadequate power supply filtering. Try a better lead orientation arrangement and/or check the capacitance of the decoupling network. Use larger capacitors at the power supply leads, increase the number of capacitors, or use of separate decoupling capacitors at each stage. Coupling capacitors that are too low in value can also cause poor low-frequency response. Poor response to high frequencies is usually caused by circuit design.

dc Coupled Amplifiers

In dc coupled amplifiers, the transistors are directly connected together without coupling capacitors. They comprise a unique troubleshooting case. Most often, when one device fails, it destroys one or more semiconductors in the circuit. If you don't find all of the bad parts, the remaining defective parts can cause the installed replacements to fail immediately. To reliably troubleshoot a dc coupled circuit, you must test every semiconductor in the circuit and replace them all at once.

Oscillators

In many circuits, a failure of the oscillator will result in complete circuit failure. If you have an internal oscillator failure, a transmitter will not transmit and a superheterodyne receiver will not receive. These symptoms do not always mean oscillator failure, however. Whenever there is weakening or complete loss of signal from a radio, check oscillator operation and frequency. There are several methods to determine how an oscillator is performing. The first method is to use a receiver with a coaxial probe to listen for the oscillator signal from the oscillator or transmitter in question. A second method is to use a device called a grid dip meter, which can be used to check oscillators. In the absorptive mode, tune the grid dip meter to within ±15 kHz of the oscillator's frequency, couple it to the circuit, and listen for a beat note in the dip meter headphones. The third method is to look at the oscillator waveform on an oscilloscope. The operating frequency cannot be determined with great accuracy with this method, but you can see if the oscillator is working at all. You must use a low-capacitance ×1 probe for oscillator observations. The fourth method to observe if an oscillator is working involves using a high-impedance voltmeter. Tube oscillators usually have negative grid bias when oscillating. Using a high-impedance voltmeter, you will need to measure grid bias. Note that the bias voltage will change slightly with frequency. The last method of determining if an oscillator is operational is based on the fact that emitter current varies slightly with frequency in transistor oscillators. So you can use a sensitive, high-impedance voltmeter across the emitter–resistor to observe the current levels, then you can use Ohm's Law to calculate the current value.

Many modern oscillators are phase-locked loops (PLLs). A PLL is a marriage of an analog oscillator and digital circuitry. To test for a failed oscillator tuned with inductors and capacitors, use a dip meter in the active mode. Set the dip meter to the oscillator frequency and couple it to the oscillator output circuit. If the oscillator is dead, the dip meter signal will take its place and temporarily restore some sembalance of normal operation. Tune the dip meter very slowly, or you may pass stations so quickly that they sound like birds.

Control Circuitry

Semiconductors have made it practical to use diodes for switching by running only a dc lead to the switching point. This eliminates problems caused by long analog leads in the circuit. Semiconductor switching usually reduces the cost and complexity of switching components. Switching speed is also increased and contact corrosion and breakage are eliminated. In exchange, troubleshooting is complicated by additional components such as voltage regulators and decoupling capacitors. The electronics hobbyist or technician must now consider many more components and symptoms when working with diode- and transistor-switched circuits.

Mechanical switches are relatively rugged. They can withstand substantial voltage and current surges. The environment does not drastically affect them, and there is usually visible damage when they fail. Semiconductor switching offers inexpensive, high-speed operation. When subjected to excess voltage or current, however, most transistors and diodes silently fail, with no visible indication. Occasionally, if the technician is lucky, a semiconductor will "smoke," indicating failure. It is well known that temperature variations can cause changes in semiconductor characteristics. A normally adequate control signal may not be effective when transistor beta is lowered by a cold environment. Heat may cause a control voltage regulator to produce an improper control signal as well.

A control signal is actually a bias for the semiconductor switch. Forward-biased diodes and transistors act as closed switches; reverse-biased components simulate open switches. If the control (bias) signal is not strong enough to completely saturate the semiconductor, conduction may not continue through a full ac cycle. Severe distortion can be the result, and failure of the circuit to operate.

When dc control leads provide unwanted feedback paths, switching transistors may become modulators or mixers. Additionally, any reverse-biased semiconductor junction is a potential source of white noise.

Microprocessor Troubles

Today, nearly every new piece of equipment or appliance is controlled by a microprocessor. Many initial microprocessor problems will result in sending the equipment back to the factory for service. Surface-mounted components are usually too difficult for most electronics enthusiasts to replace and this equipment is often sent back to the manufacturer as well. For successful repair of microprocessor-controlled circuits, you should have the knowledge and test equipment necessary for computer repair. Familiarity with machine language and other programming may also be needed.

Microprocessors have brought automation to everything from desk clocks to home appliances. In digital electronic circuits, most every aspect of their operation may be resolved to a simple 1 or 0, or tristate condition, the symptoms of their failure are far more complicated. As with other types of electronic equipment, you will need to first observe the operating characteristics of the piece of equipment that you are trying to repair. Have the schematic diagrams and manual available and study the operation so that you know how the system or circuit operates. As with other electronics, you will then move on to the testing phase of troubleshooting operation, and, finally, to replacement of the defective components.

Problems in digital circuitry usually have two basic causes. First, the circuit may give false counts because of electrical noise at the input. Second, the gates may lock in one state. False counts from noise with as little as a 15 to 20 μV voltage spike can trigger a TTL flip-flop gate into false operation. High voltage, current, or RF transients can follow the ac line or radiate directly to nearby digital equipment, causing oscillation in digital circuits that can also produce false counts.

First begin by removing the suspect equipment from nearby RF fields, high-energy switching circuits, or high-current motors. If the symptoms stop when there is no RF energy around, you need to shield the equipment from RF.

In the mid 1990s, microprocessors generally used clock speeds up to a few hundred megahertz. It may be impossible to filter RF signals from the lines when the RF is near the clock frequency. In these cases, the best, although difficult, approach is to shield the digital circuit and all lines running to it.

If digital circuitry interferes with other nearby equipment, it may be radiating spurious signals. These signals can interfere with nearby radio operation or other services. Digital circuitry can also be subject to interference from strong RF fields. Erratic operation or a complete "lock up" is often the result.

Logic Levels

To troubleshoot a digital circuit, check for the correct voltages at the pins of each IC chip. The correct voltages may not always be known, but you should be able to identify the power pins (Vcc and ground). The voltages on the other pins should be either a logic high, a logic low, or tristate. In most working digital circuitry, the logic levels are constantly changing, often at RF rates. Unfortunately, dc voltmeters may not always give reliable readings; therefore, an oscilloscope or logic analyzer is usually needed to troubleshoot digital circuitry.

Most digital circuit failures are caused by a failed logic IC. In clocked circuits, listen for the clock signal with a coax probe and a suitable receiver. If the signal is found at the clock chip, trace it to each of the other ICs to be sure that the clock system is intact. Some digital circuits use VHF clock speeds; an oscilloscope must have a bandwidth of at least twice the clock speed to be useful. If you have a suitable scope, check the pulse timing and duration against circuit specifications.

As in most circuits, failures are often catastrophic. It is unlikely that an AND gate will suddenly start functioning like an OR gate. It is more likely that the gate will have a signal at its input, and no signal at the output. In a failed device, the output in will have a steady voltage. In some cases, the voltage is steady because one of the input signals is missing. Look carefully at what is going into a digital IC to determine what should be coming out Keep manufacturers' data books on hand if possible. Data books usually can be found to describe the proper functioning of most digital devices.

Tristate Devices

Many digital devices are designed with a third logic state, commonly called tristate. In this state, the output of the device acts as if it weren't there at all. Many such devices can be connected to a common "bus," with the devices that are active at any given time selected by software or hardware control signals. A computer's data and address buses are good examples of this. If any one device on the bus fails by locking itself on in a 0 or 1 logic state, the entire bus becomes nonfunctional. These tristate devices can be locked "on" by inherent failure or a failure of the signal that controls them.

Simple Gate Tests

Logic gates, flip-flops, and counters can all be tested by manually triggering them, either with a 5 volt power supply or with a logic pulser. Diodes can be checked with an ohmmeter. Testing of more complicated ICs requires the use of a logic analyzer, multitrace scope, or a dedicated IC tester. The trick is to break the circuits down into smaller problems and test each gate separately or in small blocks, and to observe propagation of the pulses through the questionable gates or stages.

Troubleshooting Hints

Components

Once you locate a defective part, you will need to select its replacement. This is not always an easy task. Each electronic component has a function. Chapter 10 acquainted you with test procedures for resistors, capacitors, inductors, and other components. This chapter has presented functions and failure modes of circuits. These two chapters together form a solid basis for checking symptoms indicating faulty components implicated by stage-level testing. In most cases, a particular faulty component will be located by these tests. If a faulty component is not indicated, check the circuit adjustments. As a last resort, use a shotgun approach. The shotgun approach is the replacement of all parts in that problem area with components that are known to be good.

Additional Considerations

Before you install a replacement component of any type, you should make sure that another circuit defect did not cause the failure. Check the circuit voltages carefully before installing any new components. Check the potential on each trace to the bad component. The old part may have "died" as a result of a voltage increase that occurred in another part of the circuit. Measure twice, repair once!

Fuses

Usually, a fuse fails for a reason, usually because of a short circuit in the load. A fuse that has failed because of a short circuit usually shows the evidence of high current: a blackened interior with little blobs of fuse element everywhere. Fuses can also fail by fracturing the element at either end. This kind of failure is not usually visible by looking at the fuse. Check even "good" fuses with an ohmmeter. You may save hours of troubleshooting. For safety reasons, always use exact replacement fuses. Check the current and voltage ratings before replacing the fuse. Fuse timing is also very important. There are two basic fuse types—fast or normal and slow blow. You must replace the blown fuse with the original type of fuse. Never attempt to force a fuse that is not the right size into a fuse holder. The substitution of a bar, wire, or penny for a fuse invites overheating and the potential for fires.

Connectors

Connection faults are one of the most common causes of failures in electronic equipment. They can range from something as simple as the ac line cord coming out of the wall, to a connector having been put on the wrong socket, to a defective IC socket. Connectors that are plugged and unplugged frequently can wear out, becoming intermittent or noisy. Check connectors carefully when troubleshooting.

Connector failure can be hard to detect. Most connectors maintain contact as a result of spring tension that forces two conductors together. As the parts age, they become brittle and lose tension. Any connection may deteriorate because of nonconductive corrosion at the contacts. Solder helps prevent this problem but even soldered joints suffer from corrosion when exposed to weather.

The dissipated power in a defective connector usually increases. Signs of excessive heat are sometimes seen near poor connections in circuits that carry moderate current. Check for short and open circuits with an ohmmeter or continuity tester. Clean those connections that fail as a result of contamination.

Occasionally, corroded connectors maybe repaired by cleaning, but replacement of the conductor connector is usually required. Solder all connections that may be subject to harsh environments and coat them with an acrylic enamel, RTV compound, or similar coating.

Choose replacement connectors with consideration of voltage and current ratings. Use connectors with symmetrical pin arrangements only where correct insertion will not result in a safety hazard or circuit damage.

Wires

Wires seldom fail unless abused. Short circuits can be caused by physical damage to insulation or by conductive contamination. Damaged insulation is usually apparent during a close visual inspection of the conductor or connector. Look carefully where conductors come close to corners or sharp objects. Repair worn insulation by replacing the wire or securing an insulating sleeve (spaghetti) or heat-shrink tubing over the worn area.

When wires fail, the failure is usually caused by stress and flexing. Nearly everyone has broken a wire by bending it back and forth, and broken wires are usually easy to detect. Look for sharp bends or bulges in the insulation.

When replacing conductors, use the same material and size, if possible. Substitute only wire of greater cross-sectional area (smaller gauge number) or material of greater conductivity. Insulated wire should be rated at the same, or higher, temperature and voltage as the wire it replaces. Note that solid wire fails more quickly, sometimes with only a few flexings, whereas stranded wire will not easily fail with a few bendings.

Calibration and Standards

Realize that test instruments are not perfect. They age over time, lose their calibration, and eventually wear out. This means that they should be periodically checked for proper operation. This is called calibration. It is basic industry standard to calibrate instruments every year or after every repair. This takes time and can be expensive if the equipment is shipped to a calibration facility. Is it necessary? An experienced engineer will often notice variations in performance with a familiar piece of gear. Additionally, most variations are subtle and are not too likely to cause problems. It can be really serious if the instrument is used by production people for precision measurements.

A short-cut for calibration is called "confidence testing." This presents known signals to the instrument to verify that it is operating properly. A frequency counter could be tested each morning with separate and precise 250 Mhz signals. It would only take a minute. There are many simple ways to perform confidence checks on equipment. A precision resistor measured very accurately and a good crystal oscillator are required. The checks should be performed with regularity—at least every month. Any variation in performance should be noted on the instrument, using a reliable means. Any significant variation from what is expected should be a signal to repair or recalibrate the device.

Basically, confidence testing is "standards testing." You test a frequency counter with a "standard" frequency. This, of course, leads to the question of how good the frequency standard is and how it was used.

Using the techniques shown in this chapter will help you become a good troubleshooter. The ability to read schematics and mastery of the techniques outlined here, combined with common sense and some accumulated experience, will help you gain confidence and become a first-rate troubleshooting expert before too long.

Workshop
Tools

Construction of electronics projects and kits are great fun and the best place to build these projects is on your new electronic workbench. This chapter discusses tools recommended for electronics applications and their uses, as well as shop safety. Hand tools are used for so many different applications that they are discussed first, followed by some tips for proper use of power tools. At the end of this section we will also present a few ideas for obtaining tools for your electronics workbench.

Tools and Their Uses

All electronics construction projects make use of tools, from mechanical tools for chassis fabrication to the soldering tools used for circuit assembly. A good understanding of tools and their uses will enable you to perform most construction tasks. Although sophisticated and expensive tools often work better or more quickly than simple hand tools, with proper use, simple hand tools can turn out a fine piece of equipment. Table 12-1 lists the tools that are indispensable for construction of electronic equipment. These tools can be used to perform nearly any construction task. Add tools to your collection from time to time, as finances permit.

Care of Tools

The proper care of tools is more than a matter of pride. Tools that have not been cared for properly will not last long or work well. Dull or broken tools can be safety hazards. Tools that are in good condition do the work for you; tools that are misused or dull are difficult to use. Store tools in a dry place. Tools do not fit in with most living-room decor, so they are often relegated to the basement or garage. Unfortunately, many basements or garages are not good places to store tools; dampness and dust are not good for them. If your tools are stored in a damp place, use a dehumidifier. Sometimes, you can minimize rust by keeping your tools lightly oiled, but this is a second-best solution. If you oil your tools, they may not rust, but you will end up covered in oil every time you use them. Wax or silicone spray is a better alternative. Store tools neatly. A messy toolbox, with tools strewn about haphazardly, can be more than an inconvenience. You may waste a lot of time looking for the right tool and sharp edges can be dulled or nicked by tools banging into each other in the bottom of the box. As the old adage says, every tool should have a place, and every tool should be in its place. If you must search the workbench, garage, attic, and car to find the right screwdriver, you will spend more time looking for tools than building projects.

Tools Organization

The best utilization of tools is through good organization and good storage practices. One of the best ways to organize and store tools is to use pegboards hung vertically in your electronics workshop or near your electronics workbench. Pegboards and their metal hardware are readily available from Home Depot or Lowes and are inexpensive. Pegboards are generally made from composite board or pressboard material. On cement walls, furring strips or one-by-two-inch wood strips can be secured with lag bolts. On wood paneled walls, pegboard can be screwed down every 16 inches on standard two-by-four supports. Figure 12-1

Table 12-1   Electronics workshop tool list

Simple Hand Tools

Screwdrivers

   slotted, 3 inch, $\frac{1}{8}$ inch blade

   slotted, 8 inch, $\frac{1}{8}$ inch blade

   slotted, 3 inch, $\frac{3}{16}$ inch blade

   slotted, stubby, $\frac{1}{4}$ inch blade

   slotted, 4 inch, $\frac{1}{4}$ inch blade

   slotted, 6 inch, $\frac{5}{16}$ inch blade

   Phillips, $2\frac{1}{2}$ inch, #0 (pocket clip)

   Phillips, 3 inch, #1

   Phillips, stubby, #2

   Phillips, 4 inch, #2

   Phillips, 6 inch, #2

   long-shank with holding clip on blade

   jewelers set

   right-angle, slotted and Phillips

Pliers, sockets, and wrenches

   long-nose pliers, 6 and 4 inch

   diagonal cutters, 6 and 4 inch

   Channel-lock pliers, 6 inch

   locking pliers (vise grip or equivalent)

   socket nut driver set, $\frac{3}{16}$ to $\frac{1}{2}$ inch

   set of socket wrenches for hex nuts

   Allen (hex) wrench set

   wrench set

   adjustable wrenches

   6 inch and 10 inch tweezers, regular and reverse-action

   retrieval tool/parts holder, flexible claw

   retrieval tool, magnetic

Cutting and grinding tools

   file set consisting of flat, round, half-round, and triangular, large and miniature types

   burnishing tool

   wire strippers

   wire crimper

   hemostat, straight

   scissors

   tin shears, 10 inch

   hacksaw and blades

   hand nibbling tools (chassis-hole cutting)

   scratch awl or scriber

Cutting and grinding tools (*cont.*)

   knife blade set (Xacto)

   machine screw taps, #4-40, through #10-32 thread

   socket punches, $\frac{1}{2}$ in, $\frac{5}{8}$ in, $\frac{3}{4}$ in, $1\frac{1}{8}$ in, $1\frac{1}{4}$ in, and $1\frac{1}{2}$ in

   tapered reamer, T-handle, $\frac{1}{2}$ inch

   deburring tool

Miscellaneous hand tools

   combination square, 12 inch, for layout work

   hammer, ball-peen, 12 oz head

   hammer, tack

   bench vise, 4-inch jaws

   center punch

   plastic alignment tools

   mirror, inspection

   flashlight, penlight

   magnifying glass

   ruler or tape measure

   dental pick

   calipers

   brush, wire

   brush, soft

   small paintbrush

   IC-puller tool

Powered hand tools

   hand drill, $\frac{1}{4}$ inch quick chuck

   high-speed drill, hand held

   drill press

   miniature electric motor tool (Dremel)

Soldering tools and supplies

   soldering pencil iron, 30 watt, $\frac{1}{8}$ inch tip

   soldering iron, 200 watt, $\frac{5}{8}$ inch tip

   solder 60/40, resin core

   soldering gun with assorted tips

   desoldering tool

   desoldering wick

Safety devices

   safety glasses

   hearing protector

   earphones or ear plugs

   first-aid kit

Figure 12-1   Pegboards used for storing tools.

shows well-organized tools placed on pegboards for quick identification and retrieval. There are a two major options for pegboard hardware used to hold tools. One type uses magnetic strips or bars that can hold a number of tools per strip. The second type are individual metal pegs with hooks or jaw-type claws attached that hold individual tools.

Parts Organization

Practicing good tool storage organization should be carried over to good parts storage practices as well. One great way to organize electronic parts and small hardware is the use of small plastic parts-storage containers. A typical parts-storage drawer system is shown in Figure 12-2. These parts-storage containers are readily available and often on sale from large retailers such as Wal-Mart or Target. Using "see-through" plastic drawers allow quick identification of drawer contents so no time is wasted in locating small parts or hardware.

Sharpening

Many cutting tools can be sharpened. Send a tool that has been seriously dulled to a professional sharpening service. These services can resharpen saw blades, some files, drill bits and most cutting blades. Touch up the edges of cutting tools with a whetstone to extend the time between sharpenings.

Figure 12-2 Parts-storage draw system.

Sharpen drill bits frequently to minimize the amount of material that must be removed each time. Frequent sharpening also makes it easier to maintain the critical surface angles required for best cutting with least wear. Most inexpensive drill bit sharpeners available for shop use do a poor job, either due to the poor quality of the sharpening tool or inexperience of the operator. Also, drills should be sharpened at different angles for different applications. Commercial sharpening services do a much better job.

Proper Tool Use

Do not use tools for anything other than their intended purpose! If you use a pair of wire cutters to cut sheet metal, pliers as a vise, or a screwdriver as a pry bar, you will ruin a good tool. Although an experienced constructor can improvise with tools, most take pride in not abusing them.

Screwdrivers

For construction or repair, you need to have an assortment of screwdrivers. Each blade size is designed to fit a specific range of screw-head sizes. Using the wrong size blade usually damages the blade, the screw head, or both. You may also need stubby sizes to fit into tight spaces. Right-angle screwdrivers are inexpensive and can get into tight spaces that cannot otherwise be reached. Electric screwdrivers are relatively inexpensive. If you have a lot of screws to fasten, they can save a lot of time and effort. They come with a wide assortment of screwdriver and nut-driver bits. An electric drill can also function as an electric screwdriver, although it may be heavy and overpowered for some applications.

Keep screwdriver blades in good condition. If a blade becomes broken or worn out, replace the screwdriver. A screwdriver only costs a few dollars; do not use one that is not in perfect condition. Save old screwdrivers to use as pry bars and levers, but use only good ones on screws. Filing a worn blade seldom gives good results.

Pliers and Locking-Grip Pliers

Pliers and locking-grip pliers are used to hold or bend things. They are not wrenches! If pliers are used to remove a nut or bolt, the nut or the pliers is usually damaged. Pliers are not intended for heavy-duty applications. Use a metal brake to bend heavy metal; use a vise to hold a heavy component. To remove a nut, use a wrench or nut driver. There is one exception to this rule of thumb: To remove a nut that is stripped too badly for a wrench, use a pair of pliers, locking-grips, or a diagonal cutter to bite into the nut and turn it a bit. If you do this, use an old tool or one dedicated to just this purpose; this technique is not good for the tool. If a pliers' jaws or teeth become worn, replace the tool.

Needle-Nose Pliers

Needle-nose pliers of various sizes are an invaluable asset to your electronics workbench. Long- and short-nose pliers are quite useful. Sometimes, a short-nose pliers is just the right tool for a job and sometimes a long nose-pliers is needed for accessing parts and screws in tight places. Right-angle needle-nose pliers are very handy and great to have but are often overlooked because they are expensive. As your budget permits, purchase at least two different-size right-angle needle-nose pliers. Another useful tool is the hemostat. The hemostat is a physician's "clamp" used in operations. It is a type of delicate needle-nosed pliers that clamps shut. Hemostats are very useful on the electronics workbench for positioning small parts or for holding and starting screws.

Wire Cutters

Wire cutters are primarily used to cut wires or component leads. The choice of diagonal blades (sometimes called "dikes") or end-nip blades depends on the application. Diagonal blades are most often used to cut wires, whereas the end-nip blades are useful to cut off the ends of components that have been soldered into a printed-circuit board. Some delicate components can be damaged by cutting their leads with dikes. Scissors designed to cut wire can also be used. Wire strippers are handy, but you can usually strip wires using a diagonal cutter or a knife. However, wire strippers use a hole size gauge to determine wire sizes, this makes insulation easier, especially for newcomers to electronics. Do not use wire cutters or strippers on anything other than wire! If you use a cutter to trim a protruding screw head or cut a hardened-steel spring, you will usually damage the blades. Be sure to have an assortment of low-cost knives and single-edge razor blades handy.

Files

Files are used for a wide range of tasks. In addition to enlarging holes and slots, they are used to remove burrs; shape metal, wood, or plastic, and clean some surfaces in preparation for soldering. Files are especially prone to damage from rust and moisture. Keep them in a dry place. The cutting edge of the blades can also become clogged with the material you are removing. Use file brushes (also called file cards) to keep files clean. Most files cannot be sharpened easily, so when the teeth become worn, the file must be replaced. A worn file is sometimes worse than no file at all. At best, a worn file requires more effort to get the job done.

Drill Bits

Drill bits are made from carbon steel, high-speed steel, or carbide. Carbon steel is more common and is usually supplied unless a specific request is made for high-speed bits. Carbon-steel drill bits cost less than high-speed or carbide types; they are sufficient for most equipment construction work. Carbide drill bits last much longer under heavy use. One disadvantage of carbide bits is that they are brittle and break easily, especially if you are using a hand-held power drill. Drill bits are available in sets. You may not use all of the bits in a standard set, but it is nice to have a complete set on hand. You should also buy several spares of the more common sizes. Although standard bits list down to size #54, special sizes extend down to size #80. The smaller-size drills are important for printed-circuit board work.

Specialized Tools and Materials

Most constructors know how to use common tools such as screwdrivers, wrenches, and hammers. In this section we will discuss these plus other tools that are not so common.

Nibbling Tool

A hand nibbling tool is a very useful tool used to remove small "nibbles" of metal. It is easy to use and very handy to have around the electronics workbench. The nibbling tool is ideal for enlarging holes or squaring off holes for mounting square switches, meters, or LCD displays. You simply position the tool where you want to remove metal and squeeze the handle. The tool takes a small bite out of the metal. When you use a nibbler, be careful that you do not remove too much metal, clip the edge of a component mounted to the sheet metal, or grab a wire that is routed near the edge of a chassis. Fixing a broken wire is easy, but something to avoid if possible. It is easy to remove metal but nearly impossible to put it back. You have to do it right the first time!

Deburring Tool

A deburring tool is just the thing to remove the sharp edges left on a hole after most drilling or punching operations. Position the tool over the hole and rotate it around the hole edge to remove burrs or rough edges. As an alternative, select a drill bit that is somewhat larger than the hole, position it over the hole, and spin it lightly to remove the burr.

Socket Punches

Greenlee is the most widely known of the socket-punch manufacturers. Most socket punches are round, but they do come in other shapes. To use one, drill a pilot hole large enough to clear the bolt that runs through the punch. Then mount the punch with the cutter on one side of the sheet metal and the socket on the other. Tighten the nut with a wrench until the cutter cuts all the way through the sheet metal.

Crimping Tools

One important set of hand tools you will want to acquire are electronic crimping tools. There are four major types. The original crimping tool was designed for crimping spade lugs on wire ends. Spade lugs come in a number of different styles such as the closed circle or the half-round types that go underneath small screws. A new low-cost variation of the original crimping tool now incorporates a wire cutter/stripper as well as a screw-thread cleaner. This is an invaluable device for building cable harnesses for automotive or radio work.

The second type of the crimping tool that you will want to obtain is the four-pin telephone plug wire stripper crimping tool designed for putting phone plugs on four-wire phone wire in a simple two-step process. This crimping tool allows you to cut the wire, strip the four conductor wires, then crimp the wires onto the plastic phone plug in a one-minute operation. This tool will save you a lot of time and aggravation when repairing phones or rewiring phone extension cords.

The third type of crimping tool is the coaxial cable tool that is used to secure coaxial cable to a coaxial connector. This type of crimper is available for RG-59 coaxial cables for radio and instruments and Ethernet cables and BNC connectors. Another version of the coaxial crimper is also available for RG-59 cable for use with N-connectors for cable TV applications.

The fourth type of crimping tool is for connecting 10-base-T or twisted-pair Ethernet cables. One version of this crimper will allow you to strip the cable wires and permit you to crimp the eight-wire cable to the plastic connector in less than a minute. Trying to put crimp connectors on cables without the proper tool is a painful experience that results in broken connectors, shorted cables, and much frustration.

These crimping tools will make your life in electronics much more pleasant.

Heat-Shrink Tubing

Heat-shrink tubing is extremely useful around your electronics workshop or workbench, especially when splicing electronic cables. It is a much more professional approach to covering wires after splicing than trying to use any type of tape. Heat-shrink tubing can be used for all types of wire splicing from Western Union or parallel splices to end splicing. It comes in many different diameters for all sizes of wire. While you can use black electrical tape for covering a wire splice, heat-shrink tubing covers the wire more evenly, eliminating ugly bumps. When electrical tape gets old, it drys out and comes apart and makes a gooey mess everywhere.

Heat-shrink tubing can also be used for other applications in addition to normal splicing. One application for heat-shrink tubing is for covering, encasing, or protecting electronic components from outside air or chafing. Also, heat-shrink tubing can also be used to create your own special-purpose optoisolators or optocouplers. For example, you can use heat-shrink tubing to couple a neon bulb to a phototransistor to detect the presence or absence of ac line current. You can also use heat-shrink tubing to hold multiple wires together in bundle, by cutting half-inch sections of tubing and securing then periodically along a bundle of small wires as needed.

Velcro

Velcro is indeed one of most flexible and handy material known to man. It can be obtained in bulk from stores such as Home Depot and should have a prominent place in your electronics workshop. Velcro can be used in so many instances that there are too many to name. Your imagination will permit many solutions to appear from nowhere. Velcro can be used to secure wires to panels or chassis boxes or hold cables together in a bundle. It can be placed under screw terminal strips to secure them to the chassis or panels for instant placement solutions. Velcro can be used to patch holes, secure tools to walls, or support pegboards, light-weight shelves, chalk boards and white boards, as well as to hold concentric cylinders together.

Velcro can be used to secure XM "hockey puck" satellite antennas to your car rooftop. It can be used to attach headband-type earphones or headsets to walls or tables for easy storage when not in use. Velcro can be used to hold minichassis control or switch boxes to a table or wall, thus eliminating wall screws. Just when you think you have thought of all the possible applications for Velcro, then another one will come along as the need arises. Velcro knows no bounds!

Double-Sided Tape

Double-sided tape is your friend! It can be used in many applications and can be readily found in electronics and discount stores. Double-sided tape is usually white in color and is available in a number of different thicknesses from $1/32$ inch to $1/4$ inch. The 1/6 inch variety works quite well, if you were to have only one size available to you. Double-sided tape can be used to mount battery holders to circuit boards, hold terminal strips to circuit boards or chassis boxes, hold foil shields in place, hold security mounts on valuable equipment, to name just a few applications.

Electrical Tape

Electrical tape should be on your list of items to have around your electronics workshop. Black plastic electrical tape is very handy for electrical wire splices, especially if heat-shrink tubing is not available. The ability of black electrical tape to stretch is the reason for its success. When you wind plastic electrical tape around a wire or cable, you pull the tape around the item, stretch it, and pull it tight, then cut or tear it off the roll. This stretching ability is what holds the tape tight to the connection or cable. This is the reason why you use this type of tape instead of, say, masking tape or Scotch tape. Other types of tape only use the sticky quality of the tape to hold it to surfaces, whereas multiple wraps of stretched electrical tape creates a better and tighter wrap. Most people are familiar with shiny black electrical tape, but there are a few different types available, such as a thicker rubber version that is often used in outdoor applications, but they are a bit harder to find and cost more.

Scrap Circuit Boards

Do not throw away old circuit boards, because you never know when they might come in handy. Scrap circuit boards are very useful for constructing low-cost chassis boxes for small projects. They are also very useful for constructing small shielded boxes for radio circuits such as oscillators or sensitive front-ends for receivers. It is easy to create small boxes made from scrap circuit boards. All you need is a soldering gun, solder, and some scrap pieces of circuit board. You can also use scrap circuit boards for making front panels for enclosures or you can create L-shaped front and bottom open-panel chassis. Single- or double-sided circuit boards can be used for creating small chassis boxes, and any scrap flexible circuit material can be readily used for creating small shield covers or housings.

Creating small chassis boxes from scrap circuit board is very simple. Begin by cutting the scrap pieces to size using a band saw or other suitable saw. Note that you will need to use about 1/16 to 1/8 inch of circuit board for soldering near the edges, all around the inside of the new enclosure, so allow for this added dimension on all sides while initially measuring the box sides. Take a scrap board a scrap board piece that you want to use for the bottom of the box and lay it down on a flat surface. Next, hold a side piece up vertically to one edge of the bottom panel. Make sure the vertical piece is perfectly perpendicular to the bottom piece and tack solder the first side piece to the bottom in a few places. Once you have verified that the first panel is perfectly vertical, you can run a bead of solder along the entire panel. If you want the box to look professional, you should solder from the inside, so the outside surfaces are clean. You can then repeat the process for the other side panels. When finished, you will have to determine how you will secure the top panel to the box itself. You may want to just use fasteners or hinge the box top so you can open the box.

Magnifying Glass

A magnifying glass is an essential item to have at your electronics workshop or workbench for doing close-up work on small areas of a circuit or when handling small parts. Magnifying lenses come in many shapes and sizes for different applications. Some hand-held units also have an area of the lens with a greater spot magnification. The main lens could be ×3 with a spot magnifier of ×10. This type of hand lens is very useful for small-part handling or soldering. Hand lens magnifiers are also available with small light bulbs integrated into the handle, which are powered by AAA batteries. These hand magnifiers are handy in areas without bright lighting. You can also consider purchasing the low cost Radio Shack lighted, handheld microscope/magnifier (#63-851). It acts as both a magnifier lens and a microscope and is an invaluable aid in working on electronics projects, from identifying components to looking for bad solder joints on a circuit boards. You might also want to consider looking at a professional, adjustable desk- or bench-mounted lighted magnifying unit. These units mount to the side of a workbench or desk and have spring-tension-adjustable arms, so that the light can be moved over any area of the desk or workbench. These magnifiers are usually in the form of a large ring about 10 to 12 inches in diameter with a fluorescent ring lamp mounted around the magnifier lens. Higher-cost professional units will often have different replaceable magnification lenses that snap in place for higher magnification as desired.

Hot-Glue Gun

Every electronics workshop should have a low-cost hot-glue gun. The glue gun made its appearance when the craft movement began in the late 1980s. A simple but useful device, it was used to glue paper and fabric, and other applications have evolved from there. The glue gun consists of a plastic gun-shaped housing that has a small heating element placed around a cylinder. The hollow inside of the cylinder accepts a low-cost glue stick. As the heating element heats up, the solid glue stick is heated and melted and is fed through a nozzle in liquid form. The lowly glue gun has many applications around the electronics workshop, from tacking wires in place to fastening terminal strips to chassis. The glue gun can hold large electronics components in place or can be used to fasten motherboards to daughterboards. Glue guns can be used to make circuit board guides or to fasten circuit boards to a chassis directly. Glue guns only cost about $6.00, so they deserve a place at your workbench. They will serve you well for many applications.

Glue

You can never have too many different types of glue around your electronics workshop. You should consider purchasing wood glue, "super glue," and two-part epoxy cement to have at your disposal. Epoxy can be used for many applications, from fixing broken circuit boards to mending fiberglass or ceramic pieces. Two-part epoxy is available in two two basic forms. Five-minute epoxy sets up in five minutes or less. It consists of the epoxy and a hardener, generally mixed in equal parts and applied to a broken part. In five minutes, the epoxy begins to set up and harden. In a few hours, the epoxy joint becomes hard and secure. A second type of epoxy, a more costly professional, high-quality version, takes much longer to set up and harden. This industrial-grade epoxy takes a few hours to set up or begin hardening and then takes 24 hours to fully cure. The industrial-grade epoxy is generally more dense and makes a more secure bond over time. In order for epoxy to work properly, you must mix equal parts of epoxy to hardener and the surfaces must be clean and roughed up, rather than smooth.

"Super glue" is also very useful to have around your shop. It hardens instantly and works best on hard, smooth surfaces of like material. It usually comes in small bottles and is generally inexpensive. However, one caveat should be observed: this stuff is very sticky. It hardens very fast and very securely. If you get super glue between your fingers, it sticks them together very well and it is very difficult to break the bond between your fingers or between your fingers and an object. Be careful when handling super glue. Use thin rubber gloves and handle the parts carefully, trying to dispense the glue without getting your fingers glued together.

Wood glue is very useful around the shop, especially if you are building electronics circuits and housing them in wooden boxes.

Dental Picks

Dental picks may bring discomforting thoughts at the mention, but they are quite useful around the electronics workshop. Dental picks can be used for a wide range of tasks, from getting things out of tight corners to pulling wires though holes. They can be used to place small parts, wedge or nudge metal shields into place, or even bend tiny wires around corners. Dental picks come in all shapes and sizes. They usually have a tool at both ends and they are not always as sharp as knives. Some picks have spatula or flat tapered ends; some are bent at right angles and some are twisted. Dental picks can be obtained in two ways: either surplus from a dentist or from low-cost tool vendors. Many times, dental picks can be found at hamfests or flea markets for little money. Often times, a dental pick will save the day as a solution to a problem.

Knives

If you are a knife fancier, you know you can never have too many knives! Knives are very useful around your electronics workshop. A utility knife or razor-blade knife can be used in many ways from cutting insulation on wires to cutting electrical tape. Single-blade knives are also quite handy to have around, but a good set of Xacto knives are a must have! Most hobby and art/craft shops have a good selection of Xacto knives. Usually, these knife sets come in a small box with a range of blade types for different cutting applications. Xacto knives are so useful that you should have a few sets, both in your shop and in your office or home.

Xacto knives are very sharp, so care must be taken when using them. They cut paper and mylar sheets very easily and with clean, neat lines. They are the preferred knives for artists and craft people because they are so sharp cut so evenly. Xacto knives are ideal for cutting picture mats and they are also great for cutting templates for chassis front panels, making overlays, and printed circuit artwork. If you design circuit boards using the "tape and dots" method, Xacto knives will help you easily cut design-layout tape and layout pads as needed. Some circuit designers cut out parts shapes of large components and place them on their mylar layout sheets to help determine space requirements when designing circuit boards. Some technicians also use Xacto knives to scour insulation on wires to help remove it. Xacto replacement blades are low cost and readily available.

Fuses

Fuses are necessities for any well-equipped electronics workshop. You can never have enough fuses! Most electronics devices have fuses, they are safety devices designed to protect your electronic equipment as well as to help prevent fires. Never replace a low-current fuse with a high-current value; this is a sure path to damage and possible fires around both your home or workshop. Fuses values are designed specifically for the circuits they are to protect, so when replacing them be sure to install the correct value. Fuses come in many sizes and shapes, from the standard 3AG standard to the mini $\frac{1}{8}$ inch wire-pigtail types. Fuses are available in "fast-blow" and "slow-blow" types. Fast-blow fuses are used in most modern circuitry and instantly fail when their value of current is exceeded. In most instances, you want the fuse to act quickly. Some power supply or motor-drive circuits require "slow-blow" fuses, and are often required where circuits have slow start-up times or a greater surge when starting up.

Fuses were traditionally sold in small steel boxes with five fuses in them. Today, fuses are sold in blister packs or two to five in each package. A good strategy for replacing fuse stocks is to obtain a package of five fuses for the major fuse values in both "fast-blow" and "slow-blow" versions. Your stock of replacement values should contain at least the standard 3AG values from $\frac{1}{4}$ ampere up to at least 10 amperes. Try to locate a good supply of $\frac{1}{2}$ ampere, 1, $1\frac{1}{2}$, 2, 3, and 5 ampere fuses as well as 8 and 10 ampere values. You may also want to stock the next-smaller physical size fuses as money permits.

Wire Markers

At some point in your electronics career or hobby, you will likely assemble electrical and electronic cables. Building electronics projects often requires interconnecting cables between the circuits or between circuits and the front and rear panels. In large enclosures, cables may often be long and circuitous. When assembling cables, it is a good idea to label not only the ends of the cables but various points along the length. The best method for labeling wires and cables is to use electricians' wire markers. These markers come in small booklets. They are numbered pieces cloth backed with adhesive, about an inch long in groups of matching pairs. These marker labels are wrapped around wires and cables and placed along the cable at various points. The tape is very sticky and can be wrapped a number of times around each wire to secure it permanently. The object of labeling wires and cables is for ease of troubleshooting if and when the circuit or wires develop a fault condition. These labels will not only help you at a later date, but can assist others who may be called upon to troubleshoot a piece of equipment that they didn't build. In our enthusiasm to build new circuits, we often overlook the fact that our circuit can fail at a later date. The few minutes it takes to add these markers will save you enormous amounts of time when you have to service equipment at a later date.

Screw Starters

Screw starters are somewhat obscure items and many people do not even know what they do. Screw starters look much like screwdrivers from a distance. They are generally made for use with flat, slotted screws. They have a handle with a sliding bushing that moves forward and back. The bushing is attached to a sleeve that expands in two directions, with thin flat blades on either side of a center line. To use the screw starter, you adjust the sleeve so that the two flat blades are close together. Next, you insert the flat blades into a slotted screw and then adjust the sleeve to expand. In expanding, the two flat blades grip the inside of the slotted screw. You can then use the screw starter to gently begin turning the screw a few turns. Next, adjust the screw starter sleeve so that the flat blades compress, so that you can remove the screw starter and replace it with a regular screwdriver and finish driving the screw. Screw starters are very handy devices, especially if you like building equipment.

Vises

Vises are one of most useful items to have in any electronics workshop. They are available in many different shapes, sizes, and configurations. Vises for electronics applications are usually smaller than vises used in machine shops for working on large pieces of metal. Bench vises usually mount to the side of a workbench with a back mounting plate and have a long screw with a tightening handle. Bench vises are available in many sizes and head shapes, from the simple flat-sided heads to anvil or pointed and tapered head assemblies. Bench vises for electronics use generally tend to be smaller, flat-headed types. Vises are often used to hold chassis boxes and metal panels while drilling them for connectors and switches.

An alternative to bench vises are small table-top vacuum vises for electronics project use. These vacuum vises are great, since they can be positioned anywhere on the workbench at any time. Vacuum vises are easily oriented in different positions and usually have adjustment screws for tilting and rotating the head assembly. Vacuum vises are wonderful for holding connectors and other components while soldering wires to the component pins. They work quite well when the desk or bench surface is flat and clean. Occasionally, you will have to clean the bottom of the vise and the bench top with a soap and water or alcohol for best results.

Another type of vise-type item is the "extra hands device." This device can be purchased or constructed and is very helpful, since humans only have two hands that can be used at once. The "extra hands" device consists of a metal or wood base and a short vertical post set in the base. The vertical post is attached to a horizontal metal rod about 6 inches long with two to four alligator clips attached to sliding collars with set screws. The alligator clips can be positioned anywhere along the horizontal rod and can be rotated or repositioned at any angle. The "extra hands" device is often used for holding small parts while soldering wires to them or when splicing wires. This low-cost "extra hands" device can be easily moved anywhere on the workbench and is a invaluable aid.

Measuring Devices

When setting up your new electronics workbench, you will want to obtain some measuring devices. Measuring devices are needed especially by persons involved in building new circuits and projects. In constructing circuit boards and chassis boxes, you will have to locate holes for connectors and switches, etc. You should try to locate a yard stick, a 12 inch ruler, a 6 inch or longer steel rule, and at least one 12 to 24 inch right-angle steel rule. These measuring devices are the minimum requirements for measuring needs. For circuit board design work, you will should also purchase a 12 inch plastic triangle for drawing lines for artwork.

Hammers

Your electronics workshop would not be complete without a few hammers. If you already have a regular hammer, then that is one less hammer you will need to obtain. If you already have a ballpeen hammer, then you will not need to purchase one. Do you have a rubber mallet? A small rubber mallet is a very useful hammer to have, especially for inserting connector pins or panel nuts on chassis panels, marking panels prior to drilling, and so on. A mini metal hammer might also be useful for your electronics workbench. Many of these are available in low-cost tool sets.

Screwdrivers

You can never have enough screw drivers. You should try to obtain a good set of flat-blade screwdrivers with rubber handles. You should also try to purchase a good set of Phillips screwdrivers. A set of long-shaft screwdrivers, both flat blade and Phillips types, are very handy and often needed but usually overlooked. A set of small jeweler's screwdrivers is extremely useful for small electronics work. Make sure that you have a good set with a range of both flat-blade and Phillips drivers. Look for a good set of right-angle screwdrivers, which are often needed and usually forgotten when purchasing tools for the electronics workbench. A good alternative or addition to a set of flat-blade and Phillips screwdrivers is an Xcelite set of screwdrivers. An Xcelite set usually contains both flat-blade and Phillips drivers with short plastic handles. The Xcelite set also comes with a torque amplifier. Each screwdriver can be inserted into the torque amplifier handle for heavy-duty use. The Xcelite set allows you have a relatively small set of screwdrivers that becomes more substantial set when the torque amplifier is utilized. Remember to use the proper screwdriver for the job at hand.

Sources of Tools

Radio-supply houses, mail-order stores, and most hardware stores carry the tools required to build or service amateur radio equipment. Bargains are available at ham radio flea markets or local neighborhood sales. But beware! Some flea-market bargains are really shoddy imports that won't work very well or last very long. Some used tools are offered for sale because the owner is not happy with their performance.

There is no substitute for quality! A high-quality tool, though a bit more expensive, will last a lifetime. Poor-quality tools don't last long and often do a poor job even when brand new. You don't need to buy machinist-grade tools, but stay away from cheap tools; they are not the bargains they appear to be.

There are a number of sources that can be tapped for obtaining tools for your new electronics workbench. One good source for harvesting used tools are ham radio flea markets or hamfests. These get-togethers are for amateur radio operators and are held in most every state and in many towns across America. Hamfests are usually flea market style and are usually held outdoors when the weather is good. Fewer hamfests take place in colder months, and they are usually indoors. When you go to any ham flea market, you see row after row of dealers selling electronic components, tools, radio equipment, and so on. You can find out when hamfests take place by contacting local radio operators, through your local electronics parts distributors, or by looking through a current copy of one of the radio magazines such as *QST* or *CQ*. Hamfests are an excellent source for finding both new and used tools at a fraction of the cost that you would have to pay at a retail store.

Another great source for tools is to comb your local newspaper for announced tools sales or closeout sales. Jobbers and tool/equipment sellers, like Harborfreight Tools, have tools sales that move from town to town and offer great prices on new tools.

Be sure to check you local newspaper for garage sales and local fleamarkets; these can be great sources for used tools in good condition. Fleamarkets are great since you can often negotiate the final price.

Useful Parts and Shop Materials

Small stocks of useful parts and materials are handy to have when constructing electronics equipment. Most of these are available from hardware or radio supply stores. A representative list is shown in Table 12-2. Small parts, such as machine screws, nuts, washers, and soldering lugs, can be economically purchased in large quantities (but it doesn't pay to buy more than a lifetime supply). For items you don't use often, many radio supply stores or hardware stores sell small quantities and assortments.

Table 12-2   *Useful parts and materials*

Medium-weight machine oil	Contact lubricant
Contact cleaner, liquid or spray	Enameled wire
Duco cement	Tinned bare wire, bus wire
Epoxy cement	Soldering lugs, panel bearings, rubber terminal lugs,
Electrical tape	insulated tubing
Double-sided tape	Heat-shrink tubing, assorted sizes
Velcro	Heat gun
Sandpaper, assorted grits	Matches
Emery cloth	Bakelite, Lucite
Steel wool, assorted	Sheet aluminum stock 16–18 gauge
Cleaning pad, Scotchbrite	Machine screws; round and flat head, 4-40, 6-32, 8-32
Cleaners and degreasers	

Shop Safety

All the fun of building a project will be gone if you get hurt. To make sure this doesn't happen, let us first review some safety rules. Read your equipment manuals! The manual tells all you need to know about the operation and safety features of the equipment you are using. Do not work when you are tired. You will be more likely to make a mistake forget an important safety rule. Never disable any safety feature of any tool. If you do, sooner or later someone will make the mistake the safety feature was designed to prevent. Never fool around in the shop. Practical jokes and horseplay are in bad taste at social events; in a shop they are downright dangerous. A work area is a dangerous pace at all times; even hand tools can hurt someone if they are misused. Keep your shop neat and organized. A messy shop is a dangerous shop. A knife left laying in a drawer can cut someone looking for another tool; a hammer left on top of a shelf can fall down at the worst possible moment; a sharp tool left on a chair can be a dangerous surprise for the weary constructor who sits down. Wear the proper safety equipment. Wear eye-protection goggles when working with chemicals or tools. Use earplugs or ear protectors when working near noise. If you are working with dangerous chemicals, wear the proper protective clothing. Make sure your shop is well ventilated. Paints, solvents, cleaners, or other chemicals can create dangerous fumes. If you feel dizzy, get into fresh air immediately and seek medical help if you do not recover quickly. Every workshop should contain a good first-aid kit. Keep an eye-wash kit near any dangerous chemicals or power tools that can create chips. If you become injured, apply first aid and then seek medical help if you are not sure that you are okay. Even a small burn or scratch on your eye can develop into a serious problem. Respect power tools. Power tools are not forgiving. A drill can go through your hand a lot easier than metal. A power saw can remove a finger with ease. Keep away from the business end of power tools. Tuck in your shirt, roll up your sleeves, and remove your tie before using any power tool. If you have long hair, tie it back so it can't become entangled in power equipment. Don't work alone. Have someone nearby who can help if you get into trouble when working with dangerous equipment, chemicals, or voltages. Think! Pay attention to what you are doing. No list of safety rules can cover all possibilities. Safety is always *your* responsibility. You must think about what you are doing, how it relates to the tools you are using and the specific situation at hand.

First-Aid Kit

The first-aid kit is a necessity around any type of metal, wood, or electrical shop area. A first-aid kit is not a glamorous new tool or gadget but it should not be overlooked. Much like a power surge/spike protector or UPS for you computer, it is not as flashy a new piece of hardware or a new software game but it is an item you shouldn't be with out. Not only should you have a first-aid kit around your home, but you should have one close to your workshop or workbench area. If you do not have a first-aid kit by your work bench right now, go out and get one immediately. Your first-aid kit should contain different-size bandages, antiseptic creme, gauze patches, tape, absorbent cotton, triangular bandages, compress, and so on. Place the first-aid kit where you as well as other can readily see it. It should be displayed and marked as such at eye level for all to see. You know where you placed the first-aid kit, but suppose you hurt yourself badly, cried for help, and then passed out. A person coming to your rescue might not know the location of the first-aid kit.

Soldering

Everyone working in electronics needs to know how to solder well. Before you begin working on a circuit, carefully read this chapter on soldering. You will learn how to make good solder joints when soldering point-to-point wiring connections as well as PC board connections.

In all electronics work, the wiring connections must be absolutely secure. A loose connection in a radio results in noise, scratching sounds, or no sound at all. In a TV, poor connections can disrupt the sound or picture. The safe operation of airplanes and the lives of astronauts in flight depend on secure electronics connections. Soldering is not just gluing metals together. Done correctly, it unites the metals so that electrically they act as one piece of metal.

Solder

The best solder for electronics work is 60/40 rosin-core solder. It is made of 60% tin and 40% lead. This mixture melts at a lower temperature than either lead or tin alone. It makes soldering easy and provides good connections. The rosin keeps the joint clean as it is being soldered. The heat of the iron often causes a tarnish or oxide to form on the joint surface. The rosin dissolves this tarnish to make the solder cling tightly. Solders have different melting points, depending on the ratio of tin to lead. Tin melts at 450°F and lead at 621°F. Solder made from 63% tin and 37% lead melts at 361°F, the lowest melting point for a tin and lead mixture. Called 63-37 (or eutectic), this type of solder also provides the most rapid solid-to- liquid transition and the best stress resistance Solders made with different lead/tin ratios have a plastic state at some temperatures. If the solder is deformed while it is in the plastic state, the deformation remains when the solder solidifies. Any stress or motion applied to "plastic solder" causes a poor solder joint.

The 60-40 solder has the best wetting qualities. Wetting is the ability to spread rapidly and bond materials uniformly. 60-40 solder also has a low melting point. These factors make it the most commonly used solder in electronics.

Some connections that carry high current can't be made with ordinary tin–lead solder because the heat generated by the current would melt the solder. Automotive starter brushes and transmitter tank circuits are two examples. Silver-bearing solders have higher melting points, and so prevent this problem. High-temperature silver alloys become liquid in the 1100°F to 1200°F range, and a silver–manganese (85-15) alloy requires almost 1800°F. Because silver dissolves easily in tin, tin-bearing solders can leach silver plating from components. This problem can be greatly reduced by partially saturating the tin in the solder with silver or by eliminating the tin. Tin–silver or tin–lead–silver alloys become liquid at temperatures from 430°F for 96.5-3.5 (tin–silver) to 588°F for 1.0-97.5-1.5 (tin-lead-silver). A 15.0-80.0-5.0 alloy of lead–indium–silver melts at 314°F.

Never use acid-core solder for electrical work. It should be used only for plumbing or chassis work. For circuit construction, only use fluxes or solder–flux combinations that are labeled for electronic soldering. The resin or the acid is a flux. Flux removes oxide by suspending it in solution and floating it to the top. Flux is not a cleaning agent! Always clean the work before soldering. Flux is not part of a soldered connection; it merely aids the soldering process. After soldering, remove any remaining flux. Resin flux can be removed with isopropyl or denatured alcohol. A cotton swab is a good tool for applying the alcohol and scrubbing the excess flux away. Commercial flux-removal sprays are available from most electronic-part distributors.

The Soldering Iron

Soldering is used in nearly every phase of electronics construction so you will need soldering tools. A soldering tool must be hot enough to do the job and lightweight enough for agility and comfort. A temperature-controlled iron works well, although its cost is not justified for occasional projects. Get an iron with a small conical or chisel tip. Soldering is not like gluing; solder does more than bind metal together and provide an electrically conductive path between them. Soldered metals and the solder combine to form an alloy.

You may need an assortment of soldering irons to do a wide variety of soldering tasks. They range in size from 25 watt irons for delicate printed-circuit work to larger 100 to 300 watt sizes used to solder large surfaces. If you could only afford a single soldering tool when initially setting up your electronics workbench, an inexpensive to moderately priced pencil-type soldering iron with between 25 and 40 watt capacity is the best for PC board electronics work. A 100 watt soldering gun is overkill for printed-circuit work, since it often gets too hot, cooking solder into a brittle mess or damaging small parts of a circuit. Soldering guns are best used for point-to-point soldering jobs, large-mass soldering joints, or large components. Small "pencil" butane torches are also available, with replaceable soldering-iron tips. A small butane torch is available from the Solder-It Company. Butane soldering irons are ideal for field-service problems and will allow you to solder where there is no 110 volt power source. This company also sells a soldering kit that contains paste solders (in syringes) for electronics, pot metal, and plumbing. See the Address List in the References section for the address information.

Keep soldering tools in good condition by keeping the tips well tinned with solder. Do not run them at full temperature for long periods when not in use. After use, remove the tip and clean off any scale that may have accumulated. Clean an oxidized tip by dipping the hot tip in sal ammoniac (ammonium chloride) and then wiping it clean with a rag. Sal ammoniac is somewhat corrosive, so if you don't wipe the tip thoroughly, it can contaminate electronic solder joints. If a copper tip becomes pitted, file it smooth and bright and then tin it immediately with solder. Modern soldering iron tips are nickel or iron clad and should not be filed.

The secret of good soldering is to use the right amount of heat. Many people who have not soldered before use too little heat, dabbing at the joint to be soldered and making little solder blobs that cause unintended short circuits. Always use caution when soldering. A hot soldering iron can burn your hand badly or ruin a tabletop. It's a good idea to buy or make a soldering iron holder.

Soldering Station

Often, when building or repairing a circuit, your soldering iron is kept switched on for unnecessarily long periods, consuming energy and allowing the tip to burn and develop a buildup of oxide. Using this temperature-controlled soldering station, you will avoid destroying sensitive components when soldering. Buying a lower wattage iron may solve some of the problems, but new problems arise when you want to solder some heavy-duty component, setting the stage for creating a "cold" connection. If you've ever tried to troubleshoot some instrument in which a cold solder joint was at the root of the problem, you know how difficult such defects are to locate. Therefore, the best way to satisfy all your needs is to build or buy a temperature-controlled station for your electronics workbench.

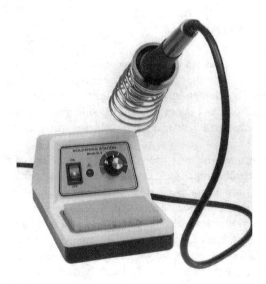

Figure 13-1 Soldering station.

A soldering station usually consists of a temperature-controlled soldering iron with an adjustable heat or temperature control and a soldering iron holder and cleaning pad. If you are serious about your electronics hobby or if you have been involved with electronics building and repair for any length of time you will eventually want to build or invest in a soldering station at some point in time. There are low-cost soldering stations for hobbyists available for under $30, but it makes more sense to purchase a moderately priced soldering station such as the quality Weller series. A typical soldering station is shown in Figure 13-1.

Soldering Gun

An electronics workbench would not be complete without a soldering gun. Soldering guns are useful for soldering large components to terminal strips, splicing wires together, or putting connectors on coaxial cable. There are many instances in which more heat is needed than a soldering iron can supply. For example, a large connector mass cannot be heated with a small soldering iron, so you would never be able to "tin" a connector with it. A soldering gun is a heavy-duty soldering device that does in fact look like a gun. Numerous tips are available for soldering guns and they are easily replaceable using two small nuts on the side arm. Soldering guns are available in two main heat ranges. Most have a two-step "trigger " switch that enable you to select two heat ranges for different soldering jobs. The most common soldering guns provide both a 100 watt setting when the trigger is pressed to its first setting. As the trigger is advanced to the next step, the soldering gun will provide 150 watts. A larger or heavy-duty, 200 to 250 watt soldering gun is also available but is a little harder to locate. The first trigger switch position provides 200 watts and the second provides 250 watts. When splicing wires together, either using the "Western Union" or parallel splice or the end splice, a soldering gun should be used, especially if the wire gauge is below size 22. Otherwise, the solder may not melt properly and result in a "cold" solder joint and, therefore, a poor or noisy splice. Soldering wires to binding post connections should be performed with a soldering gun to ensure proper heating. Most larger connectors should be soldered or pretinned using a soldering gun for even solder flow.

Preparing and Using the Soldering Iron

If your iron is new, read the instructions about preparing it for use. If there are no instructions, follow this procedure. The iron should be hot enough to melt solder applied to its tip quickly (half a second when dry, instantly when wet with solder). Apply a little solder directly to the tip so that the surface is shiny. This process is called "tinning" the tool. The solder coating helps conduct heat from the tip to the joint face, the tip in contact with one side of the joint.

When soldering, you can place the tip on the underside of the joint, do so. With the tool below the joint, convection helps transfer heat to the joint. Place the solder against the joint directly opposite the soldering tool. It should melt within a second for normal PC connections and within two seconds for most other connections. If it takes longer to melt, there is not enough heat for the job at hand. Keep the tool against the joint until the solder flows freely throughout the joint. When it flows freely, solder tends to form concave shapes between the conductors. With insufficient heat, solder does not flow freely and forms convex blobs. Once solder shape changes from convex to concave, remove the tool from the joint. Let the joint cool without movement at room temperature. It usually takes no more than a few seconds. If the joint is moved before it is cool, it may take on a dull, satin look that is characteristic of a cold solder joint. Reheat cold joints until the solder flows freely and hold them still until cool. When the iron is set aside, or if it loses its shiny appearance, wipe away any dirt with a wet cloth or sponge. If it remains dull after cleaning, tin it again.

Overheating a transistor or diode while soldering can cause permanent damage. Use a small heat sink when you solder transistors, diodes, or components with plastic parts that can melt. Grip the component lead with a pair of pliers close to the unit so that the heat is conducted away You will need to be careful, since it is easy to damage delicate component leads. A small alligator clip also makes a good heat sink to dissipate heat from the component. Mechanical stress can damage components, too. Mount components so there is no appreciable mechanical strain on the leads.

Soldering the pins of coil forms or male cable plugs can be difficult. Use a suitable small twist drill to clean the inside of the pin and then tin it with resin-core solder. While it is still liquid, clear the surplus solder from each pin with a whipping motion or by blowing through the pin from the inside of the form or plug. Watch out for flying hot solder; you can get severe burns. Next, file the nickel tip, insert the wire, and solder it. After soldering, remove excess solder with a file, if necessary. When soldering the pins of plastic coil forms, hold the pin to be soldered with a pair of heavy pliers to form a heat sink. Do not allow the pin to overheat; it will loosen and become misaligned.

Preparing Work for Soldering

If you use old junk parts, be sure to completely clean all wires or surfaces before applying solder. Remove all enamel, dirt, scale, or oxidation by sanding or scraping the parts down to bare metal. Use fine sandpaper or emery paper to clean flat surfaces or wire. (Note: no amount of cleaning will allow you to solder to aluminum. When making a connection to a sheet of aluminum, you must connect the wire by a solder lug or a screw.)

When preparing wires, remove the insulation with wire strippers or a pocket knife. If using a knife, do not cut straight into the insulation; you might nick the wire and weaken it. Instead, hold the knife as if you were sharpening a pencil, taking care not to nick the wire as you remove the insulation. For enameled wire, use the back of the knife blade to scrape the wire until it is clean and bright. Next, tin the clean end of the wire. Now, hold the heated soldering-iron tip against the under surface of the wire and place the end of the rosin-core solder against the upper surface. As the solder melts, it flows on the clean end of the wire. Hold the hot tip of the soldering iron against the under surface of the tinned wire and remove the excess solder by letting it flow down on the tip. When properly tinned, the exposed surface of the wire should be covered with a thin, even coating of solder.

How to Solder

The two key factors in quality soldering are time and temperature. Generally, rapid heating is desired, although most unsuccessful solder jobs fail because insufficient heat has been applied. Be careful; if heat is applied too long, the components or PC board can be damaged, the flux may be used up, and surface oxidation can become a problem. The soldering-iron tips should be hot enough to readily melt the solder without burning, charring, or discoloring components, PC boards, or wires. Usually, a tip temperature about 100°F above the solder melting point is about right for mounting components on PC boards. Also, use solder that is sized appropriately for the job. As the cross section of the solder decreases, so does the amount of heat required to melt it. Diameters from 0.025 to 0.040 inches are good for nearly all circuit wiring.

A well-soldered joint depends on:

1. Soldering with a clean, well-tinned tip
2. Cleaning the wires or parts to be soldered
3. Making a good mechanical joint before soldering
4. Allowing the joint to get hot enough before applying solder
5. Allowing the solder to set before handling or moving soldered parts

Making a Good Mechanical Joint

Unless you are creating a temporary joint, the next step is to make a good mechanical connection between the parts to be soldered. For instance, wrap the wire carefully and tightly around a soldering terminal or soldering lug, as shown in Figure 13-2(a). Bend the wire and make connections with long-nosed pliers. When connecting two wires together, make a tight splice before soldering. Once you have made a good mechanical contact, you are ready for the actual soldering.

The next step is to apply the soldering iron to the connection as shown in Figure 13-2(a). When soldering a wire splice, hold the iron below the splice and apply solder to the top of the splice. If the tip of the iron has a bit of melted solder on the side held against the splice, heat is transferred more readily to the splice and the soldering is done more easily. Do not try to solder by applying solder to the joint and then pressing down on it with the iron. Be sure not to disturb the soldered joint until the solder has set. It may

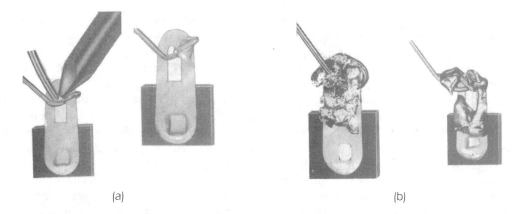

(a)   (b)

Figure 13-2   *Soldering to a terminal. (a) Good connection. (b) Poor connection.*

take a few seconds for the solder to set, depending upon the amount of solder used in making the joint. Now, take a good look at the joint. It should have a shiny, smooth appearance—not pitted or grainy. If it does have a pitted, granular appearance as seen in Figure 13-2(b), reheat the joint, scrape off the solder, and clean the connection. Then start over again. After the solder is well set, pull on the wire to see if the connection is good and tight. If you find that you did a poor soldering job, don't get upset, be thankful that you found it and do it over again. A quick-reference soldering checklist is found in Table 13-1.

Soldering Printed Circuit Boards

Most electronic devices use one or more printed circuit (PC) boards. A PC board is a thin sheet of fiberglass or phenolic resin that has a pattern of foil conductors "printed" on it. You insert component leads

Table 13-1   *Soldering checklist*

1. Prepare the joint. Clean all surfaces and conductors thoroughly with fine steel wool. First clean the circuit traces, then clean the component leads.
2. Prepare the soldering iron or gun. The soldering device should be hot enough to melt solder applied to the tip. Apply a small amount of solder directly to the tip, so that the surface is shiny.
3. Place the tip in contact with one side of the joint. If possible, place the tip below the joint.
4. Place the solder against the joint directly opposite the soldering tool. The solder should melt within two seconds. If it takes longer use a larger iron.
5. Keep the soldering tool against the joint until the solder flows freely throughout the joint. When it flows freely, the joint should form a concave shape; insufficient heat will form a convex shape.
6. Let the joint cool without any movement. The joint should cool and set up within a few seconds. If the joint is moved before it cools, it will look dull instead of shiny, and you will likely have a cold solder joint. Reheat the joint and begin anew.
7. Once the iron is set aside, or if it loses its shiny appearance, wipe away any dirt with a wet cloth or wet sponge. When the iron is clean, the tip should look clean and shiny. After cleaning the tip, apply some solder.

into holes in the board and solder them to the foil. This method of assembly is widely used and you will probably encounter it if you choose to build from a kit. Printed circuit boards make assembly easy. First, insert component leads through the correct holes in the circuit board. Mount parts tightly against the circuit board unless otherwise directed. After inserting a lead into the board, bend it slightly outward to hold the part in place.

Touch the tip of your soldering iron to both the component lead and the foil until they are hot enough to melt solder; see Figure 13-3(a). Apply a small amount of solder to both the tip and the connection as shown. Remove the iron and let the solder harden before moving the wire or the board. The finished connection should be smooth and bright. Reheat any cloudy or grainy-looking connections. Finally, clip off the excess wire length, as shown in Figure 13-3(b).

Occasionally, a solder "bridge" will form between two adjacent foil conductors. You must remove these bridges; otherwise, a short circuit will exist between the two conductors. Remove a solder bridge by heating the bridge and quickly wiping away the melted solder with a soft cloth. Often, you will find a hole on the board plugged by solder from a previous connection. Clear the hole by heating the solder while pushing a component lead through the hole from the other side of the board.

How to Unsolder

In order to remove components, you need to learn the art of unsoldering. You might accidentally make a wrong connection or have to move a component that you put in an incorrect location. Take great care while unsoldering to avoid breaking or destroying good parts. The leads on components such as resistors or transistors and the lugs on other parts may sometimes break off when you are unsoldering a good, tight joint. To avoid heat damage, you must use as much care in unsoldering delicate parts as you do in soldering them. There are three basic ways of un-soldering. The first method is to heat the joint and "flick" the wet solder off. The second method is to use a metal wick or braid is available at most electronics parts stores to remove the melted solder. Place the braid against the joint that you want to unsolder. Use the

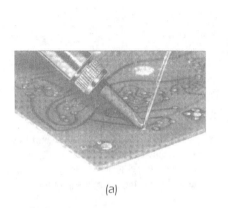

(a)

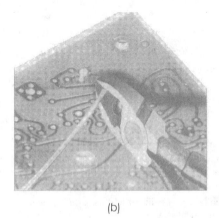

(b)

Figure 13-3   Soldering a PC board. (a) Applying solder to the iron tip and the connection. (b) Clipping off excess wire.

heated soldering iron to gently press the braid against the joint. As the solder melts, it is pulled into the braid. By repeating this process, you can remove virtually all the solder from the joint. Then reheat the joint and lift off the component leads. Another useful tool is an air-suction solder remover. Most electronics parts stores have these devices. Before using a desolder squeeze bulb, use your soldering iron to heat the joint you want to unsolder until the solder melts. Then squeeze the bulb to create a vacuum inside. Touch the tip of the bulb against the melted solder. Release the bulb to suck up the molten solder. Repeat the process until you have removed most of the solder from the joint. Then reheat the joint and gently pry off the wires. This third method is easy and is the preferred method, since it is fast and clean. You can also use a vacuum device to suck up molten solder. There are many new styles of solder vacuum devices on the market that are much better than the older squeeze bulb types. The new vacuum desoldering tools are about 8 to 12 inches long with a hollow Teflon tip. You draw the vacuum with a push handle and set it. As you reheat the solder around the component to be removed, you push a button on the device to suck the solder into the chamber of the desoldering tool.

Desoldering Station

A desoldering station is a very useful addition to your electronics workshop or workbench but in most cases they cost too much for most hobbyists. Desoldering stations are often used in production environments or as rework stations, when production changes warrant changes to many circuit boards in production. Some repair shops use desoldering stations to quickly and efficiently remove components. A professional desoldering station is shown in Figure 13-4.

Another useful desoldering tool is one made specifically for removing integrated circuits. The specially designed desoldering tip is made the same size as the integrated circuit, so that all IC pins can be desoldered at once. This tool is often combined with a vacuum suction device to remove the solder as all the IC

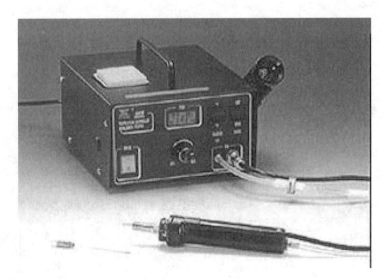

Figure 13-4   Desoldering station.

pins are heated. The IC desoldering tips are made in various sizes. There are 8 pin, 14 pin, and 16 pin versions that are used to uniformly desolder all IC pins quickly and evenly, so as not to destroy the circuit board. The specialized soldering tips are often used in conjunction with vacuum systems to remove the solder at the same time.

Remember these things when unsoldering:

1. Be sure that there is a little melted solder on the tip of your iron so that the joint will heat quickly.
2. Work quickly and carefully to avoid heat damage to parts. Use long-nosed pliers to hold the leads of components, just as you did while soldering.
3. When loosening a wire lead, be careful not to bend the lug or tie point to which it is attached.
4. Practice unsoldering connections on some old parts until you are satisfied that you can do them quickly and without breaking wires or lugs.

Caring for Your Soldering iron

To get the best service from your soldering iron, keep it clean and well tinned. Keep a damp cloth on the bench as you work. Before soldering a connection, wipe the tip of the iron across the cloth, then touch some fresh solder to the tip. The tip will eventually become worn or pitted. You can repair minor wear by filing the tip back into shape. Be sure to tin the tip immediately after filing it. If the tip is badly worn or pitted, replace it. Replacement tips can be found at most electronics parts stores. Remember that oxidation develops more rapidly when the iron is hot. Therefore, do not keep the iron heated for long periods unless you are using it. Do not try to cool an iron rapidly with ice or water. If you do, the heating element may be damaged and need to be replaced or water might get into the barrel and cause rust. Take care of your soldering iron and it will give you many years of useful service.

Remember, soldering equipment gets hot! Be careful. Treat a soldering burn as you would any other. Handling lead or breathing soldering fumes is also hazardous. Observe these precautions to protect yourself and others!

Ventilation

Properly ventilate the work area where you will be soldering. If you can smell fumes, you are breathing them. Often, when building a new circuit or repairing a "vintage" circuit, you may be soldering continuously for a few hours at a time. This can mean you will be breathing solder fumes for many hours and the fumes can cause you to get dizzy or light-headed. This is dangerous because you could fall down and possibly hurt yourself in the process. Many people who are highly allergic are also allergic to solder fumes. Solder fumes can also cause sensitive people to get sinus infections, so ventilating solder fumes is important. There a few different ways to handle this problem. One method is to purchase a small fan unit housed with a carbon filter that sucks the solder fumes into the carbon filter to eliminate them. This is the most simple method of reducing or eliminating solder fumes from the immediate area. If there is a window near your soldering area, be sure to open it to reduce exposure to solder fumes. Another method of reducing or eliminating solder fumes is to buy or build a solder-smoke removal system. You can purchase one of these systems but they tend to be quite expensive. You can create you own solder-smoke removal system by locating or purchasing an eight to ten foot piece of two-inch diameter flexible hose, similar to your vacuum cleaner hose. At the solder station end of the hose, you can affix the hose to a wooden stand in front of your work area. The other end of the hose is funneled into a small square "muffin" type fan placed near a window. Finally, be sure to wash your hands after soldering, especially before handling food, since solder contains lead; also try to minimize direct contact with flux and flux solvents.

Build Your Own Solder Station

Since the prices of commercially available units are rather high for the beginner or even the advanced hobbyist, the next-best solution is to build your own temperature controlled soldering station. Figure 13-5 is the schematic diagram of the basic temperature-controlled soldering station. The operation of the circuit is very simple. Once the temperature-controlled soldering station is connected to the ac line, capacitor C1 starts to charge through a variable resistor R1. The diac–triac combination that forms what is known as a quadrac contained in a single TO-220 package can be purchased from electronic parts supply houses. When the voltage across C1 reaches the break-over voltage of the diac (about 30 to 40 volts), the diac conducts, dumping C1's charge across the gate of the Triac and triggering it into conduction. The time constant for charging the capacitor is determined by the capacitor and R1 (a 200,000 ohm potentiometer used as a rheostat).

Once the Triac is turned on, it continues to conduct until the ac current applied to its two main terminals (MT1 and MT2) falls below the minimum holding current. When the polarity of the ac input reverses, the cycle starts again, but with reversed polarity across C1. It must be noted that the triac does not conduct until the amplitude of the gate voltage reaches the break-over point, even when R1 is at minimum resistance. The triac does not conduct unless the RC time constant is lower than the time required to

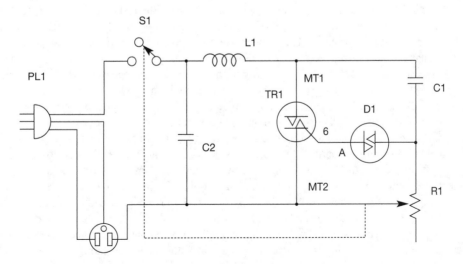

Figure 13-5   Temperature-controlled soldering station circuit.

change the amplitude of the mains below the break-over voltage of the diac. Note that the triac is either off or on, but not in an intermediate state, and therefore dissipates very little power.

Although power loss is negligible in either end position of R1, it is maximum in the middle position. In that position, the difference of the voltage being switched is maximum, which leads to maximum power dissipation across the triac and increased generation of RF interference. The circuit can be simplified by using only a triac, diac, C1, and R1. Capacitor C2 and the inductor (L1) decrease the RF interference caused by the switching action of the triac.

Generally, the triac has to have about twice the current handling capability of the highest-wattage iron that will be connected to the project. Note that there are two kinds of triac's: two-mode and four-mode. The four-mode device can have its main terminals in any order with respect to the gate, whereas two-mode units won't operate with their main terminals reversed. Be sure to check the data sheet before soldering into position.

Usually, triacs have their gates referenced to the main terminal 1 (MT1) and all conduct when the voltage between MT1 and the gate reaches about 0.9 to 1.4 volts It is important to know the lead configuration, so check the data sheet. If no data sheet is available, the gate can be distinguished from the main terminals with the aid of an ohmmeter.

Switch the ohmmeter (VOM or DMM) to the lowest resistance range to promote sufficient current flow (which must be greater than the holding current) through the device under test. Measure the resistance between all leads of the triac. If the resistance between any two leads is low (regardless of polarity), you have located the gate and MT1, making the final lead (assuming a good unit) MT2. Check the resistance between the first two leads (the gate and MT1) and MT2. The resistance between MT2 and either MT1 or the gate should be at or near infinity (without respect to polarity). Now, with one test lead connected to MT2, connect the other test lead to one of the leads (which we now know have to be MT1 and the gate) and touch MT2 to the last lead. If the meter reading drops to some value substantially lower value than that obtained during your preliminary measurements, the lead shorted to MT2 is the gate. If not, with MT2 still connected, switch the connection of the other two leads, and again short MT2 to the remaining lead.

Now, while holding the test lead in contact with MT2, release the test lead from the gate. If the resistance does not stay low in any one of the test situations, you probably do not have sufficient current flowing in the device to keep it turned on. This may be true for light current devices. In that case, try to use another VOM that provides a higher current for resistance measurements. Also, it helps to have a fresh battery installed in the VOM. Change the polarity of the test leads and repeat the whole procedure in order to assure yourself that you are dealing with the triac and not with an SCR, which behaves just like the triac, except only when its cathode and anode are connected to negative and positive terminals (respectively) and its gate being triggered with positive voltage only (referenced to the cathode).

Assembling the temperature-controlled solder station is quite simple and straightforward. The prototype was built on a small PC board and mounted inside a small plastic box. If a perfboard is used, mount all components to the board. Then, using insulated wire, connect the PC board leads to the case-mounted components. For safety reasons, it should be remembered that the "hot" side of the line cord must be connected to the unit through switch S1. A three-prong line cord must be used, with the third wire connecting the ground of the outlet with the ground of SO1 (to which the soldering iron is connected).

It is advisable to use a plastic or some other nonmetallic box to house the project, thereby avoiding any possible shock hazard. Since the circuit described here does not use any isolation transformer, all components must be assumed to be "hot, " conducting 117 volts ac. Therefore, it is advisable not to operate the unit without its case.

Once the unit is assembled, the testing procedure is to test for any short circuits with an ohmmeter connected between line and neutral cables of the line cord. Calibrating the unit is a straightforward task. Simply connect a load to the output and measure the voltage in the "through" (100% V) position. Then flip S1 to the "on" position and calibrate the R1 setting according to the voltage readings across the output terminals. A parts list for the unit follows.

R1	200 kohm ½ watt potentiometer (panel mount)
R2	22 kohm ¼ watt resistor
C1, C2	0.1 μF, 200 volt ceramic disk capacitor
L1	100 μH coil
TR1	2N6073 triac
D1	1N5760 diac
S1	SPST switch
S2	SPDT switch
NL	NE2H neon lamp
PL1	115 volt ac power plug and cable
SO1	115 volt, three-prong, grounded ac power outlet
Miscellaneous	PC board, metal enclosure wire, mounting lugs, screws, etc.

Circuit
Fabrication

Electronic Circuits

Most construction projects started by electronics enthusiasts involve some kind of electronic circuitry. The circuit is the "heart" most electronics equipment. In order for you to build a reliable electronic circuit, the circuit design has to work! Before fabricating an electronic circuit board, you will want to make sure that the circuit design is without error. You cannot always assume that a "cookbook" circuit design that you saw in an electronics magazine or applications note is flawless. Often, these are design examples that have not always been thoroughly debugged. Many home-construction projects are "one-time" projects created by an author who has designed and built the original prototype and got the circuit to work. In some cases, component tolerances or minor layout changes might make it difficult to get a second unit to work. It is advisable to build a "test" circuit before actually committing a design to a circuit board. A prudent approach to circuit design is to construct the intended circuit on a protoboard circuit layout device, where you push the components into the holes and run jumpers between ICs and components. These protoboards allow you to build and "test" a circuit before committing it to copper. In this way, you can ensure that the circuit design is sound before taking time to build it. It is no fun building a circuit board that doesn't work because the design was not finalized.

Circuit boards are often the preferred method of creating electronic circuits, but there are other methods. Building a circuit using breadboard, ground-plane, or wire-wrap techniques are other alternatives to using a circuit board and they do have some advantages, such as fluid design, which can be changed as the circuit evolves or if design criteria change, or if mistakes in the design become apparent later. Alternative circuit-building techniques are useful for one-of-a-kind designs that will not be duplicated. If you are designing a circuit that will be distributed to others or if the design is meant for production of hundreds or thousands of circuits, then circuit boards are the way to go. We will discuss these alternative techniques later in the chapter.

Electrostatic Discharge Protection

Many components, especially high-impedance components such as FETs, MOSFETs, and CMOS gates, can be readily damaged by electrostatic discharge (ESD). Protecting these components from static charges is an important concern when handling these sensitive components. Most people are familiar with the static charge that builds up when walking across a carpet and then is discharged by touching metallic objects. On a dry day, walking across a carpet can generate up to 30 or 40 kilovolts! A person sitting at an electronics workbench can generate voltages up to 6 kV, depending on conditions, especially if relative humidity is less than 20%.

Sensitive electronic components can be damaged with as little as 30 V, and damage is not always catastrophic. A MOSFET device can often become noisy, or lose gain properties; an integrated circuit can suffer damage that causes early failure. To prevent damage to integrated circuits, you need to take some precautions while handling these components.

The energy from a spark can travel inside a piece of equipment to affect internal components. Protection of sensitive electronic components involves the prevention of static build-up together with the removal of any existing charges by dissipating any energy that does build up. Several techniques can be used to minimize static build-up. The first step in reducing the risk of ESD damage is to remove any carpets in your work area, or treat them with antistatic spray, which is available from Chemtronics, GC

Thorsen, and other lines carried by electronics wholesalers. You can also replace carpets in your work area with special antistatic carpets, but this is expensive.

Even the choice of clothing you wear can affect the amount of ESD. Polyester has a much greater ESD potential than cotton. Many builders who have their bench work table on a concrete floor use a rubber mat to minimize the risk of electric shocks from the ac line. Unfortunately, the rubber mat increases the risk of ESD. An antistatic rubber mat can serve both purposes.

Many components are shipped in antistatic packaging. Be sure to leave components in their conductive packaging until you are ready to solder or place them into IC sockets. Other components, such as MOSFETs, are shipped with a small metal ring that temporarily shorts all of the leads together. Leave this ring in place until the device is fully installed in the circuit. These precautions help reduce the build-up of electrostatic charges. Other techniques offer a slow discharge path for the charges or keep the components and the person handling them at the same ground potential.

One of the best techniques is to connect the circuit builder and the devices being handled to an earth ground or a common reference point. It is not a good idea to directly ground yourself while working on electronic equipment, since there is a risk of shock. However, if you are grounded through a high-value resistor, ESD protection is still offered but there is no risk of shock. You can "ground" yourself through a conductive wrist strap. The 3M company makes a low-cost grounding wrist band. This wrist band is equipped with a snap-on ground lead, with a 1 megohm ohm resistor built into the snap handle. The conductive wrist strap can be used in conjunction with a resistive bench mat. The mat is an insulator that has been impregnated with a resistance material. Suitable mats and wrist straps are made by 3M, GC Electronic, and others; they are available from most electronics supply houses.

The work area should also be grounded via an electrical antistatic mat. The mat can also be connected to the conductive wrist band for more protection. Use a soldering iron with a grounded tip to solder sensitive components. Most irons that have three-wire power cords are properly grounded. When soldering static-sensitive devices, use two or three jumpers to ground you, the work, and the iron. If the iron does not have a ground wire in the power cord clip, run a jumper from the metal part of the iron near the handle to the metal box that houses the temperature control. Another jumper connects the box to the work. Finally, a jumper goes from the box to an elastic wrist band to the conductive mat for optimum grounding.

Use anti-static bags to transport susceptible components or equipment. Keep your bench free of objects such as paper, plastic, and other static-generating materials. Use conductive containers with a dissipative surface coating for equipment storage. All of the anti-static products described above are available from Newark Electronics and other suppliers. See the Address List in the References.

Electronics Construction Techniques

There are a number of different techniques that can be used to construct electronic circuits. In this chapter, we will look at a number of alternative types of circuit construction methods. Protoboard circuit construction is used for initial prototyping or early circuit design to work out the circuit "bugs." Point-to-point wiring techniques were often used in early tube-type equipment where terminal lugs were used to connect components and tube sockets. This technique is still used where multiple circuit construction techniques are combined. Wire-wrap techniques were used extensively in the 1980s and 1990s for connecting digital integrated circuit packages together. This technique is also used for prototyping projects in the early design stages as well as one-of-a-kind designs. Perf-board construction, uses general-purpose circuit boards with predrilled holes and preplaced copper pads. Perf-board construction is ideal for quick circuits or prototype circuits using preexisting copper layouts. Printed-circuit boards are used to construct complex electronic circuits. They are often used to increase reliability and reduce circuit noise, and when large numbers of boards are needed. Many electronic projects use a combination of techniques. The selection of techniques used depends on many different factors and builder preferences.

Point-to-Point Techniques

Point-to-point techniques include all circuit construction techniques that rely on tie points, terminal lugs, terminal posts, component leads, and point-to-point wiring. This is the technique used in most homemade construction projects. It is sometimes used in commercial construction, such as old vacuum-tube receivers and modern tube amplifiers. Point-to-point techniques are also used to connect the "off-board" components used in printed-circuit projects. It can be used to interconnect the various modules and printed-circuit boards used in more complex electronic systems. Most pieces of electronic equipment have at least some point-to-point wiring.

Solderless Protoboard Construction

Protoboard construction techniques work very well for audio and digital circuits. A solderless prototype board or protoboard is shown in Figure 14-1. Protoboards are usually not suitable for RF circuits but work well for most low-frequency prototype building. A protoboard consists of grooves with spring-loaded metal strips inside. Wires and components are inserted into the grooves making contact with the metal

Figure 14-1   Components on a protoboard.

strips. Components that are inserted into the same groove rows are connected together. Circuits are easy to construct using a protoboard and component/connection changes are quite easy to make at a moment's notice. Protoboards have some minor disadvantages. The boards are not good for building RF circuits; the metal strips add too much stray capacitance to the high-frequency circuit. Large and/or thick component leads can deform the metal strips and ruin the protoboard very quickly. In order to avoid damaging your protoboard, you must avoid trying to insert thick or oversized components leads, such as those on 1 watt resistors, into the grooves. If you plan on using components with large leads, then solder a thinner wire to the thick wire and insert the thin wire into the protoboard.

Perf-Board Construction

A simple approach to circuit building uses something called perf-board. Perf-board is available with many different hole patterns. Choose the one that best suits your needs. Perf-board is usually unclad, although it is made with pads that facilitate soldering. Circuit construction on perf-board is easy. Start by placing the components loosely on the board and moving them around until a satisfactory layout is obtained. Most of the construction techniques described in this chapter can be applied to perf-board. The audio amplifier shown in Figure 14-2 was constructed with the perf-board technique. Perf-board and accessories are widely available. Accessories include special mounting hardware and a variety of solder and solderless connection terminals. Perf-board construction is a quick way to construct a permanent circuit, without the headache of designing a circuit board. Perf-board construction is easy and fluid, which means you can make changes and additions quite easily.

Ground-Plane Construction

Ground-plane construction is a point-to-point construction technique that uses the leads of the components as tie points for electrical connections. It is also known as "dead-bug" or "ugly" construction. Dead-

Figure 14-2   Audio amplifier constructed on perf-board.

bug construction gets its name from the appearance of an IC with its leads sticking up in the air. In most cases, this technique uses copper-clad circuit board material as a foundation and a ground plane on which to build a circuit using point-to-point wiring, hence the name "ground-plane construction." An example is shown in Figure 14-3.

Ground-plane construction is quick and simple. You can construct a circuit on an unetched piece of copper-clad circuit board. Wherever a component connects to ground, you solder it to the copper board. Ungrounded connections between components are made point to point. Once you learn how to build with a ground-plane board, you can grab a piece of circuit board and start building any time you see an interesting circuit.

A PC board has defined size limits, and all the components must fit in the allotted space. Ground-plane construction is a bit more flexible since it allows you to use the components on hand. The circuit can be easily changed; this is a big help you are designing an experimental circuit. The greatest virtue of ground-plane construction is that the circuit can be constructed very quickly.

Ground-plane construction is much like model building, connecting parts using solder instead of glue. In ground-plane construction, you build the circuit directly from the schematic, so it can help you become familiar with a circuit and how it works. You can build subsections of a large circuit on small ground-plane modules and combine into a larger design.

Circuit connections are made directly, minimizing component lead length. Short lead lengths and a low-impedance ground conductor help prevent circuit instability. There is usually less intercomponent capacitive coupling than would be found between PC-board traces, so it is often better than PC-board construction for RF high-gain or sensitive circuits.

Circuit components are used to support other circuit components. You generally start by mounting components onto the ground plane and building from there. A two-handed technique is used to mount a component to the ground plane. Bend one of the component leads at a 90 degree angle, then trim off the excess. Solder a blob of solder to the board surface, perhaps about 0.1 inch in diameter. Using one hand, hold the component in place on top of the soldered spot and reheat the component and the solder. It should flow nicely, soldering the component securely. Next, remove the soldering iron tip and hold the component perfectly still until the solder cools.

Connections should be mechanically secure before soldering. Bend a small hook in the lead of a component, then crimp it to the next component. Do not rely only on the solder connections to provide mechanical strength; sooner or later, one of these connections will fail, resulting in a dead circuit.

Figure 14-3   An example of ground-plane construction.

In most instances, each circuit has enough grounded components to support all of the components in the circuit, but this may not always be possible. In some circuits, high-value resistors can be used as stand-off insulators. One resistor lead is soldered to the copper ground plane and the other lead is used as a circuit connection point. You can use ¼ or ½ watt resistors in values from 1 to 10 megohms. Such high-value resistors permit almost no current to flow, and in low-impedance circuits they act more like insulators than resistors. As a rule of thumb, resistors used as stand-off insulators should have a value that is at least 10 times the circuit impedance at that circuit point.

The Lazy Builder's PC Board

If you already have a PC-board design but don't want to copy the entire circuit or you don't want to make a double-sided PC board, then the easiest construction technique is to use a bare board or perf-board and hard-wire the traces. Drill the necessary holes in a piece of single-sided board, remove the copper ground plane from around the holes, and then wire up the back using component leads and bits of wire instead of etched traces, as shown in Figure 14-4.

To transfer an existing board layout, make a 1:1 photocopy and tape it to your piece of PC board. Prick through the centers of the holes with an automatic (one-handed) punch, aligning the two sheets. Holes for ground leads are optional and, generally, you get a better RF ground by bending the component lead flat to the board and soldering it down. Remove the copper around the rest of the holes by pressing a drill bit lightly against the hole and twisting it between your fingers. A drill press can also be used, but be careful not to remove too much board material. Lastly, wire up the circuit beneath the board.

Circuits that contain components originally designed for PC-board mounting are good candidates for this technique. Wired traces would also be suitable for circuits involving multipin integrated circuits designed for radio frequencies, double-balanced mixers, and similar components. To bypass the pins of these components to ground, connect a miniature ceramic capacitor on the bottom of the board directly from the bypassed pin to the ground plane. A wired-trace board is fairly sturdy even though many of the components are only held in by their bent leads and blob of solder. A drop of "super glue" can hold down

Figure 14-4 *The lazy builder's PC-board technique.*

some of the larger components, components with fragile leads, or any long leads or wires that might move. This technique is quite similar to the ground-plane technique.

"Ready-Made" or "Utility" PC Boards

"Ready-made" or "utility" PC boards are an alternative to custom-designed etched PC boards. They offer the flexibility of perf-board construction and the mechanical and electrical advantages of etched circuit connection pads. Utility PC boards can be used to build anything from simple passive filter circuits to computers. Circuits can be built on boards on which the copper cladding has been divided into connection pads. Power supply voltages can be distributed on bus strips.

The audio amplifier circuit shown in Figure 14-5 was constructed on a utility PC board. Component leads are inserted into the board and soldered to the etched pads. Wire jumpers connect the pads together to complete the circuit.

Utility boards with one or more etched plugs for use in computer-bus, interface, and general-purpose applications are widely available. Connectors, mounting hardware, and other accessories are also available.

Wire-Wrap Construction

Wire-wrap techniques can be used to quickly construct a circuit without solder. Low- and medium-speed digital circuits are often assembled on a wire-wrap board. The technique is not limited to digital circuits, however. Figure 14-6 shows an audio amplifier built using wire-wrap techniques. Circuit changes are easy to make, yet the method is suitable for permanent assemblies.

Wire wrapping is done by wrapping a wire around a small square post to make each connection. A wrapping tool resembles a thick pencil. Electric wire-wrap guns are convenient when many connections

Figure 14-5   Audio amplifier constructed on a utility PC board.

Figure 14-6   Audio amplifier built with the wire-wrap technique.

must be made. The wire is almost always #30 wire with thin insulation. The traditional or standard wire-wrap method and the incorrect wire-wrap method are shown in Figure 14-7. The wrapping post terminals are square for best results. Note that wire wrapping only works with sharp cornered posts, and the posts should be long enough for at least two connections. Be sure to mount small components on an IC header.

Surface-Mount Construction

Surface-mount construction techniques have been around for quite some time. These techniques are particularly suitable for PC-board construction, although they can be applied to many other construction

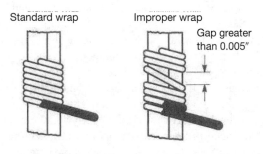

Figure 14-7

techniques. Surface-mounted components take up very little space on a board. Automated manufacturing techniques and surface-mount technology have evolved together; most modern ICs are being made specifically for this technique.

Surface-mount techniques are not limited only to "surface-mount" integrated circuit chips. This technique can be used to mount standard resistors and capacitors. Chip resistors and capacitors are common in UHF and microwave designs, since they have low stray inductance and capacitance, making them excellent components to use in the high-frequency range. Other components, such as transistors and diode arrays are also available in this space-saving format.

There are two different ground-plane construction techniques that work well for surface mounting components. One method is to cut out or etch small insulated islands in the PC board. There are now "island" cutter tools, which are special drills for this purpose. The technique works with either single or double-sided board. Cut out the islands with a small hobby knife, making parallel cuts spaced as needed. Peel away the copper with the point of a hot soldering iron. An alternative is to use a hand-held hobbyist's grinder with a cutting bit. You can also design a surface-mount PC-board pattern to streamline the design process.

In the second method of surface mounting, you cut small patches of single-sided board and glue them onto the copper ground plane. Although very effective, this technique is tedious for all but the simplest circuits. The surface-mount ICs used in industry are not easy for experimenters to use. They have tiny pins designed for precision PC boards. Most electronic enthusiasts avoid these in home-created designs. Sooner or later, you may need to replace one, though. If you do, don't try to get the old IC out in one piece! This will badly damage the IC and may also damage the PC board.

Although it requires a delicate touch and small tools, it is possible to change a surface-mount IC at your electronics bench. To remove the old one, use small, sharp wire cutters to cut the IC pins flush with the IC. This usually leaves just enough old pin to grab with a tiny pair of needle nose pliers, tweezers, or a hemostat. Heat the soldered connection with a small iron and use the pliers to gently pull the pin from the PC board.

To install a new part, apply a small blob of solder to one of the pads. Position the IC on the PC board and press it down while applying heat to the pin over the solder blob. With only that one pin soldered, inspect the position of each other pin relative to its pad. Reheat the first pin and reposition the part until all pins are in place. Then solder the remaining pins to their pads. Watch out for solder bridges; they are easy to make on traces with such small spacing. When you are done, inspect your work carefully.

Small "chip" surface-mount components are a bit trickier to deal with because they are too small to hold safely during soldering. Special care must be used when soldering chip components. All surfaces should be clean and lightly tinned before you solder the chip. Position the chip in place and hold it down with a toothpick. Do not use a screwdriver or tweezers, because the metal can easily damage the ceramic base of the component. Lay the chisel tip of a 15 to 20 watt soldering pencil on the surface to which the chip is to be soldered, with the tip of the chisel just touching the chip component. Be careful not to use too much heat in one spot for too long, in order to avoid component and circuit-board damage.

Touch and flow a minimum amount of solder between the chip and the soldering pencil tip and pull the tip away at a low angle to the surface. Repeat the procedure on the other side of the chip. If you don't use too much heat while soldering the second end, you may not have to hold it in place. Inspect the solder and chip with a magnifier. The solder should have flowed properly and there should be no fractures or cracks. Don't overheat the chip or the metallization may separate from the ceramic and ruin the component. This technique requires a little practice but can be mastered with some practice.

When are PC Boards the Best Choice?

PC boards are in all types of consumer electronics. They are also used in most kits and construction projects. A newcomer to electronics might think that there is some unwritten law against building equipment in any other way! The misconception that everything needs to be built on a printed-circuit board is often a stumbling block in project construction for first-time builders. In fact, a PC board is probably the worst choice for a one-time project. In actuality, a moderately complex project can be built in much less time using other techniques. The additional design, layout, and manufacturing usually entails much more work than it would take to build the project by hand. In most cases, if a ready-made board is not available, ground-plane construction is a lot less work than designing, debugging, and then making a PC board. In this section, we will explain how to turn a schematic into a working circuit. It is not as simple as laying out the PC board just like the circuit is drawn on the schematic.

So, why does everyone use PC boards? The most important reason for using PC boards is that they lend themselves to reproducibility. They allow many units to be introduced with exactly the same layout, reducing the time and work of conventional wiring and minimizing the possibilities of wiring errors. Often, friends or club members want to build the same circuit and the PC board makes the job much easier for all. If you can buy a ready-made PC board or kit for your project, it can save a lot of circuit design and circuit board construction time. If you want to build a project and a ready-made PC board is available, you can assemble the project quickly and expect it to work. This is true because someone else has done most of the real work involved in designing the PC board layout and fixing any bugs caused by intertrace capacitive coupling, ground loops, and similar problems.

Using a PC board usually makes project construction easier by minimizing the risk of wiring errors or other construction blunders. Inexperienced constructors usually feel more confident when construction has been simplified to the assembly of components onto a PC board. One of the best ways to get started with home construction is to start by assembling a few kits using PC boards.

Schematic to PC Board

Many people do not know how to turn a schematic into a working circuit; it is more of an art rather than of science. One thing is usually true: you cannot build the circuit the way it looks on the schematic. Many design and layout considerations that apply in the real world of practical electronics do not appear on a schematic. The skill of turning a schematic into a circuit board design comes with some practice and experience. A circuit diagram is a poor guide to a proper layout. Circuit diagrams are drawn to look attractive on paper. They follow drafting conventions that have very little to do with the way the circuit works. On a schematic, ground and supply voltage symbols are scattered all over the place. The first rule of RF layout is to not wire RF circuits as they are drawn. How a circuit works in practice depends on the layout. Poor layout can ruin the performance of even a well-designed circuit. Basically, there are two major methods of designing a circuit board. The original method of converting a circuit schematic to a PC board design is the manual method of transferring the design directly to circuit board stock for single quantity designs or a mylar sheet for multiple PC board designs. The second method of designing a circuit board it to use one of the many available circuit-board design software packages.

Rough Layout—Manual Method

Start by drawing a rough-scale pictorial diagram of the layout. Draw the interconnecting leads to represent the traces that are needed on the board. Rearrange the layout as necessary to find an arrangement that completes all of the circuit traces with a minimum number of jumper-wire connections. In some cases, however, it is not possible to complete a design without at least a few jumpers.

After you have completed a rough layout, redraw the physical layout on a grid. Graph paper works well for this. Most IC pins are on 0.1-inch centers. Use graph paper that has 10 lines per inch to draw artwork at 1:1 and estimate the distance halfway between lines for 0.05 inch spacing. Drafting templates are helpful in the layout stage. Local drafting supply stores should be able to supply them. The templates usually come in either full scale or twice normal size.

To lay out a double-sided board, ensure that the lines on both sides of the paper line up; you can hold the paper up to the light or use a light table to check this. When using graph paper for a PC board layout, include bolt holes, notches for wires, and other mechanical considerations. Fit the circuit into and around these, maintaining clearance between parts.

Most modern components have leads on 0.1 inch centers. The rows of dual-inline package (DIP) IC pins are spaced at 0.3 or 0.4 inch. Measure the spacing for other components. Transfer the dimensions to the graph paper. It is useful to draw a schematic symbol of the component on the layout. Most IC specification sheets show a top view of the pin locations. If you are designing the "foil" side of a PC board, be sure to invert the pin-out. This is the cause of the most the common errors found *after* a circuit board has been designed and is about to be built, by newcomers and sometimes by veterans alike. Be very careful and get it right the first time.

Draw the traces and pads the way they will look. Using dots and lines is confusing. It's okay to connect more than one lead per pad or run a lead through a pad, although using more than two creates a complicated layout. In that case, there may be problems with solder bridges that form short circuits. Traces can run under some components; it is possible to put two or three between 0.4-inch centers for a ¼ watt resistor.

Leave power supply and other dc paths for last. These can usually run just about anywhere, and jumper wires are fine for these noncritical paths. Do not use traces less than 0.010 inch (10 mil) wide. If 1 ounce board stock is used, a 10 mil circuit trace can safely carry up to 500 mA. To carry higher current, increase the width of the traces proportionally. A circuit trace should be 0.2 inch to carry 10 A. Allow 0.1 inch between traces for each kilovolt in the circuit.

When making a double-sided board, use pads on both sides of the board to connect traces through the board. PC boards for home construction generally do not use plated-through holes (a manufacturing technique that has copper and tin plating inside all of the holes to form electrical connections). Use a through-hole and solder the associated component to both sides of the board. Make other through-hole connections with a small piece of bus wire providing the connection through the board; solder it on both sides. This serves the same purpose as the plated-through holes found in commercially manufactured boards.

After you have planned the physical design of the board, decide on the best way to complete the design. For one or two simple boards, draw the design directly on the board, using a resist pen, paint, or rub-on resist materials. To transfer the design to the PC board, draw light, accurate pencil lines at 0.1 or 0.05 inch centers on the PC board. Draw both horizontal and vertical lines, forming a grid. You only need lines on one side. For single-sided boards, use this grid to transfer the layout directly to the board surface. To make drilling easier, use a center punch to punch the centers of holes accurately. Do this before applying the resist so the grid is visible. When drawing a pad with plenty of room around it, use a pad about 0.05 to 0. 1 inch in diameter. For ICs or other close quarters, make the pad as small as 0.03 inch or so. A "ring" that is too narrow invites soldering problems; the copper may delaminate because of the heat. Pads need not be round. It's okay to shave one or more edges if necessary, to allow a trace to pass nearby.

Draw the traces next. A drafting triangle can help. It should be spaced about 0.1 inch above the table, to avoid smudging the artwork. Use a large triangle with a rubber grommet taped to each corner (to hold it off the table). Select a sturdy triangle that doesn't bend easily.

Align the triangle with the grid lines by eye and make straight, even traces similar to the layout drawing. The triangle can help with angled lines, too. Practice on a few pieces of scrap circuit board. Make sure that the resist adheres well to the PC board. Most problems can be seen by eye; there can be weak areas or bar spots. If necessary, touch up problems with additional resist. If the board is not clean, the resist will not adhere properly. If necessary, remove the resist, clean the board, and start from the beginning.

Resist pens dry out quickly. Discard troublesome pens. Keep a few on hand and switch back and forth, putting the cap back on each for a bit to give the pen a chance to recover. Once all of the artwork on the board is drawn, check it against the original artwork. It is easy to leave out a trace and it is not easy to put copper back after a board is etched. In a pinch, replace the missing trace with a small wire. Applied resist takes about an hour to dry at room temperature. Fifteen minutes in a 200°F oven is also adequate.

There are several basic techniques that can be used to make PC boards. They usually start with a PC board "pattern" or artwork. All of the techniques have one thing in common: the pattern needs to be transferred to the copper surface of the PC board. Unwanted copper is then removed by chemical or mechanical means. Most variations in the PC-board manufacturing technique involve differences in resist or etchant materials or techniques.

PC Board Design Software

If you want your circuit or project to look its best and function without trouble, then you may want to consider designing your circuit board using PCB design software, which will greatly ease the burden of PC board design.

Using this type of inexpensive software to design a circuit board can be little more than a connect-the-dots game. Inexpensive PCB layout software does not have some of the nice features of more elaborate design packages, or, if it does have the features, you are limited on how complex your PCB can be.

Some prototype printed circuit board companies will actually give you rudimentary PCB layout software for free. Point your browser to www.expresspcb.com for an example or see www.mentala.com for some low-cost design packages. There are several companies that offer PCB layout programs for a few hundred dollars.

More expensive PCB layout software will allow you to enter your schematic, automatically place the components, automatically route (connect the dots), and then check your PCB layout against the schematic. These packages also have built-in design rule checkers with which you can check your design against the generic rules or enter your own design rules. For instance, your contract manufacturer may not want any components closer than 0.060″ to the edge of the PCB because the component may get impacted when the PCBs are separated. You can set up a design rule such that you get an error message when you violate this constraint. There are plenty of other design constraints that you may want to consider to make sure your PCB design is robust and can be manufactured in volume for commercial applications.

Printed-Circuit Boards

Many builders prefer the neatness and miniaturization made possible by the use of etched printed-circuit boards. Once designed, a PC board is easily duplicated, making them ideal for group projects. To make a PC board, resist material is applied to a copper-clad bare PC board, which is then immersed into an acid etching bath to remove selected areas of copper. In a finished board, the conductive copper is formed into a pattern of conductors or "traces" that form the actual wiring of the circuit.

PC Board Stock

PC board stock consists of a sheet made from insulating material, usually glass epoxy or phenolic, coated with conductive copper. Copper-clad stock is manufactured with phenolic, FR-4 fiberglass, and Teflon-based materials in thicknesses up to ⅛ inch. The copper thickness varies. It is usually plated from 1 to 2 ounces per square foot of bare stock.

Resists

A resist is a material that is applied to a PC board to prevent the acid etchant from eating away the copper on those areas of the board that are to be used as conductors. There are several different types of resist materials, both commercial and home use. When resist is applied to those areas of the board that are to remain as copper traces, it "resists" the acid action of the etchant. The PC board stock must be clean before any resist is applied. This is discussed later in the chapter. After you have applied resist, by whatever means, protect the board by handling it only at its edges. Do not let it get scraped. Etch the board as soon as possible, to minimize the likelihood of oxidation, moisture, or oils contaminating the resist or bare board. Resists are available in many different forms, including tape, resist pens, and rub-on transfers, as discussed below.

Tape. If you are making a single PC board, Scotch, adhesive, or masking tape, securely applied, makes a good resist. (Don't use drafting tape; its glue may be too weak to hold in the etching bath.) Apply the tape to the entire board, transfer the circuit pattern by means of carbon paper, then cut out and remove the sections of tape where the copper is to be etched away. An Xacto hobby knife is excellent for this purpose. This method is crude and tedious. In order to streamline the direct-tape method, use commercial tape and dots (see below).

Resist Pens. Several electronics suppliers sell resist pens. Use a resist pen to draw PC-board artwork directly onto a bare board. Commercially available resist pens work well. Several types of permanent markers also function as resist pens, especially the Sharpie brand. They come in fine-point and regular sizes; keep two of each on hand.

Paint. Some paints are good resists. Exterior enamel works well. Nail polish is also good, although it tends to dry quickly so you must work fast. Paint the pattern onto the copper surface of the board to be etched. Use an artist's brush to duplicate the PC board pattern on bare PC-board stock. Tape a piece of carbon paper to the PC-board stock. Tape the PC-board pattern to the carbon paper. Trace over the original layout with a ballpoint pen. The carbon paper transfers the outline of the pattern onto the bare board. Fill in the outline with the resist paint. After paint has been applied, allow it to dry thoroughly before etching.

Rub-on Transfer. Several companies, (Kepro Circuit Systems, DATAK Corp., GC Electronics) produce rub-on transfer material that can also be used as a resist. Patterns are made with various width traces and for most components, including ICs. As the name implies, the pad or trace is positioned on the bare board and rubbed to adhere to the board. This method is called the "tape and dots" method of circuit trace layout. It is a manual process but works very well and produces nice clean lines. There is a wide variety of tape widths and dot sizes as well as finger contacts pads and integrated circuit transfer for all types of IC packaging. This method tends to be expensive, since you may want to have a variety of tape widths and dot sizes in your inventory. This process produces uniform circuit traces that are reliable, stick to the board quite well, and will not come off the board when least expected. The "tape and dots" are usually laid out on a thin sheet of mylar and used as an artwork master for producing a number of boards.

Photographic Process

Many magazine articles feature printed circuit layouts. Some of these patterns are difficult to duplicate accurately by hand. A photographic process is the most efficient way to transfer a layout from a magazine page to a circuit board. A copper board coated with a light-sensitive chemical is at the heart of the photographic process. In a sense, this board becomes your photographic film.

Make a contact print of the desired pattern by transferring the printed-circuit artwork to special copy film. This film is attached to the copper side of the board and both are exposed to intense light. The areas of the board that are exposed to the light (those areas not shielded by the black portions of the artwork) undergo a chemical change. This creates a transparent image of the artwork on the copper surface.

Your Own Artwork

Artwork can be applied directly to the circuit board if you are only making a single board. If you anticipate making more than a single circuit board, then your artwork should be created on a sheet of mylar film, which is readily available through an art supply store. Once the artwork and design process has been completed, the next step is to transfer the design to the bare circuit board. Circuit-board stock is available in three different types: blank with no photosensitive layer; and with positive-acting or negative-acting photo-resist. You can purchase blank boards and then apply a spray-on photo-resist material layer yourself. The blank boards are less expensive than presensitized boards; however, you must do the extra work of applying your own resist and ensure that the applied layer is even and free of contamination. In the long run, it is much easier to purchase presensitized boards. Many circuit builders transfer their mylar artwork directly to a positive presensitized board, since this eliminates the extra photographic process of using a negative presensitized board. If you choose the negative presensitized boards, then your mylar artwork must be turned into a photographic negative before exposing the boards to light in the development process.

Iron-On Resist

Another popular circuit-board processing technique is the iron-on resist method. Meadowlake Corporation makes an iron-on resist. Their products make an artwork positive using a standard photocopier. A clothes iron transfers the printed resist pattern to the bare PC board.

Some experimenters have reported satisfactory results using standard photo copier paper or the output from a laser printer. Apparently, the toner makes a reasonable resist. Note that the artwork for this method must be reversed with respect to a normal etching pattern because the print must be placed with the toner against the copper.

To transfer the resist pattern onto the board, place the pattern on the board (image side toward the copper), then firmly press a hot iron onto the entire surface. Use plenty of heat and even pressure. This melts the resist, which then sticks to the bare PC board. This is not a perfect process; there will probably be bad areas on the resist. The amount of heat, the cleanliness of the bare board, and the skill of the operator may affect the outcome.

The key to making high-quality boards with the photocopy techniques is to be good at retouching the transferred resist. Fortunately, the problems are usually easy to retouch, if you have a bit of patience. A resist pen does a good job of reinforcing any spotty areas in large areas of copper.

Once the artwork is ready and your board type is chosen, then you can move on to exposing the artwork to the presensitized circuit board. Your artwork is placed over the light-sensitive side of the presensitized circuit board. Be careful when placing the artwork over the circuit board to ensure that the original art side is on the top side, facing you. Once the artwork and presensitized board are sandwiched together, you will need to expose the combination to ultraviolet light for the specified time period recommended by the manufacturer, which is usually 3 to 8 minutes. Once the exposure time has been completed, the circuit board must be placed in a developer solution in relative darkness for a few minutes. The circuit board is gently agitated back and forth in a Pyrex glass dish for the duration. Gently lift the board out of the developer, handling the boards by the ends. Run the circuit board under cool water to wash off the developer solution. The circuit board is now ready for the etching process, which removes the unwanted copper from the board.

Etchant

Etchant is an acid solution that is designed to remove the unwanted copper areas on PC-board stock, leaving the areas that will function as conductors. Almost any strong acid bath can serve as an etchant, but some acids are too strong to be safe for general use. Two different etchants are commonly used to fabricate prototype PC boards: ammonium persulphate and ferric chloride. The latter is the more common of the two.

Ferric chloride etchant is usually sold ready-mixed. It is made from one part ferric chloride crystals and two parts water by volume. No catalyst is required.

Etchant solutions become exhausted as they are used, so keep a supply on hand. Dispose of the used solution safely; follow the instructions of your local environmental protection authority. Most etchants work better if they are hot. Etching a PC board at room temperature takes about 45 minutes, but will only take 10 to 15 minutes if the etchant is hot. Use a heat lamp or rubber pad heater under a Pyrex cake dish to warm the etchant to the desired temperature. A darkroom thermometer is handy for monitoring the temperature of the bath. Do not heat your etchant above the recommended temperature, which is typically 160°F. If it gets too hot, it will probably damage the resist. Hot or boiling etchant is also a safety hazard.

Insert the board to be etched into the solution and agitate it continuously to keep fresh chemicals near the board surface This speeds up the etching process. Normally, the circuit board should be placed in the bath with the copper side facing up. After the etching process is completed, remove the board from the tray and wash it thoroughly with water. Use medium-grade steel wool to rub off the resist. Most etchants will react with a metal container, so use only glass, ceramic, or a nonreactive container to hold etching chemicals. Do not attempt to use an aluminum pie dish. Etchant is caustic and can burn eyes or skin easily. Use rubber gloves and wear old clothing or a lab smock when working with any chemicals. If you get some on your skin, wash it with soap and cold water. Wear safety goggles (the kind that fit snugly on your face) when working with any dangerous chemicals. Read the safety labels and follow them carefully. If you get etchant in your eyes, wash immediately with large amounts of cold water and seek immediate medical help. Even a small chemical burn on your eye can develop into a serious problem. Ferric chloride stains clothes permanently, so be forewarned.

Once the circuit board has been etched, which usually takes about 15 to 20 minutes, you will need to rinse the board clean. Finally, the photo resist that remains on the board must be removed. Take a small square of steel wool or Brillo and rub the circuit board until the board appears very shinny and clean. Now the circuit board is ready to be cut and drilled. No matter what technique you use to produce your circuit boards, you should determine the required size of the PC board, then cut the board to size before installing components. Trimming off excess PC-board material can be difficult after the components are installed.

Kepro and Meadowlake Corporation sell materials and supplies for all types of PC-board manufacturing, such as printed-circuit board kits, chemicals, tools and other materials. They are available through distributors such as Ocean State Electronics or Newark Electronics.

Double-Sided PC Boards

All of the examples used to describe the above techniques were for single-sided PC boards, with traces on one side of the board and either a bare board or a ground plane on the other side. PC boards can also have patterns etched onto both sides, or even have multiple layers. Most home-construction projects use single-sided boards, although some kits supply double-sided boards. Multilayer boards are rare in electronics hobbyist construction projects. Before attempting to create a double-sided PC board, consult with an experienced mentor.

Plating

Most commercial PC boards are tin plated, to make them easier to solder. Commercial tin-plating techniques require electroplating equipment not readily available to the home constructor. Immersion tin plating solutions can deposit a thin layer of tin onto a copper PC board. The technique is easy; put some of the solution into a plastic container and immerse the board in the solution for a few minutes. The chemical action of the tin-plating solution replaces some of the copper on the board with tin. The result looks nearly as good as a commercially made board. Agitate the board or solution from time to time. When the tinning is complete, take the board out of the solution and rinse it for five minutes under running water. If you do not remove all of the residue, solder may not adhere well to the surface. Kepro sells an immersion plating solution.

Complex Printed Circuit Boards

More complex, multilayer or commercial printed circuit boards are made up of several components. These components are composed of the laminate, conductive layers, lands, holes, solder mask, and silk screen. A PC board is often made up of different layers laminated together. The laminate material is the foundation of the PCB. These various materials such as FR2, FR4, or CEM-1 are made of such things as epoxy and fiberglass. For flexible printed circuit boards, polyimide film laminates are used. Laminates are electrical insulator materials. The choice of substrate material greatly affects the cost of an individual PCB.

The solder mask is the material that is placed over the areas to which you do not want solder to adhere. There are a variety of different types of solder masks and processes with which to apply them. Screening the solder mask is an inexpensive but inexact way to apply the mask material. Liquid photoimageable (LPI) material is a slightly more expensive but more precise way to apply the mask. Solder mask the stuff that is usually all over the surface of the PCB, except the component leads.

A silk screen is a print, in ink, of characters or graphics on the surface of the PCB. The manufacturing department may require a silk screen to show the locator and reference designators of the components for inspection and rework purposes. It is possible to print characters in copper, but it is often advantageous to use a nonconductive method of printing, especially in high-density designs.

Copper is what makes the electrical connections on the PCB, and strips of copper connecting components are called traces or tracks. A layer of the PCB that contains traces or a continuous sheet of copper (like a ground plane) is called a conductive layer. A PCB can have one or more conductive layers. If you hear someone say they have a four-layer board, it means they have a PCB with four conductive layers. An odd number of conductive layers is generally not used because of warping effects on the PCB.

There are holes in most PCBs for mechanical mounting and soldering of components and interconnecting the conductive layers. A "land" is the area around a hole to which a component is soldered. "Via holes" are what connect conductive layers together. Vias are conductively plated on their interior surface and traces on the appropriate layers are connected to the vias. Vias can be all the way through a PCB or they can be buried. Buried means that the via interconnects internal layers and cannot be seen from the exterior of the PCB. A higher density of conductors is possible with buried via holes since some layers do not have space taken up by the via.

Design Considerations

PCB layout is important in circuits that operate at very high frequencies (VHF) and beyond because it can reduce or increase the amount of stray inductance and capacitance that is present. This stray inductance and capacitance makes up a significant part of the total inductance and capacitance in the circuit at these frequencies. Wide and short PCB traces will help reduce the stray inductance. It is also good practice to use a ground plane (a layer of your PCB that is entirely at ground potential) instead of having the ground connection fragmented among the various layers.

Even if your circuit does not operate at VHF and beyond, it may be sensitive to interference from transmitters operating at those frequencies. A 100 MHz signal can disrupt the functioning of a circuit operating at a few Hz. If your circuit ever has to go through electromagnetic compatibility (EMC) testing for applications such as automotive or aerospace, a proper layout will be critical. EMC testing subjects even low-frequency devices to varying frequency ranges to make sure they function properly and are not damaged.

You will have to decide what type of component packages you will need to use. If they are standard component packages, you will be able to download them from a menu on your design software and pop them onto the PCB. If they are not standard packages, you will have to draw each different type of package. You will need to determine if you are going to use all through-hole components ("old-fashioned" components with wire leads), all surface-mount technology (SMT) components, or a mixture of the two types. There are other types of component packages such as ball grid array (BGA), which rest on a multitude of solder balls that get reflowed in an oven, and chip-on-board (COB) devices, which are bare semiconductor dies without a plastic package, but these are beyond the scope of easy prototyping. It is even difficult to build and troubleshoot prototypes with fairly large SMT devices. Your best bet for prototype building and troubleshooting is to use through-hole package components. However, keep in mind that your circuit might work differently if you change to SMT, BGA, or COB later on. This is especially true if it involves radio frequencies.

Board Production

If you are considering producing many circuit boards for a large club project or commercial production, then you will need to get your circuit design to a PCB production company, in the format that they require, in order to get your prototype produced. The format that PCB suppliers require is generally a Gerber file for copper artwork or an Excellon for the hole locations and sizes.

You must provide a Gerber file for each layer in your design: for each conductor layer, the silk screen layer, the solder mask layers, and so on. The PCB design software will usually give different file extensions for each layer's Gerber file.

An aperture list is a file that dictates the size of a specific entity on the artwork. For instance, the aperture list might call out a 0.060″ round shape that is used to form a component solder pad (land). RS274X format Gerber files do not have to have a separate aperture list file since they are built into the Gerber files for each layer. However, RS274D format files do require a separate aperture list that must be provided to the PCB manufacturer with the Gerber files for each layer. There should be one comprehensive aperture file that covers all the different-layer Gerber files. Most modern PCB layout software packages input RS274X format Gerber files.

There will be a couple of Excellon files. One will call out the software codes for the drill bit sizes for all the different hole sizes in your design. The other will call out the X-Y location of each hole and the corresponding drill bit code. The X-Y locations are referenced to an origin somewhere on your PCB (probably one of the comers). All the files can be combined in a zip file to make transfer to the PCB supplier easier. However, I have seen the zipping process corrupt some of the Gerber files. Therefore, you may want to zip, unzip, and view the files to make sure you get want you think you are getting. When you view the Gerber files, you are seeing exactly what you will get. Gerber files can be viewed in various software programs such as GoPrevue. GoPrevue can be downloaded free from www.graphicode.com.

Commerical Production Costs

Commerical printed circuit boards are usually constructed from FR4, CEM-I, or CEM-3 substrate material. The fiberglass materials such as FR4 typically cost more than the epoxy-type materials such as CEM-I. The materials vary greatly in temperature characteristics and other physical properties such as stiffness. The application of your design may dictate what type of substrate material is used. Not all materials will be available from all PCB suppliers.

The thicker the copper, the more expensive the PCBs. Copper thickness is specified in ounces per square foot. For instance, three-ounce copper is thicker than two-ounce copper and will cost more. Your choice of copper thickness may also affect other cost-impacting aspects of your board such as solder-mask type. The thickness of the copper also dictates how much current a given copper trace can handle. For example, a 0.020″ wide trace can handle more current if it is on a PCB using three-ounce copper than if on a PCB using two-once copper.

The more layers your design uses, the higher the cost. Every layer you add may add 30% to the cost of each PCB. The more complicated your circuit and the smaller your PCB length and width, the more layers you will need. A silk screen layer gets graphics such as a part number, reference designators, or notes printed on the surface of the PCB. If you specify a silk screen, the cost will be slightly higher. This is true for both prototypes and production PCBs.

If you want an electrical test before the PCBs is shipped, there will be a substantial charge. An electrical test verifies that all the connections are present. You can specify that this test not be done and save substantial money. However, you are taking a risk. If your design is simple enough, it may be worth the risk. Some PCB suppliers will do the test even if you do not pay for it.

If you want your PCBs in a hurry (within 24 hours), you will pay dearly. If, for example you want ten, 10 × 10″, FR4, single-layer, two-ounce copper PCBs in 24 hours, you may have to pay $2,000.00. If you can wait two weeks for the same PCBs, you may only pay $800.00. With this type of order, there may also be a tooling charge of $250.00. However, if you only need a few PCBs, there are cheaper options.

There are a few PCB suppliers that offer PCBs for $30.00 to $50.00 each with no tooling charges. Check out www.PCBexpress.com and www.4PCB.com for these prices. The lead time on these types of offers is usually one week. The number of layers is usually limited to two and the copper is one ounce. There is also a limit on the size of the PCB. The suppliers are usually strict on the constraints on these types of bargain PCBs. Search for "PCBs" on the Internet to find more of these kinds of offers. As stated before, requiring an electrical test to make sure all the connections are there will substantially increase the cost, but may be worth it if your design is complex.

Drilling Your PC Board

After you make a PC board using one of the above techniques, you need to drill holes in the board for the components. Use a drill press, or at least improvise one. Boards can be drilled entirely "free hand" with a

Table 14-1 *Printed circuit drill sizes*

Drill size	Decimal size	Drill size	Decimal size
80	0.0135	69	0.0292
79	0.0145	68	0.0310
78	0.0160	67	0.0320
77	0.0180	66	0.0330
76	0.0200	65	0.0350
75	0.0210	64	0.0360
74	0.0225	63	0.0370
73	0.0240	62	0.0380
72	0.0250	61	0.0390
71	0.0260	60	0.0400
70	0.0280		

hand-held drill but the potential for error is great. A drill press or a small motor tool in an accessory drill press makes the job a lot easier. A single-sided board should be drilled after it is etched; the easiest way to do a double-sided board is to do it before the resist is applied.

To drill in straight lines, build a small guide for the drill press so you can slide one edge of the board against it and line up all of the holes on one grid line at a time. This is similar to the "rip fence" setup used by most woodworkers to cut accurately and repeatedly with a table saw.

The drill-bit sizes available in hardware stores are too big for PC boards. You can use high-speed steel bits, but glass epoxy stock tends to dull these after a few hundred holes. (When your drill bit becomes warm, it makes a little "hill" around each drilled hole as the worn bit pushes and pulls the copper rather than drilling it.) A PC-board drill bit, available from many electronics suppliers, will last for thousands of holes. If you are doing a lot of boards, it is clearly worth the investment. Small drill bits are usually ordered by number, as can be seen in Table 14-1. Use high RPM and light pressure to make good holes. Count the holes on both the board and your layout drawing to ensure that none are missed. Use a larger-size drill bit, lightly spun between your fingers, to remove any burrs. Do not use too much pressure; remove only the burr.

PC-Board Assembly Techniques

Once you have etched and drilled a PC board, you are ready to use it in a project. Several tools will come in handy: needle-nose pliers, diagonal cutters, pocket knife, wire strippers, clip leads, and soldering iron.

Cleanliness

Make sure your PC board and component leads are clean. Clean the entire PC board before assembly; clean each component lead before you install it. Corroded areas look dark instead of bright and shiny. Do not use sandpaper to clean your board. Use a piece of fine steel wool or a Scotchbrite cleaning pad to clean component leads or the PC board before you solder them together.

Installing Components

In a construction project that uses a PC board, most of the components are installed on the board. Installing components is easy. Place the components in the right circuit-board holes, solder the leads, and cut off the extra lead length. Most construction projects and kits have a parts-placement diagram that shows you where each component is installed.

Getting the components in the right holes is called "stuffing" the circuit board. Inserting and soldering one component at a time takes too long. Some people like to put the components in all at once, and then turn the board over and solder all the leads. If you bend the leads a bit, about 20° from the bottom side, after you push them through the board, the components are not likely to fall out when you turn the board over.

A better approach is to install about five or six components, bending each lead about 20° off center to hold them in place, then turn the PC board over and solder the leads of all the recently installed components. Then begin again by inserting another five or six components. Many people place some ESD or electrostatic foam under the circuit board while soldering components, to hold the components in place while soldering.

Start with the shortest components such as horizontally mounted diodes and resistors. Larger components sometimes cover smaller components, so these smaller parts must be installed first. Use adhesive tape to temporarily hold difficult components in place while you solder.

PC-Board Soldering

To solder components to a PC board, bend the leads at a slight angle, apply the soldering iron to one side of the lead, and flow the solder in from the other side of the lead. Too little heat causes a bad or "cold" solder joint; too much heat can damage the PC board. Practice a bit on some spare copper stock before you tackle your first PC board project. After the connection is soldered properly, clip the lead flush with the solder.

Special concerns

Make sure you have the components in the right holes before you solder them. Components that have polarity, such a diodes, ICs, and some capacitors, must be oriented as shown on the parts-placement diagram.

Off-Board Wiring

Selecting the interconnecting wire used in electronic equipment is accomplished by considering the maximum current it must carry as well as the voltage its insulation must withstand. To minimize EMI or electromagnetic interference, the power wiring of transmitters should use shielded wire. Receiver and audio circuits may also require the use of shielded wire at some points for stability or the elimination of hum. Coaxial cable is recommended for all 50 ohm circuits. Use it for short runs of high-impedance audio wiring.

When choosing wire, consider how much current it will carry. The current-handling capabilities of common wire sizes are listed in Table 14-2. Stranded wire is usually preferred over solid wire because it better withstands the inevitable bending that is part of building and troubleshooting a circuit. Solid wire is more rigid than stranded wire; use it where mechanical rigidity is needed or desired.

Wire with typical plastic insulation is good for voltages up to about 500 V. Teflon-insulated wire should be used for higher voltages. Teflon insulation does not melt when a soldering iron is applied. This makes it particularly helpful in tight places or large wiring harnesses. Although Teflon-insulated wire is more expensive, it is often available from industrial surplus houses. Inexpensive wire strippers make the removal of insulation from hookup wire an easy job. Bare soft-drawn, solid #22 to #12 gauge wire is often used to wire HF radio and power circuits. Avoid kinks by stretching a piece 10 or 15 ft long and then cutting it into short, convenient lengths. Note that solid wire will only permit a few bends before the wire breaks, whereas stranded wire can be bent numerous times before breaking. Run RF wiring directly from point to point with a minimum of sharp bends and keep the wire well spaced from the chassis or other grounded metal surfaces. Where the wiring must pass through the chassis or a partition, cut a clearance hole and line it with a rubber grommet. If insulation is necessary, slip spaghetti insulation or heat-shrink tubing over the wire. For power-supply leads, bring the wire through the chassis via a feed through the capacitors.

In transmitters where the peak voltage does not exceed 500 V, shielded wire is satisfactory for power circuits. Shielded wire is not readily available for higher voltages; use point-to-point wiring instead. When wiring filament circuits carrying heavy current tube equipment, it is necessary to use #10 or #12 bare or

Table 14-2   Current-handling capacity of wire

Wire size, AWG	Copper wire (ohms/ft)	Maximum current, wiring in air	Maximum current, wiring confined
32	188	0.53	0.32
30	116	0.86	0.52
28	72.0	1.4	0.83
26	45.2	2.2	1.3
24	28.4	3.5	5.0
22	22.0	7.0	5.0
20	13.7	11.0	7.5
18	6.5	16	10
16	5.15	22	13
14	3.20	32	17
12	2.02	41	23
10	1.31	55	33
8	.734	73	46
6	.459	101	60
4	.290	135	80
2	.185	181	100
1	.151	211	125
0	.117	245	150
00	.092	283	175
000	.074	328	200

enameled wire. Slip the bare wire through spaghetti, then cover it with copper braid pulled tightly over the spaghetti. Slide the shielding back over the insulation and flow solder into the end of the braid; the braid will stay in place, making it unnecessary to cut it back or secure it in place. Clean the braid first so solder will take with a minimum of heat.

For receivers, RF wiring follows the methods described above. At RF, most of the current flows on the surface of the wire, a phenomenon called the "skin effect." Hollow tubing is just as good a conductor at RF as solid wire.

A Final Check

No matter what construction technique you use for constructing your electronic circuits, do a final check before applying power to the circuit. Things do and can go wrong—remember Murphy's Law! Double check that each component is installed in the proper holes on the board and that the orientation is correct. Make sure that no component leads or transistor tabs are touching other components or PC board connections. Sometimes, components get installed in the wrong places, and sometimes ICs get orientated wrongly, so careful inspection minimizes the risk of a project beginning and ending its short life in a puff of black smoke! Check your wiring carefully. Make a photocopy of the schematic and mark each lead on the schematic with a red X when you have verified that it is connected to the right spot in the circuit.

Inspect solder connections. A bad solder joint is much easier to find before the PC board is mounted to a chassis. Look for solder "bridges" between adjacent circuit-board traces. Solder bridges occur when solder accidentally connects two or more conductors that are supposed to be isolated. It is often difficult to distinguish a solder bridge from a conductive trace on a tin-plated board. If you find a bridge, remelt it and the adjacent trace or traces to allow the solder's surface tension to absorb it. Cold solder joints often look "gritty," dark, and unsmooth. Check the circuit voltages before installing ICs in their sockets. Ensure that the ICs are oriented properly and installed in the correct sockets.

Chassis Fabrication

Most electronics projects end up in some sort of an enclosure, and most hobbyists choose to purchase a ready-made chassis for small projects, but some projects require a custom enclosure. Even a ready-made chassis may require a fabricated sheet metal shield or bracket, so it is good to learn something about sheet-metal and metal fabrication techniques.

Most often, you can buy a suitable enclosure. These are sold by Radio Shack and most electronics distributors. See the supplier list in the Appendix. Select an enclosure that has plenty of room. A removable cover or front panel can make future troubleshooting or later modifications a simple matter. A project enclosure should be strong enough to hold all of the components without bending or sagging; it should also be strong enough to stand up to expected use and abuse. A general rule of thumb is to use a metal chassis for high-frequency oscillators, transmitters, and VHF/UHF receivers. Amplifier circuits, slow-speed digital circuits, and some sensing circuits can be housed in plastic enclosures.

Cutting and Bending Sheet Metal

Enclosures, mounting brackets, and shields are usually made of sheet metal. Most sheet metal is sold in large sheets 4 × 8 ft or larger. It must be cut to the size needed. Most sheet metal is thin enough to cut with metal shears or a hacksaw but a jigsaw or bandsaw makes the task easier. If you use any kind of saw, select a blade that has teeth fine enough so that at least two teeth are in contact with the metal at all times.

If a metal sheet is too large to cut conveniently with a hacksaw, it can be scored and broken. Make scratches as deep at possible along the line of the cut on both sides of the sheet. Then, clamp it in a vise and work it back and forth until the sheet breaks at the line. Do not bend it too far before the break begins to weaken, or the edge of the sheet might bend. A pair of flat bars, slightly longer than the sheet being bent, makes it easier to hold a sheet firmly in a vise. Use C clamps to keep the bars from spreading at the ends.

Smooth rough edges with a file or by sanding with a large piece of emery cloth or sandpaper wrapped around a flat block.

Finishing Aluminum

Give the aluminum chassis, panels, and parts a sheen finish by treating them in a caustic bath. Use a plastic container to hold the solution. Ordinary household lye can be dissolved in water to make a bath solution. Follow the directions on the container. A strong solution will do the job more rapidly.

Stir the solution with a stick of wood until the lye crystals are completely dissolved. If the lye solution gets on your skin, wash with plenty of water. If you get any in your eyes, immediately rinse with plenty of clean, room-temperature water and seek medical help. It can also damage your clothing, so wear something old. Prepare sufficient solution to cover the piece completely. When the aluminum is immersed, a very pronounced bubbling takes place. Provide ventilation to disperse the escaping gas. A half hour to two hours in the bath is sufficient, depending on the strength of the solution and the desired surface characteristics.

Chassis Working

With a few essential tools and proper procedure, building electronic gear on a metal chassis is a relatively simple matter. Aluminum is better than steel, not only because it is a superior shielding material, but also because it is much easier to work and provides good chassis contact when used with secure fasteners.

Spend some time planning your project enclosure, the circuit board placement, and layout of the controls; this will save you time and energy later on and avoid undue surprises. The actual construction is much simpler when all details are worked out beforehand. Here, we discuss a large chassis-and-cabinet project, such as a high-power amplifier. The techniques are applicable to small projects as well.

Cover the top of the chassis with a piece of wrapping paper or graph paper. Fold the edges down over the sides of the chassis and fasten them with adhesive tape. Place the front panel against the chassis front and draw a line there to indicate the chassis top edge.

Assemble the parts to be mounted on the chassis top and move them about to find a satisfactory arrangement. Consider that some will be mounted underneath the chassis and ensure that the two groups of components won't interfere with each other.

Be sure to place all controls with shafts that extend through the cabinet first, and arrange them so that the knobs will form the desired pattern on the panel. Position the shafts perpendicular to the panel.

Locate any partition shields and panel brackets next, then sockets and any other parts. Mark the mounting-hole centers of each part accurately on the paper. Watch out for capacitors with off-center-shafts that do not line up with the mounting holes. Do not forget to mark the centers of socket holes and holes for wiring leads. Make the large center hole for a socket before the small mounting hole, then use the socket itself as a template to mark the centers of the mounting holes. With all chassis holes marked, center-punch and drill each hole.

Next, mount on the chassis the capacitors and any other parts with shafts extending to the panel. Fasten the front panel to the chassis temporarily. Use a machinist's square to extend the line (vertical axis) of any control shaft to the chassis front and mark the location on the front panel at the chassis line. If the layout is complex, label each mark with an identifier. Also mark the back of the front panel with the locations of any holes in the chassis front that must go through the front panel. Remove the front panel.

PC-Board Chassis Boxes

An alternative to aluminum chassis box project construction is the use of printed circuit board material for the construction of enclosures. You can eliminate tedious sheet-metal work by fabricating chassis boxes and enclosures from copper-clad printed-circuit board material. Although it is manufactured in large sheets for industrial use, some hobby electronics stores and surplus outlets sell usable scraps at reasonable prices. PC-board stock cuts easily with a small hacksaw or bandsaw. The nonmetalic base material is not malleable so it cannot be bent. Corners are easily formed by holding two pieces at right angles ant soldering the seam. This technique make excellent RF-tight enclosures. If mechanical rigidity is required of a large copper-clad surface, solder stiffening ribs at right angles to the sheet. Start by laying the bottom piece on a workbench, then placing one of the sides in place at right angles. Tack solder the second piece in two or three places, then start at one end and run a bead of solder down the entire seam. Use plenty of solder and plenty of heat. Continue with the rest of the pieces until all but the top cover is in place. In most cases, it is better to drill all needed holes in advance. It can sometimes be difficult to drill holes after the enclosure is soldered together.

You can use this technique to build enclosures, subassemblies, or shields. This technique is easy with practice; hone your skills on a few scrap pieces of PC-board stock.

Drilling Techniques

Before drilling holes in metal with a hand drill, indent the hole centers with a center punch. This prevents the drill bit from "walking" away from the center when stating the hole. Predrill holes greater than $\frac{1}{2}$ inch in diameter with a smaller bit large enough to contain the flat spot at the large bit's tip. When the metal being drilled is thinner than the depth of the drill-bit tip, back up the metal with a wood block to smooth the drilling process. Table 14-3 lists machine screw tap sizes and clearance drilling holes dimensions.

Table 14-3   Machine screw tap and clearance drill sizes

Type	Tap drill	Clearance	Type	Tap drill	Clearance
2-56	50	42	10-24	25	13/64
2-64	50	42	10-32	21	13/64
3-48	47	36	12-24	16	7/32
3-56	45	36	12-28	14	7/32
4-40	43	31	¼-20	7	17/64
4-48	42	31	¼-28	3	17/64
4-48	42	31	⁵/₁₆-18	F	21/64
6-32	36	25	⁵/₁₆-24	1	21/64
6-40	33	25	³/₈-16	5/16	25/64
8-32	29	16	⁷/₁₆-20	25/64	29/64
8-36	29	16	½-20	29/64	33/64

The chuck on the common hand drill is limited to ⅜ inch bits. Some bits are much larger, with a ⅜ inch shank. If necessary, enlarge holes with a reamer or round file. For very large or odd-shaped holes, drill a series of closely spaced small holes just inside of the desired opening. Cut the metal remaining between the holes with a cold chisel and file or grind the hole to its finished shape. A nibbling tool also works well for such holes.

Use socket-hole punches to make socket holes and other large holes in an aluminum chassis. Drill a guide hole for the punch center bolt, assemble the punch with the bolt through the guide hole and tighten the bolt to cut the desired hole. Oil the threads of the bolt occasionally. Cut large circular holes in steel panels or chassis with an adjustable circle cutter or "fly-cutter."

Occasionally apply a small amount of machine oil to the cutting groove to speed the job and to keep the tool from getting overheated. Test the cutter's diameter setting by cutting a block of wood or scrap material first. Remove bumps or rough edges that result from drilling or cutting with a burr remover, round or half-round file, sharp knife, or chisel. Keep an old chisel sharpened and available for this purpose.

Rectangular Holes

Square or rectangular holes can be cut with a nibbling tool or a row of small holes, as previously described. Large square or rectangular holes or openings can be cut easily using socket-hole punches.

Construction Notes

If a control shaft must be extended or insulated, a flexible shaft coupling with adequate insulation should be used. Satisfactory support for the shaft extension, as well as electrical contact for safety, can be provided by means of a metal panel bushing made for the purpose. These can be obtained singly for use with existing shafts, or they can be bought with a captive extension shaft included. In either case the panel bushing gives a solid feel to the control. The use of fiber washers between ceramic insulation and metal brackets, screws, or nuts will prevent the ceramic parts from breaking.

Painting

Painting is an art, but, like most arts, successful techniques are based on skills that can be learned. The surfaces to be painted must be clean to ensure that the paint will adhere properly. In most cases, you can wash the item to be painted with soap, water, and a mild scrub brush, then rinse thoroughly. When it is dry, it is ready for painting. Avoid touching it with your bare hands after it has been cleaned. Your skin oils will interfere with paint adhesion. Wear rubber or clean cotton gloves.

Sheet metal can be prepared for painting by abrading the surface with medium to fine grade sandpaper, making certain the strokes are applied in the same direction, that is, with a noncircular motion. This process will create tiny grooves on the otherwise smooth surface. As a result, paint or lacquer will adhere well. On aluminum, one or two coats of zinc chromate primer applied before the finish paint will ensure good adhesion.

Keep work areas clean and the air free of dust. Any loose dirt or dust particles will probably find their way onto a freshly painted project. Even water-based paints produce some fumes, so properly ventilate work areas.

Select a paint suitable to the task. Some paints are best for metal, others for wood and plastics. Some dry quickly, with no fumes; others dry slowly and need to be thoroughly ventilated. You may want to select a rust-preventative paint for metal surfaces that might be subjected to high moisture or salts. Most metal surfaces are painted with some sort of spray, either from a spray gun or from spray cans of paint. Either way, follow the manufacturer's instructions for a high-quality job.

Buying Equipment, Components, and Tools

Buying New Equipment

Buying new test equipment is always preferable to buying used equipment if possible. Budget permitting, who wouldn't want to have new equipment. The benefits of buying new test equipment are numerous. One of the benefits of purchasing new equipment is that it will come with a warranty for a period of time, usually a full 90 days on parts and labor. Often, warranties are 90 days on labor and one year on parts. When purchasing a new piece of test equipment, be sure to understand the warranty features completely. Another benefit of buying new test equipment is that you can usually extend the warranty for a longer period and not have to worry about service or breakdowns for a long period of time. Electronic circuits, if they are going to fail, will usually do so within the first few weeks of operation and after that period will operate trouble free for a long time. That is the general rule, but there are always exceptions to the rule. When buying new test equipment, you will always have the benefit of having an operating manual and sometimes a service manual. Another benefit of purchasing new equipment is that you can get company sales people to call on you to demonstrate the new equipment before purchasing it. You can have the salesman fully demonstrate all the features to you and often they will leave the equipment for a period of time for you to test it or to become familiar with the equipment before purchasing it. You can actually have a number of different representatives call on you to demonstrate their product before making a final decision. In this way, you can see the benefits of the different makes and models of equipment before parting with your hard-earned money. When purchasing more than one piece of equipment, always try to negotiate a discount. It never hurts to ask, and you may receive 10% or more off additional items simply by saying that you will buy immediately if you receive a discount.

New or Used Equipment?

The topic of used equipment needs to be discussed before delving into specific requirements. Obviously, used equipment is much less expensive than new. And "used" doesn't necessarily mean "used up." Good-quality used equipment is readily available. The question is how to determine what is "good" and what is not.

Generally, look to the used market before buying any major piece of test gear. However, there are a few exceptions to this. Used digital logic analyzers and computerized equipment is often available new for comparable prices. Also, used mechanical tools and equipment can simply wear out and be very difficult to repair. There is also one very important rule to follow when buying previously owned gear. If you do not know exactly what you are buying, do not buy it. There are many products that have similar part numbers that are quite different in performance. There are often "A" versions, "B" versions, and so forth. Additionally, some oscilloscopes need "plug-in" subassemblies and are not functional without them. Obviously, this can come as a big surprise if you don't know exactly what you are buying. Try to find a 5- or 10-year-old catalog to old specifications. Another point to remember is that a name-brand piece of used gear will often outlast a new piece of off-brand gear. This is because name brands (Tektronix, Hewlett-Packard, and Fluke) are designed for professional use and tend to age well. They also have a large user base, so special parts are relatively easy to get. Off-brands (Heath, Radio Shack, and Eico) are usually built for the hobbyist. They are fine for that purpose. However, in a business (or serious amateur) setting, these are simply not reliable enough. This is why used-equipment companies don't carry off-brands.

Finally, *always* get a service/operator's manual when you buy the equipment. Often, the seller can provide you with one. If not, then go to the manufacturer or try third-party documentation vendors like Sam's Photofacts. It is hard to repair, calibrate, or use equipment without a service manual. If you don`t get one when you are thinking about it, you won't have one when you really need one. What's more, as time goes by, manuals become harder to locate.

There are several classes of used equipment dealers: salvagers, resellers, calibrators, and leasers. Salvagers are companies that buy equipment, literally by the ton. They may have no background in electronics. To them, an oscilloscope works if you turn it on and a trace is seen. Usually, the equipment is quite old and no warranty is available. The prices are great—often a few pennies on the dollar. But the risk is also great. This is a gambler's market, since usually all sales are final. If an individual is selling an item on the Internet, it should probably be considered in this class.

Resellers are a step up from the salvagers. They understand test equipment better. They also know how to perform basic tests to see if a unit is functional. They usually warrant functionality but not performance. This means that the oscilloscope works reasonably well, but the time-base may be off slightly. Companies selling surplus equipment on the Web usually fall into this class. Prices can be 10% to 25% of the original price. Usually, there is a warranty for functionality and some type of basic return policy (perhaps 30 days). This choice may be acceptable for a hobbyist, but not for a business. The equipment is still not reliable enough for that.

Calibrators sell equipment that is guaranteed to meet original factory specifications. This means that it operates as if it was new. Their prices are slightly higher than the resellers, but this class of equipment is suitable for any purpose. Individuals and companies selling surplus equipment do not fall into this class. This means that anything on eBay is probably not in this class, either. Only companies that specialize in selling calibrated used equipment fall into this class. Look for the "Fully calibrated to factory specifications" phrase. If it does not specifically say it is fully calibrated it probably isn't. These companies provide a warranty for performance and a return policy. This is a good choice for older equipment (5 to 10 years old).

Leasers are companies that buy new equipment and lease it out. After a couple of years, they have recouped their investment and made a profit, so they then sell off the equipment for cash. This equipment is newer than that from the calibrators. In fact, it may still be under the original factory warranty. The equipment is generally calibrated. At worst, you can pay a nominal fee for calibration. Prices are usually 70% to 90% of the original price, but, if you need a $20,000.00 spectrum analyzer and only have $15,000.00, this is the class to examine. This class is suitable for any purpose, as well.

Always be sure to ask about the warranty and how long it will be honored. Also ask the seller if you can return the equipment within the warranty period. If a dealer is reputable, then he will give you some leeway on this. If a seller says all sales are final, with or without a warranty, then you should be hesitant about the sale. Do not be pressured into buying a piece of test equipment in a few minutes from a fast-talking salesman. Take your time and make a careful decision when buying test equipment, since you want it to serve you well and last many years.

If an equipment reseller is local to your area or in a nearby town, consider driving to their location and inspecting what they have for sale. Often, holding a piece of equipment and looking it over as well as looking a salesperson in the eye and asking questions about the equipment is very desirable. A local dealer or lease agent can be very helpful in the event that your equipment develops problems at a later date. It is a lot easier to deal with a local seller, rather than sending equipment back to the manufacturer for repair.

If you are looking to purchase a used piece of test equipment from a local or regional dealer, make sure that you physically pick up the equipment that you are interested in and hold it in your hands. Look to see how dirty or dusty the piece of equipment is. If it is dusty, then it has been sitting around for a long time.

If the equipment is dirty, then the you can tell that the dealer has not cleaned it up, that it has had lots of use, or that its owner did not take good care of it. When looking over the prospective piece of used test gear, also look for broken knobs and bent shafts on the controls, this is a sign of rough handling or old age. When looking over equipment, also be sure to see if there is any rust on the back, front, and bottom of the case. If you see rust, that is an indication that the instrument was stored or used in a damp area, and not a good sign. If there is rust on the outside it means that it is likely that the inside of the equipment has lots more rust. Also look to see that the control shafts will rotate. If the control shafts are very stiff or will not turn, then stay away from that particular piece of equipment. The best motto is to "Buy the best test equipment that you can afford."

Guidelines for Buying Used Equipment

In a tight economy, many small businesses and hobbyists are discovering the benefits of buying used instead of new. It's one thing, of course, to buy used chairs or desks. It's quite another to buy used test equipment or computers. It can be a somewhat risky proposition to purchase used high-tech equipment! You will need to read and become familiar with technical terms, functions, and features and have a sense of familiarity with the high-tech equipment you wish to purchase before parting with your hard-earned cash. Talk to a friend or mentor and make comparisons before purchasing used test equipment or computers. Many people buying a used computer, for example, have made poor choices because they were not properly aware of terms and features prior to their purchase. Advanced knowledge is the key to making sure that you buy reliable, long-lasting equipment. You can save yourself a bundle in the process if you follow certain purchasing guidelines. Look first to the company offering a warranty; second, look at the warranty itself.

Most used equipment comes with a warranty, 30 days being most common. This means little, however, if you can't find reliable technicians to repair the equipment if any glitches show up during this time. It is usually best to purchase from a company offering in-house repairs. Before buying, ask about their experience with the particular type of equipment and brand that you are buying. Another option is buying from a company that guarantees to send the equipment back to the manufacturer for repair.

Understand the difference between "refurbished," "reconditioned," and "remanufactured." These terms are commonly used to describe how the used equipment has been brought back into good working condition. "Refurbished" usually means that equipment has simply been cleaned, possibly with minor repairs performed. "Reconditioned" means that all repairs have been made to bring the equipment back into near-perfect or perfect working order. "Remanufactured" means that the equipment has gone back to the plant and has been worked on by expert technicians. Remanufactured equipment is often available only through authorized dealers or resellers. Always ask for extended warranty options. Ask for a warranty that includes free parts and labor for 120 days. The seller may agree, just to make the sale. When possible, find out where others have had success buying used equipment.

Ask friends, business associates, repair shops and retail equipment sellers about the equipment and repair service of the particular company you are thinking of buying from. You might find out that the company's technicians are not skilled or the company does not honor its warranties. You might also find out about another company that offers much better deals.

Consider saving money by buying new but not state-of-the-art equipment. Most retailers of new computer equipment and peripherals, copiers, telephone systems, and other electronic devices sell brands that offer varying levels of features and pricing. Included in these may be models with fewer features, slower performance, or more basic controls, but they still may be all that you need. Such equipment is new out of the box (even though the box may have been around for awhile) and comes with the company's new equipment warranty/replacement guarantees. The cost of this equipment may be considerably less than the most advanced models but will not be as low as used equipment. Even so, the added peace of mind can be worth it.

Hidden Costs

Although the price of used test and measurement equipment can be low, the risks can be high. Hidden costs often emerge when used equipment arrives without, for example, the necessary accessories or options. The cost to bring the instrument up to the needed performance can cost more than a brand new instrument. A number of considerations must be taken into account before purchasing a used piece of test equipment. Does the test equipment come with any accessories? What options or plug-ins come with your purchase? Oscilloscope and function generator require calibration for maximum performance. How much will it cost to have the equipment calibrated after the purchase?

When purchasing used test equipment, you will need to know if the equipment you wish to purchase comes "as is" or is fully refurbished. If you are purchasing a function generator or especially an oscilloscope as an individual, you will be farther ahead if you purchase the equipment reconditioned or refurbished and recently calibrated. Ask the seller as many questions as possible: does the oscilloscope come with probes and all plug-ins; has it been calibrated recently? Purchasing a bargain oscilloscope and then having to buy additional accessories and probes and then having it recalibrated may add a significant cost to your purchase and may prove to be false economy.

If you are purchasing test equipment for a large company that has a calibration lab, then it does not matter much if your purchase includes calibration. If the buyer is purchasing many pieces of test equipment all at the same time, then they have bargaining power for buying accessories if needed. Companies will also have the option of evaluating the cost of ownership, and may want to consider leasing equipment rather than purchasing it. Individuals, however will likely want to purchase their test equipment outright. When purchasing used test equipment you will want to make sure that you are well aware of the return policy and the length of the warranty before making your final purchase.

To summarize, the upside of buying new equipment includes full warranty, new components, new calibration, and instructions and manual. The downside of buying new equipment is the higher cost. The upside of buying used test equipment includes lower cost, excellent or good condition, and years of service left. The downside of buying used test equipment includes the possible lack of instructions or a manual, surface wear (scratches, etc.), possible need for calibration, and component aging.

Additional Thoughts on Buying Used Test Equipment

Used Oscilloscopes

You can readily locate used test equipment for your electronics workbench, with a little detective work. The two most popular brands of used test gear are Hewlett Packard (Agilent) and Tektronix. For oscilloscopes, the only real choice is Tektronix. "Service grade" equipment such as B&K, Sencore, Simpson, Eico, or Triplett are not generally very desirable.

Needless to say, clean working scopes with probes and manuals will command much higher prices than untested junk. The oldest scopes are those Tektronix vacuum tube "doghouses." These are of interest mostly to hard core collectors and are useless for general lab work due to their outrageous size and weight. Of these, the 515, 545A, and 561 are probably the best of the lot. The 570 vacuum tube curve tracers still can bring outrageously high prices at market. Stan Griffiths is a recognized authority on older Tektronix restorations, having published *Oscilloscopes—Restoring a Classic*. His website includes definitive resources. Somewhat newer are the Tektronix 5000 mainframes. These have virtually zero demand due to poor performance. Next are the 7000 series which still has modest popularity, especially the higher-speed horizontal plug-ins. The industry-standard workhorse scope for years and years was the Tektronix 465 or its related 455 plastic or military ruggedized versions. These were the first "modern" scopes, being portable, reasonably lightweight, fast, and largely solid state. Although they remain eminently useful, they are getting somewhat long in the tooth. The best and most desirable Tektronix scopes are the 2213 and 2215 for individuals and students, or the 2246 for serious professionals.

Check eBay for the current model-by-model going rate. Rebranded scopes such as Tektronix/Telequipment or Tektronix/Sony generally do not sell well. Tektronix scopes whose model numbers begin with "T" are typical of this breed. Most Tektronix spectrum analyzers do command premium prices, but replacing the sometimes burned-out mixer diodes may require advanced skills or a lot of money. Note that the newest spectrum analyzers use wildly improved FFT computer techniques, leaving all of the older versions in the dust. As mentioned, awkward and heavier scope carts are not usually worth bothering over, but there can be some real surprises hidden in their drawers. Some carts may also have valuable TM-series mainframes or plugins that can easily and profitably be removed and sold separately. You can visit the Tektronix website, but support for older equipment is limited and the prices for service parts is several orders of magnitude beyond outrageous. One useful source for older Tektronix parts is Deane Kidd.

The closest thing to a "Blue Book" are all those free catalogs from Test Equipment Connection. Divide these prices by six to get the best possible eBay selling price and by thirty to get the absolute maximum you should ever pay for anything. A search on eBay should give you their current price spreads of more popular items. Note that these vary widely with how clean the instrument is, which accessories are provided, and presence or absence of manuals and documentation, warranties, calibration, or inspection privileges. Supply and demand plus the competence of the seller also makes a huge difference in your final price. Be sure to check the completed auctions, not the current ones. Final prices are typically determined by last-second snipers. National Serial Numbers, or NSNs, sometimes will tell you what the government paid for any piece of test gear, along with how old it is and how long it was popular. These are available free from Surplus Bid, but normally only on their current items. Other NSNs can sometimes be gotten for a fee through USA Info. Be certain to divide any NSN price by 100 to get the absolute maximum you should pay.

Equipment companies such as Tucker, Test Equity, and Megahertz Electronics, and Fair Radio Sales offer older military surplus electronics. Naturally, a good way to establish price is to try to buy the item on the Web. A surprising number of scientific or other more specialized instrument manufacturers may still have current or upgraded versions in production. Sometimes, you can simply guess at their website as www.theirname.com. Do not forget to try Hotbot, Google, and the Thomas Register. Also visit newsgroups such as sci.electronics.equipment or sci.electronics.repair.

Hamfests have long been a useful source for electronic bargains, but these have gotten a lot more junky. Dot.bomb bankruptcies can offer exceptional bargains as can related distress auctions. I've found the best results on auctions that list only a very few electronic or techie items in an otherwise large sale. Auction Advisory is a superb source to pin down these events. At an auction, either start off with a ridiculously lowball bid and quit early, or bid your maximum once, very late in the process. Note that you can cut your bid increment in half by waving your hand across your chest palm down. Most Web truckload "palletized salvage" outfits are best avoided. Cherry picking and scams are the norms. Avoid anything excessively heavy, certainly anything that cannot be shipped by UPS.

Equipment Manuals

Do not attempt serious repairs without full manuals and proper test gear! By far the easiest way to pick up manuals is to use eBay as a lending library. Popular and some obscure manuals are often found in the $12 to $30 range, and CD collections of military electronics stuff are readily available. You can often resell the manual on eBay for more than you paid for it. At the least, you have added value to the equipment

Of the commercial manual vendors, check out HK Porter. His competitors include Tannebaum, Manual Man, and Manuals Plus. There are hundreds more, easily found through the Boatanchor Circle. Besides the previously mentioned links, chances are also good that somebody has posted something to a random news group, so be sure to try Deja News, which recently has become a part of Google.

Equipment Cleaning

In many instances, cleaning up used equipment is a matter of cleaning years of grime, labels, and gunk, and tightening knobs, cleaning switches, and replacing line cords. Generally, it is wise to check over the electrolytic capacitors, since aged capacitors often need replacement. Serious repairs however, should not be attempted without the proper manuals, tools, and test gear on hand. Your two most important refurbishing tools are a 0.050 hex Allen wrench and a can of Radio Shack tuner cleaner for dealing with noisy switches.

Other items high on the "must have" list are an ac power pen detector (Radio Shack model 1LAC-A VoltAlert or equivalent), a tube of super glue, fuses, some silicon rubber adhesive/sealant, a strong household cleaner such as Simple Green, plus glop removers such as full strength Citra-solv or Armorall. Tire cleaner works surprisingly well to renew black Bakelite cases. Soak stuck-on labels or whatever you are trying to remove overnight. Removing the "top half" of a label makes getting the sticky part off easier. A small sponge bungee-corded in place helps a lot. Plastic spatulas and toothbrushes can also be a great help. Steel wool may be too harsh. Try to match missing screws. Be sure you get all knobs in the correct position after tightening! Repainting is usually a bad idea and should always be avoided. Sharpie permanent marker "retouch" pens can cover multiple sins, especially on Wavetek knobs. Solder damage on plastic bezels can sometimes be fixed by a little filing or sanding. A good rule on cosmetics is to "take half and leave half" when improving appearance. Should parts be needed, search the web for collectors or specialists, then try Questlink, Chipcenter, or similar resources (see the Appendix). For obsolete semiconductors, check into Rochester Electronics, Luke Systems, or Electronic Expediters International.

Buying Versus Building Test Equipment

Should you buy test equipment or build it yourself? Gaining knowledge through understanding and building equipment is not only educational but fun as well. The knowledge that you gain by building your equipment is invaluable in obtaining practical knowledge in the field of electronics. You will learn how to read schematics, how circuits operate, as well as how to solder. Building test equipment yourself will also save you considerable money if you are just starting out in electronics and will also give you a great sense of personal satisfaction in creating something and then actually using it.

Start by building smaller pieces of test equipment, such as continuity testers, logic pulsers or tracers, or signal injectors. You will be learning about electronics and test equipment but, additionally, you will save money that you can use to purchase more critical pieces of test equipment such as an oscilloscope or an arbitrary waveform generator. These more expensive and what might be called "critical" pieces of equipment that require calibration should be purchased rather than built. The oscilloscope is usually the workhorse of the electronics workbench or laboratory and is used most every day to perform critical and important measurements. It is, therefore, important to buy an oscilloscope rather than attempt to build one. Saving money by building less-critical measurement devices will allow you to save money in order to buy a good oscilloscope. Once again, the best policy is to buy the best oscilloscope, spectrum analyzer, or arbitrary waveform generator that you can afford.

Another approach to obtaining test equipment is to build a kit. Electronics kits are a great way to learn about electronics in general and the about a specific piece of test gear that you can build. Electronics kits were very popular in the 1950s and 1960s from companies like Heathkit and Knight Kits. These companies provided many types of kits during the post World War II boom in electronics equipment building. Many electronics enthusiasts became technicians and engineers during this period. With advances and price decreases in technology and microelectronics, the interest in building electronics circuits waned during the 1970s and 1980s. Fortunately, beginning in the mid 1990s there has been an upsurge in electronic kit building. Marketers have recognized the tremendous demand for good electronics kits. This is a fortunate occurrence, since there is now a new cycle of young electronics enthusiasts who will benefit from the joy and knowledge gained from building kits. It is now possible to locate and build numerous types of kits including electronic test equipment, and it is just plain fun for young and old alike.

Buying Electronic Components

One of the most common and perplexing problems faced by new electronics enthusiasts is where to purchase parts. Sometimes, you get lucky when reading electronics articles and the author makes a parts list available, but not every project has a parts list. If you want to expand your construction horizons, you will need to become your own "purchasing agent." That means you will have to search out parts sources on your own and develop an understanding of how this is done.

In reality, it is not all that difficult to find most parts. Unfortunately, though, the days of the local electronic parts supplier seem to have disappeared. Years ago, an electronics supplier had to stock a relatively small number of electronic components: resistors, capacitors, tube sockets, a few relays, and variable resistors. You could walk into a local parts supplier and show them your parts list and they would likely have most of the components and know how to substitute for components that they did not have. Today, that service is much more difficult to obtain. Most parts houses today only have order takers, who are not familiar with electronics or components. In many instances, companies like Digi-Key and Mouser have technicians available who can help, but you have to ask for them specifically. The technicians can often help you if you need to substitute one part for another. Technology has increased so much in recent years that the number of components has increased by a few orders of magnitude. The number of integrated circuits alone is enough to fill a multivolume book. No single electronics supplier could possibly stock them all. It has become a mail-order world; electronics is no exception.

Although it is no longer always possible to purchase all of your electronic needs from a local electronics supplier, and fortunately you do not have to. For a few dollars, mail-order companies are willing to supply whatever you need. You only need do two things to obtain nearly any electronic component: make a phone call and either to write a check or use your credit card. You will need to become an electronic catalog collector, establishing your own reference library at your electronics workbench. The Appendix has a list of electronic-component suppliers. Write to these companies and request their catalogs. Many electronics suppliers will gladly supply you with their latest parts catalog, which might be in the form of a CDROM, or on the Web. Electronic suppliers also advertise in magazines that cater to electronics enthusiasts. If you are lucky, you may have a local source of electronics parts. Look in the Yellow Pages under "Electronic Equipment Suppliers" to find the local outlets. Radio Shack is one local source found nearly everywhere. They carry an assortment of the more common electronic parts. You will probably need to order from more than one mail-order company. No matter how wide a selection you find in one mail-order catalog, there's always at least one part you must buy somewhere else!

While you are waiting for your catalogs, look at the parts list for the project you want to build. Unfortunately, you cannot just photocopy the list and send it off to a mail-order company with a note that says "please send me these parts." You need to convert the parts list into a part-order list that shows the order number and quantity required of each component. This may require a similar list for each parts supplier if one supplier does not have all the parts. Sometimes, you will have to use more than one supplier to locate all the parts for a particular project.

Check the type, tolerance, power rating, and other key characteristics of the parts. Group the parts by those parameters before grouping them by value. If all of the circuit's components are already grouped by value on the parts list, you can just count the number of each value. Each time you add parts to the order list, check them off the published parts list. Sometimes, the parts list does not include common components like resistors and capacitors. If this is the case, make a copy of the schematic and check off the parts as you build your shopping list.

Although you will probably be able to order exactly the right number of each part for a project, buy a few extras of some parts for your junk box. It's always good to have a few extra parts on hand; you may break a component lead during assembly, damage a solid-state component by applying too much heat, or wire a component in backwards. If you do not have extra parts, you will often wish you did, especially if it is a critical or hard-to-find component. Even if you do not need the extras for this project, they may come in handy later, and with those extra parts around you might be encouraged to build another project!

When constructing projects, you will have to decide whether you are going to construct your projects with ground-plane construction, PC boards, or perf-boards. If you are going to use ground-plane construction, buy a good-sized piece of single-sided, copper-clad glass–epoxy board if you can. Phenolic board is inferior because it is brittle and deteriorates rapidly with soldering heat. If you construct many projects using circuit boards, purchase them in packages of five boards. This saves money and also provides you with boards on hand if you decide to build another project in the near future. Companies such as Newark Electronics or Digi-key Electronics are great sources for circuit boards. Circuit boards come in two basic forms: the blank copper-clad board with 1 or 2 ounce thickness of copper, or the photoresist layered or covered circuit board. The photoresist boards come in both negative and positive resist formats. The blank copper boards with no photoresist are designed as quick single boards where the circuit board design will be created manually and the circuit trace lines created with a wide, black "Sharpie." This is the quick, one-time design board. If you will be creating multiple circuit boards, then you will need to create "circuit artwork," either with software on your PC or on a mylar plastic sheet using the "tape and dots" method of laying out your circuit board. The artwork is then placed over your photoresist circuit board and developed with UV light. If your artwork is used place directly over the circuit board, then you will need to use the positive photoresist circuit boards. If your artwork is first converted to a photographic negative before developing your circuit board, then you would have to use a negative photoresist PC board. PC boards serve many purposes. They provide a professional appearance, a means to replicate your design to pass it on to others, and a reliable design free of glitches and noise problems due to loose or poor wiring. If you need a PC board for popular projects found in electronics or radio magazines, you should contact FAR Circuits. Using these prefabricated PCBs will save you many hours of work.

There are almost always a few items you can't get from one company and most parts suppliers have minimum orders. You may need to distribute your order between two or more companies to meet minimum-order requirements. Some companies put out beautiful catalogs, but their minimum order is $25 or they charge $5 for shipping if you place a small order. One electronics supplier that does not have minimum orders is Mouser Electronics; they are on-line and have a good selection of parts at reasonable prices.

If you order enough parts, you will soon find out which companies you like to deal with, which have fast service, and which have slow service. It is frustrating to receive most of an order, then wait months for parts that are on back-order. If you do not want the company to back-order your parts, write clearly on the order form, "Do not back-order parts." They will then ship the parts they have and leave you to order the rest from somewhere else. If you are in a hurry, call the company to inquire about the availability of the parts in your order. Most electronics parts suppliers are now on-line and, take credit-card orders. You generally have the option of ordering over the Internet or over the telephone. Some companies hold orders a few weeks to allow personal checks to clear. If you are familiar with the catalogs and policies of electronic-component suppliers, you will find that getting parts is not difficult. Concentrate on the fun part building the circuit and getting it working.

After creating an electronics project or electronics kit, you will have to enclose it in order to protect the components in the circuit and give your project a professional look. This is often overlooked in parts lists for most projects, because different builders like different types of enclosures. Enclosures come in many different types, shapes, and sizes as well as materials. There are metal chassis boxes, cast-aluminum boxes, as well as many different types of plastic boxes in which to house your projects. Plastic boxes are easier to drill and cut and are often cheaper than metal boxes; however, sometimes there is no substitute for a metal enclosure. If you are constructing a radio project, for example, and the circuit is a sensitive receiver or transmitter, then you should use a metal enclosure. The general rule is that if a circuit receives or radiates, then the enclosure probably should be metal. When constructing a new project, make sure there is room in the box for all of the components. Some people like to cram projects into the smallest possible box, but miniaturization can be extremely frustrating if you are not good at it. Miniaturization takes lots of practice to get right, so until you have some experience under your belt, do not try to make the smallest possible enclosure.

Buying Tools—New and Used

The availability and selection of new tools boggles the mind. If you want the highest quality tools, you will definitely have to pay for it. Sears still offers their Craftsman series of tools with the full replacement guarantee, which in the long run are an excellent value. Make sure the tool has the Craftsman name stamped into the metal. Be aware that Sears also offers the Companion tool series; this series does not offer the same guarantee as the Craftsman series.

All types of lower-priced tools from Asia have become ubiquitous in recent years, and tool companies such as Harbor Freight Tools have brought them to towns across America. The quality is usually not as good as high-quality American tools, but most of these tools will perform quite adequately for most applications. Look over the tools carefully and the differences will become apparent. The lower-cost Asian tools often have lower tolerances and poorer finishes than high-quality tools. Many early Asian tools were made of pot metal or low-quality castings, but in recent years have become considerably better. The Asian tool makers and sellers discovered that in order to have repeat buyers in the United States, they had to improve quality, so many of their tools are now of higher quality and now compete with other high-quality manufacturers. Most people are now familiar with the Japan syndrome. Toys made in Japan in the early 1960s were cheap and plentiful but mostly junk. The Japanese discovered after awhile that the market was filled with junk and no one wanted it any more. They then discovered that if they made quality things, maybe even better than those from other countries, that consumers would beat a path to their door, and companies like Sony and Panasonic became world leaders in consumer products. The Chinese repeated the Japanese mistakes.

Dollar stores offer tools that are still generally inferior to most other offerings, so *Caveat Emptor* (buyer beware). You will have to weigh the options and try to make intelligent decisions when purchasing tools to equip your electronics workshop.

A good general rule to follow is that tools that get the most use should be of the highest quality to ensure a long life. Tools that get one-time or infrequent use or tools that do not get regular use can be tools of lower quality. You will need to assess a particular tool's purpose before purchasing it, as to whether you should buy a quality or higher-priced tool. Tools that you might use every day should be of the best quality, since they will get a lot of constant use. A good Xcelite screwdriver or nut driver set that you will every day on your electronics workbench will serve you well and last a lifetime. Many tools are often available through mail-order suppliers, check the brand names to ensure that you know what you are buying. Technitools and Jensen, for example, offer high-quality tools, and the prices reflect their quality. Look for new tools at your local retail outlets, electronics stores, and Sears, or through mail-order suppliers such as Newark Electronics. The Appendix lists a number of new tool suppliers as well as specialty tool companies. Look for new tools at lower or below retail costs at tool shows that travel from town to town, local and regional hamfests, and through eBay offerings.

Used tools are also readily available, generally at reasonable prices. Good used tools can generally have many more years of useful life left in them. Before purchasing used tools, be sure to fully inspect and handle them if you can, before laying down your money. Used tools can be obtained through local classified advertisements, local flea markets, garage sales, and at local and regional hamfests as well as through eBay. Buying used tools from an individual will give you a sense of who used them. You can also ask about the history of the tools before purchasing them. The important thing to remember is to know what you are looking at, what you are looking for, and what you are willing to spend, in order to make an intelligent choice. Craftsmen and hobbyists like and usually appreciate tools. It may take you a number of years to obtain a full compliment of tools for different purposes, but for most people, the process is never ending and fun.

Building Your Own Test Equipment

Now that you have constructed your own electronics workbench and acquired some basic tools and troubleshooting techniques, you will need to acquire some test equipment. One way to obtain test equipment is to build it yourself. This chapter presents a number of useful, low-cost test equipment projects that you can build yourself, in order to save money. You can then spend the money saved on some of the more expensive equipment, like a good oscilloscope or function generator.

Continuity Tester

The continuity tester is one of the simplest but most used pieces of test gear. You will find that it is indispensable around your electronics workshop. The continuity tester shown in Figure 16-1 uses an audible indicator to indicate a continuous circuit, which will allow you to keep your eyes on the circuit or wires being tested. The continuity tester can be built on a small circuit board and runs on a 9 volt transistor-radio battery.

The continuity tester begins at the probe at (A). When a continuous or "short" circuit is present between the probe and ground, transistor Q1 is biased. When Q1 is turned on, it also turns on transistors Q2 and Q3, starting the multivibrator or oscillator circuit composed of Q4 and Q5. The oscillator drives the small radio speaker at SP. The on–off switch circuit composed of Q1 through Q3 negates the use of a conventional power switch. The battery life is about two years using the tester intermittently. The continuity

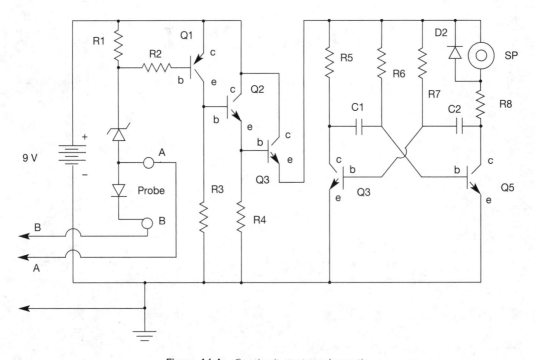

Figure 16-1 Continuity tester schematic.

tester can be assembled on a small 2⅞ by 1¼ inch circuit board and placed in a small "magic marker" type housing. The parts list for the tester follows.

R1	1 kohm, ¼ watt resistor	D1, D2	1N4148 silicon diodes
R2	2.2 kohm, ¼ watt resistor	Z1	8.2 volt, ¼ watt zener diode
R3, R4	22 kohm, ¼ watt resistor	Q1	2N3905 (PNP) transistor
R5	2.7 kohm, ¼ watt resistor	Q2, Q3, Q4, Q5	2N3904 (NPN) transistor
R6, R7	56 kohm, ¼ watt resistor	B1	9 volt transistor radio battery
R8	240 ohm, ¼ watt resistor	Miscellaneous	Battery clip, PC board, wire, probe,
C1, C2	22 nF, 35 volt ceramic capacitors		housing

Operation of the continuity tester is simple. First, connect the 9 volt battery to the tester and "short" the ground lead probe and either probe lead (A) or (B) to ground and the you should begin to hear sound from the continuity tester. If the circuit doesn't work immediately, check the polarity of the diodes and the transistors.

Logic Probe

One of the most difficult problems of testing circuits is how to test "fast" circuits without the use of a triggered oscilloscope. You can trace logic gates easily and without the use of a scope by using the logic probe circuit shown in Figure 16-2. The logic probe has three state indicators, a yellow LED that indicates short-

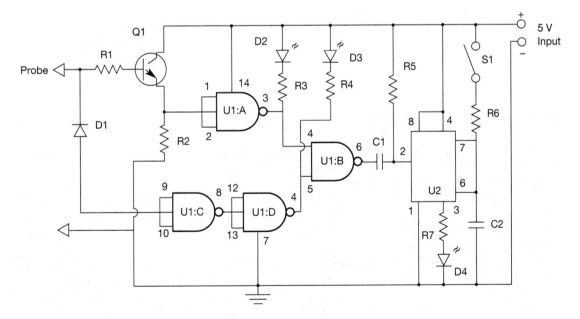

Figure 16-2 Logic probe circuit.

duration pulses, a green logic (0) State indicator, and a red LED or logic (1) state indicator. The heart of the logic probe is the LM 555 timer integrated circuit, a 7400 NAND gate IC, and a 2N2219 transistor. An input pulse is fed through R1 into the base of transistor Q1. Transistor Q1 in turn activates U1:A or U1:C, depending upon the pulse levels detected. Diode D1 is used to protect U1 and the LEDs from large reverse-input voltages. The LM555 is used to catch "fast" pulses through the probe's circuitry. The logic probe is powered from a 5 volt power supply. No power switch is needed in this circuit, since it is just powered up when it is being used.

In operation, the yellow LED comes on for approximately 200 milliseconds to indicate a pulse without regard to the pulses width. This feature enables you to observe very short duration pulses that might not be seen by on the logic 1 or logic 0 indicators. A small switch at S1 is used to keep the pulse LED on permanently after a pulse occurs. For a logic 0 input signal, the logic 0 green LED will come on and the pulse LED will light up, but after 200 milliseconds the yellow LED will go off. For a logic (1) state the red LD will come on. With switch S1 closed, the circuit will indicate whether a negative or positive going pulse has occurred. If the incoming pulse is positive-going the logic state (0) LED and the yellow pulse LED will turn on. If the pulse is negative-going the logic state (1) red and the yellow pulse LED will be lit.

The logic probe circuit can be built into a small "magic marker" housing if desired; any other suitable enclosure could be used. A small probe pin can be placed at the end of the housing to be used to "probe" the circuit under test. Any low-current 5 volt power supply can be used to power the circuit. The parts list for the logic probe follows.

R1	1 kohm, ¼ watt resistor	U1	7400 NAND gate
R2	390 ohm, ¼ watt resistor	U2	LM555 timer IC
R3, R4, R7	150 ohm, ¼ watt resistor	D1	1N914
R5	3.3 kohm, ¼ watt resistor	D2	red LED
R6	20 kohm, ¼ watt resistor	D3	green LED
C1, C2	0.01 µf, 35 volt disc capacitor	D4	yellow LED
Q1	2N2219 (A) transistor or NTE 123	S1	on–off switch (slide)
		Miscellaneous	PC board, wire, probe, clip leads

Wire-Tracer Circuit

The wire-tracer circuit can assist you in tracing telephone wires or computer wiring through walls and, in addition, the circuit can be used to inject audio signals into an audio amplifier to trace the amplifier stages when troubleshooting inoperative equipment. The operation of the wire tracer is straightforward. The transmitter portion (the circuit shown in Figure 16-3) is connected to one end of the wire line to be tested and a small AM transistor radio is placed near or connected to the other end of the wire under test. The ferrite loop antenna picks up the signal wire tracer's RF signal and produces a tone through the radio's speaker when in proximity to the tracer's signal.

The heart of the signal tracer is the 4001 CMOS gate. The tracer circuit uses all four sections of the 4001 IC. The 1 MHz crystal forms an oscillator that generates a 100 KHz RF signal at the output of U1:D. The output of the 4001 on pin 11 is fed to a 10 kohm audio taper potentiometer used to attenuate the output signal. The potentiometer is coupled to a 0.001 µF capacitor connected to a probe or clip lead, which is attached to the wire or circuit under test. The wire tracer is powered by a 9 volt transistor radio battery.

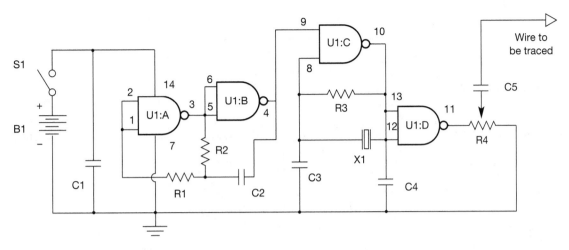

Figure 16-3 Wire-tracer circuit.

The battery should last for a long time since the circuit is only used intermittently. The wire tracer circuit can be built on a small printed circuit board and placed in a small metal or plastic enclosure.

After building the wire tracer, take a few moments to look over the circuit to make sure you have wired the components in correctly. The only critical component is the 4001 integrated circuit. It is advisable to use an IC socket on the printed circuit in case the IC ever gets damaged. The parts list for the wire tracer follows.

R1, R2	100 kohm, ¼ watt resistor	C5	0.001 µF, 35 volt ceramic disc capacitor
R3	10 megohm, ¼ watt resistor		
R4	10 k audio taper potentiometer, PC mount	U1	4001 CMOS IC
		S1	SPST toggle switch
C1, C2	0.001 µF, 35 volt ceramic disc capacitor	B1	9 volt transistor radio battery
		Miscellaneous	PC board, wire, probe/clip, battery leads, sockets, box, etc.
C3, C4	22 pF, 35 volt ceramic disc capacitor		

The operation of the wire tracer is simple. Install the battery in the wire tracer and switch the circuit on via S1. Place the tracer's signal output lead near the loop antenna of a small AM transistor radio. Tune the radio until the signal from the tracer is loudest. Then back down potentiometer R4 until the signal from the wire tracer is just barely heard on the radio. Your wire tracer is now ready for operation.

Zener Diode Tester

The zener diode tester shown in Figure 16-4 can save you time and money when testing new or used zener diodes. The heart of the zener diode tester is the LM555 timer integrated circuit at U1. The LM555 is

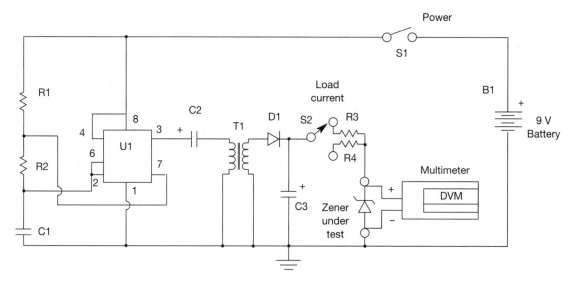

Figure 16-4   Zener diode tester.

set up to operate in the astable mode. Resistor R2 and capacitor C1 set the oscillation frequency of the timer IC. The output of U1 is coupled through a 4.7 μF capacitor to the input of the mini transformer. The transformer has a primary impedance of 1 kohms and a secondary impedance of 8 ohms. Used in reverse, the unloaded ac voltage is about 120 volts ac. The output of the transformer is next rectified by diode D1 and smoothed by capacitor C3, a 2.2 μF capacitor. Switch S2 is used to simulate two different load current characteristics that can be inserted into the circuit to evaluate the zener diode under test. The load current switch enables the zener diode to be tested at either 1 or 2 mA dc. The zener under test is measured with a multimeter set to dc as shown. A good zener diode should maintain a constant voltage reading on the dc voltmeter as the load current switch is toggled between both settings. The zener diode tester is powered from a 9 volt transistor radio battery.

The zener diode tester can be housed in one of those small plastic enclosures that have a small 9 volt battery compartment underneath. The power switch S1 and the load current switch S2 were both mounted on the top of the mini box. Two pin jacks are also mounted on the top of the box. The pin jacks are used to insert the zener diode leads for testing. An RCA jack was mounted on the side of the plastic box. A mating RCA plug can later be wired to your multimeter. The zener tester is simple to use and makes a nice addition to your electronics workbench. The parts list for the zener diode tester follows.

R1	2.2 kohm, ¼ watt resistor	D1	1N4004 silicon diode
R2	82.0 kohm, ¼ watt resistor	T1	1 kohm to 8 ohm mini inter-stage
R3	22 kohm, ¼ watt resistor		transformer (LT700)
R4	10 kohm, ¼ watt resistor	U1	LM555 timer IC
C1	4.7nF, 35 volt capacitor	S1, S2	SPST toggle switch
C2	4.7 μF, 35 volt electrolytic capacitor	B1	9 volt transistor radio battery
C3	2.2 μF, 150 volt electrolytic capacitor	Miscellaneous	PC board, pin jacks, battery clips,
			wire, etc.

Visual Diode/Transistor Curve Tracer

A low-cost diode/transistor curve tracer is shown in Figure 16-5. It is a very useful device for quickly analyzing diode and transistor characteristics. The curve tracer will help you determine if a diode or transistor is good or leaky and it will also test for open or shorted conditions as well. Figure 16-6 illustrates the curve tracer waveforms displayed on an oscilloscope screen.

The low-cost curve tracer consists of only two basic components: a low-voltage 110 V ac to 6.3 V ac step-down transformer and a 270 ohm, ½ watt resistor. The primary of the transformer is plugged into the wall outlet. One end of the secondary of the transformer (point A) is connected to the horizontal input of an oscilloscope. The other secondary transformer lead (point B) on the input of the 270 ohm resistor is connected to the vertical input of the oscilloscope. Point C is connected to the ground terminal of the oscilloscope as shown.

The diode/transistor curve tracer can be mounted in a small aluminum chassis box, which will also house the transformer and resistor. A ¼ amp fuse (F1) was mounted on the rear panel of the chassis, along

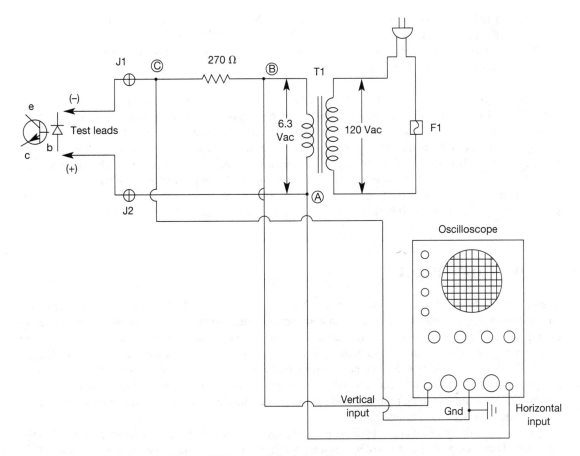

Figure 16-5 Visual diode/transistor curve tracer.

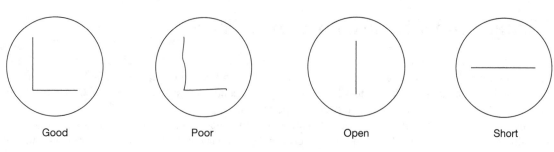

Good   Poor   Open   Short

Figure 16-6 Curve tracer waveforms.

with the line cord. The front panel of the chassis can be used to mount two terminal or binding posts at J1 and J2, which are connected to a pair of multimeter probes. The three leads from points A, B, and C can all be brought out to three-pin jacks on the front or rear panel of the chassis box as desired. Pin plugs can be used to connect the curve tracer to the oscilloscope. The parts list follows.

T1	6 volt transformer, 500 mA	J1, J2, J3, J4	"pin jacks" for DUT
D1	1N4001 silicon diode	J5, J6	RCA jacks to scope
R1, R2	470 ohm, ½ watt resistors	Miscellaneous	wire, chassis box, etc.
R3	5 ohm potentiometer		

Once the diode/curve tracer is completed, you can begin testing diodes and transistors. The test lead at (point C) or J1 is the ground or minus lead, and the probe lead at J2 is the plus lead. Connect a diode between the plus and minus test probe leads and the oscilloscope should display one of the waveforms shown in Figure 16-6. Remember, a transistor is two diodes back to back. In order to test a transistor, you will have to make two test configurations. Place the minus test lead on the base lead of the transistor and then place the plus test lead on the collector of the transistor and observe the oscilloscope. Finally, place the minus lead on the base of the transistor and the plus led on the emitter lead of the transistor and observe the oscilloscope screen. If both tests indicate a good condition, then the transistor is good.

Transistor Tester

Even if you are fairly adept at electronic troubleshooting, it can still be difficult to zero in on a defective circuit component. Often, you are faced with confirming your suspicion that a transistor is defective. To do this, you will most likely have to unsolder and remove the transistor or diode from a crowded PC area and test it with a transistor tester or an ohmmeter. If your guess is wrong, you will have to resolder the device to the PC board, risking excessive-heat damage to the transistor and possibly to the circuit board.

A much better way to troubleshoot transistorized circuits is provided by an in-circuit transistor tester, which eliminates the foregoing problems and saves a lot of time and frustration. You can build the one described below for about $25 or less.

The transistor tester's small test clip leads make it easy to connect it directly to the leads of any bipolar transistor. The in-circuit transistor tester uses a pair of light emitting diodes, one each for NPN and PNP transistors, to indicate whether a suspect transistor is good or bad and simultaneously identify it by type.

If either LED lights, the transistor is good, and if the two LEDs alternately flash or do not flash at all during a test, the transistor is bad.

The transistor tester, shown in Figures 16-7 and Figure 16-8, is simple to build, straightforward in design, and makes use of a single CMOS NAND-gate Schmitt trigger integrated circuit, identified as U1. The 4093 IC is a bistable device that does not respond directly to an input signal. Rather, its snap action response, known as "hysteresis," creates a dead band that is useful for cleaning up slow and noisy digital signals.

The dead band is the result of the fact that a Schmitt trigger's input voltage must rise above a certain level, known as the high trip point, before the output voltage can change. Similarly, the input voltage must drop below a certain level, which is lower than the high trip point, before the output can change once again. Any change that occurs within the dead band itself has no effect on the output.

Built-in hysteresis makes it possible to design and build an astable oscillator with just a single gate, as shown. Frequency of oscillation is calculated using the formula $F = 1/(1.4\ RC)$. With the component values shown in the schematic diagram and specified in the parts list for R1 and C1, oscillator frequency is approximately 7 Hz. All remaining gates in U1 are used simply for inversion and buffering. The first of these gates following the oscillator isolates the latter from the rest of the circuit to prevent frequency drift as a result of loading. The remaining two gates are used to produce the complementary outputs needed to enable testing of both NPN and PNP transistors.

Note that both inputs of each of the four gates that make up U1 are tied together. This is done because only the Schmitt trigger inversion, not the normal NAND function, is required in this project. The complementary outputs produced by the final two gates in U1 are connected to light emitting diodes D5 and D6 through current-limiting resistor R4. Only one current-limiting resistor is needed in this arrangement because only one of the LEDs will normally be on at any given time.

The LEDs are connected in a reverse condition as shown, that is, the anode of one is connected to the cathode of the other and the remaining two leads are tied together. With this arrangement, when a logic

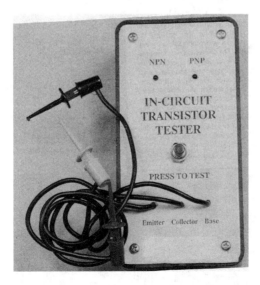

Figure 16-7　In-circuit transistor tester.

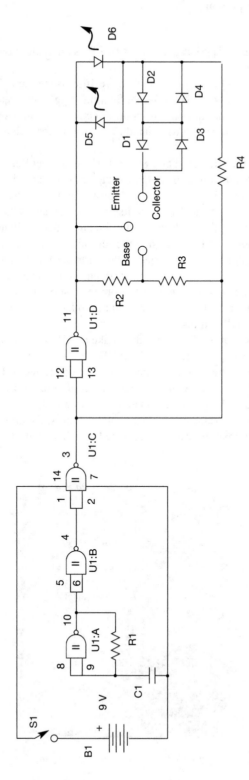

Figure 16-8 In-circuit transistor tester schematic.

high voltage appears at pin 11 of U1 and the voltage at pin 3 is at a logic low voltage, D5 is held in reverse bias while D6 conducts and the LED lights. With no transistor connected to the circuit, the outputs at pins 3 and 11 of U1 change at a frequency of about 7 Hz, allowing the LEDs to blink on and off alternately.

In addition to being connected to the LEDs, the complementary outputs at pins 3 and 11 of U1 are applied to the resistor network composed of R2 and R3. The junction of these two resistors is brought out as a test point and is connected to the base of the transistor under test. The emitter of the transistor connects directly to pin 11 of U1, and the collector connects to the D1 through D4 parallel diode arrangement and D5 and D6 antiparallel LED arrangement.

An important purpose is served by this strange diode arrangement. When a transistor under test has an internal short circuit between its collector base or base-emitter junctions, the good half of the transistor acts like an ordinary diode and will normally conduct and indicate a "good" transistor. When either D1 and D2 or D3 and D4 are conducting, a drop of about 1.2 volts appears across the operating pair. This voltage adds to the voltage dropped across the transistor being tested. If the transistor is good, the drop will be about 0.1 volt, and the total drop across the LEDs will be 1.3 volts for the half cycle that the transistor is conducting.

On other hand, if the transistor being tested has a base-emitter or base-collector short, the 1.2-volt drop across the diodes adds to another 0.6-volt drop across the bad transistor to produce a drop of 1.8 volts. This is enough to turn on the LED. Therefore, internal short circuits will cause both LEDs to alternately flash.

Construction

Because the circuitry for this project is very simple, just about any wiring technique can be used to build the project. For example, you can build your own printed-circuit board or you can assemble the circuit on perf-board with holes on 0.1 inch centers.

With the PC board component-side up and properly oriented, start populating it by installing and soldering into place a 14 pin DIP IC socket in the U1 location. An IC socket is optional but highly recommended should the 4093 IC ever have to be replaced. Next, install diodes D1 through D4 on the lower left of the board, taking care to properly orient each, and solder all leads to the copper pads.

Install and solder into place the resistors, clipping off excess lead lengths. Make sure to properly polarize the electrolytic capacitor as shown and plug its leads into the specified holes in the board.

Prepare three 18 inch lengths of miniature test-lead wire by stripping ¼ inch of insulation from both ends of each piece. Tightly twist together the fine wires at both ends of each wire and sparingly tin with solder. Plug one end of these wires into the holes labeled E, B, and C in the upper left of the board and solder into place.

You can use any suitably sized project box as an enclosure for the tester. An ideal enclosure to use has an aluminum front panel and measures 5 inches long by 2.5 inches wide by 1.5 inches deep. Only the front panel requires machining to prepare it for housing the tester. A parts list for the transistor tester follows.

R1	100 kohm, ¼ watt resistor	U1	4093 CMOS, quad two-input
R2, R3, R4	330 ohm, ¼ watt resistor		NAND Schmitt trigger
C1	1 µF, 25 volt electrolytic	S1	SPST power switch
	capacitor	B1	9 volt transistor radio battery
D1, D2, D3, D4	1N914 silicon diodes	Miscellaneous	PC board, pin jacks, IC socket,
D5, D6	Red LEDs		plastic enclosure, wire, etc.

Checkout

Snap a fresh 9 volt battery onto the connector. Touch the common lead of a dc voltmeter, set to indicate at least 10 volts, to the pin 7 connector and the "hot" lead to the pin 14 connector of U1's socket. Press the switch's pushbutton and note the reading on the meter. If you do not obtain a reading of approximately 9 volts, release the switch and carefully check all components for proper orientation, soldering bridges, and poor connections. Correct the problem before proceeding. Once you obtain the proper 9 volt dc reading between pins 14 and 7 of the IC socket, release the switch's pushbutton and install the 4093 in the IC socket. Make sure you orient the IC as shown and take care to avoid having any pins overhang the socket or fold under between the socket and the IC body as you push the 4093 home. Exercise the usual precautions for MOS devices when handling the IC. If you are soldering the IC directly on the board, use a soldering iron with a grounded tip.

With the IC installed, press and hold down the switch's pushbutton and observe the activity of the LEDs. If everything is working properly, two LEDs will alternately flash on and off. If both LEDs flash in step with each other, the two are not properly connected. In this event, power down the circuit and transpose the leads of one LED. If neither LED flashes, you have an error in construction and must recheck your wiring to correct the problem. Having obtained proper flashing, release the switch's pushbutton and connect to the test leads a good transistor. You can use either a PNP or an NPN transistor. Once again, press and hold the switch's pushbutton and note which LED lights. Mount this LED in the appropriate hole in the front panel of the enclosure. That is, if you are using an NPN transistor for this test, the LED that lights goes into the NPN hole in the panel. The other LED then goes in the remaining LED hole. If the LEDs tend to fall out of their holes, secure them in place with either clear nail enamel or fast-set epoxy cement.

Using the in-circuit transistor tester is almost self-explanatory. You simply determine the base, emitter, and collector leads of the transistor to be tested, clip the test-lead connectors to them accordingly and press the "Press To Test" pushbutton while observing the LEDs. If the transistor is good, either the NPN or the PNP LED will light up, simultaneously identifying the type of transistor under test and verifying that it is good. If both LEDs or neither LED lights, the transistor is almost certainly bad, regardless of type, and should be replaced.

Capacitance Leakage Meter

A capacitance leakage meter will help you find leaky capacitors that most capacitor checkers miss (see Figure 16-9). The new low-cost digital capacitance testers make capacitor checking quick and easy, but they have shortcomings because they do not check capacitors for leakage and series resistance.

After spending a few hours troubleshooting a piece of electronic equipment, you find some defective components and replace some defective capacitors with units from your "junk box." After hours of additional troubleshooting, you discover that those replacement capacitors are no good after all, even though they checked out OK on your new digital capacitor tester. That's because they are leaky, and your meter is one of the many that has no leakage-test function. That "missing feature" wound up costing you much time and aggravation.

Perhaps you just bought some electrolytic capacitors from your dealer, installed them, and watched as the power supply fuse blew. You later discover that the cause was those "new" capacitors; they had be-

Figure 16-9 Capacitance leakage tester.

come leaky after sitting on your dealer's shelf for the better part of a decade. If you had only known that, those capacitors could have been rejuvinated with a shot of current from the proper source.

Don't trash your capacitance tester just because it doesn't have a leakage-test function. Instead, supplement it with the leakage tester described. Our tester checks the all-important leakage parameter quickly and easily, weeding out defective capacitors that otherwise test good. It can also bring those elderly "new" capacitors back to life fast.

Besides checking capacitor leakage, the circuit has many other uses on the bench and in the field. For example, you could use it to test insulation resistances on power tools and appliances. If you find just one tool or appliance with a dangerous fault, you'll be very glad you took the time to build and use the tester. You can also test suspected lossy cables, as well as high-voltage diodes, rectifiers, neon lamps, and other high-voltage components that are often difficult to troubleshoot with conventional equipment.

How it Works

The leakage tester is basically a regulated dc power supply with a metering circuit to indicate leakage current, as shown schematically in Figure 16-10. In operation, 9 volts from a plug-in transformer passes through switch Sl-a and is rectified by diodes D1–D4. If field operation is required, it is possible to substitute a 12 volt battery for the plug-in transformer. The switch is a DPST unit that selects either power on or capacitor discharge. From the rectifier, the output is filtered by capacitor C1 and is then used to power the rest of the circuitry.

Op-amp U1 serves as an error amplifier. Basically, that amplifier is set up as an inverting gain-of-100 unit. Resistors R3 and R5 set the gain value. The op-amp's job is to ensure that the voltage applied to the capacitor under test is regulated. It does that by "sampling" the voltage at capacitor C5 through the range switch. The range switch is nothing more than a resistive attenuator network reducing the output voltage to about 1.2 volts. The op-amp simply adjusts the power supply until its minus (inverting) input equals 1.2

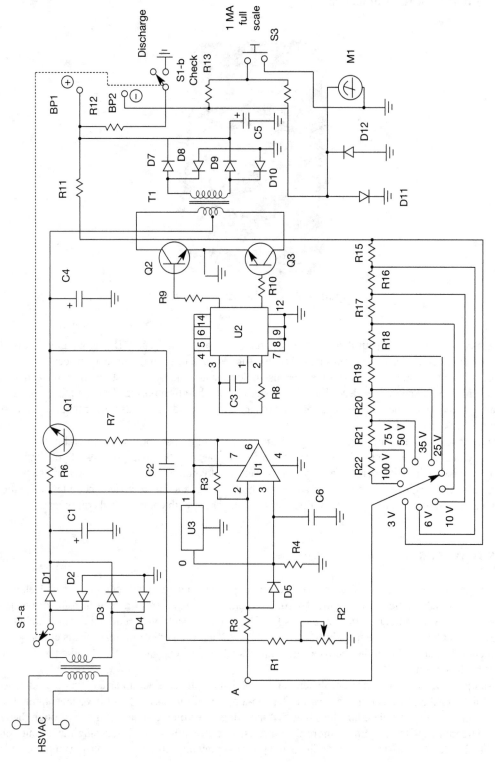

Figure 16-10 Capacitance leakage tester schematic.

volts, which regulates the output voltage. Diode D5 serves as overload protection, preventing excessive voltage from range switch S2 from damaging U1. That condition might occur if you were to switch rapidly from 100 volts to 3 volts. And finally, capacitor C2 provides some ac feedback, ensuring stable operation over a wide range of capacitor loads.

Transistor Q1 serves as a control element for the rest of the power supply. Since the op-amp can't provide enough current to do the job directly, a Darlington power transistor is used here to boost the current. Resistor R6 limits current to the rest of the circuitry, preventing transformer damage. Moving on, the dc output from capacitor C1 powers U2, which is a CMOS, programmable one-shot wired as a 100 Hz oscillator with complimentary outputs. The outputs from U2 alternately drive transistors Q2 and Q3, which serve as switches. They alternately switch each side of the transformer T1 winding to ground, generating current pulses.

Transformer T1 performs two purposes. First, it steps up the current pulses so that they can be rectified. Second, under heavy load (as from a charging capacitor) it saturates, limiting the output current to the capacitor to about 20 mA. That forms a constant-current type power supply, which is especially effective in charging test capacitors.

The output from transformer T1 is rectified by diodes D7 to D10 and filtered by capacitor C5. From that point, the dc output feeds back to the range switch, which is used along with op-amp UI to set the output voltage. The dc output also drives the test capacitor, which is connected to the project through binding posts.

Meter M1 is included in the minus leg of the test capacitor for monitoring the charging and discharging currents. Note that resistor R12 and switch S1-b are included to discharge the test capacitor when the power is turned off.

Next, look at integrated circuit, U2. Resistor R8 and capacitor C3 set the operating frequency to 100 Hz. That frequency, while not critical, was chosen to prevent "beats" with the 60 Hz power line and permits increased output from transformer T1.

Finally, look at the metering circuit. Diodes D11 and D12 are included to protect the meter from harmful overloads, especially when a large capacitor is being discharged. A 10 mA current shunt consisting of resistors Rl3 and R14 is also included for measuring currents in noncapacitor testing applications. That shunt can be selected via pushbutton switch S3.

Construction

Transformer T1 can be any 12.6 volt, 1.2 ampere, center-tapped transformer, although it likely will have to be mounted off the board.

The capacitor leakage tester is constructed on a PC board to eliminate noise and wiring problems later. Once you have created your PC board, you can start construction. Start by placing the board in front of you with the foil side down. First, locate and install sockets for the ICs; this is a recommended in the event of a circuit problem either after the circuit is built or at a later date. Next locate and identify all the resistors that are needed. Pay particular attention to the resistors that have 1% values. Be sure to use 1% resistors where needed for accuracy in measurements that you will make. Once the resistors have been identified, they can be soldered to the circuit board. Next, locate the capacitors need for the circuit. Be sure to pay attention to polarity when installing the electrolytic capacitors. Locate all the diodes need for the circuit; again, be sure to observe the correct polarity when installing the diodes on the PC board. Install TIP120 transistors at Q3 to Q1 next. Note that the leads are bent back 90 degrees, allowing the transistors to be mounted with the metal tabs flush against the board. When installing the transistors, make sure that you know the pin-outs of the device before installing it into the board. Usually, the tab on the transistor is the collector, but if you are not sure, look up the transistor specification in a data book.

Place the PC board in the bottom of your chosen chassis box. Drill three mounting holes for the board, plus a ¼ inch hole in the top side for the power cord. Drill another ¼ inch hole in the bottom side for access to pot R2. Drill mounting holes for T1 where you will locate it. Note that transformer T1 is not mounted inside on the PC board, but on the chassis itself. Next, wire the transformer to the appropriate points on the PC board.

Complete the mechanical work by installing the switches, binding posts, and meter on the front panel. Connect the switches to the appropriate points on the board. Install the board in the box using 4-40 × 1 inch screws and nuts. Use a nut as a spacer between the board and box on each screw. Feed the power cord (from T2) through the hole that has been drilled for it and connect the cord to the appropriate pads on the board. Finish up the assembly by installing a CD4047 at U2 and an LF356 at UI. When installing the ICs in their respective sockets, be sure to orient them correctly to avoid damaging them when power is applied. A parts list for the device follows.

R1	8.2 kohm, ¼ watt, 5% resistor	C5	2 μF, 16 volt electrolytic capacitor
R2	5 kohm potentiometer	U1	LF356N op-amp
R3	100 kohm, ¼ watt, 5% resistor	U2	CD4047 CMOS, one-shot IC
R4	270 ohm, ¼ watt, 5% resistor	U3	LM317LE adjustable voltage
R5	10 megohm, ¼ watt, 5% resistor		regulator
R6	10 ohm, 2 watt resistor	Q1, Q2, Q3	2N3904, TIP120 NPN or RS
R7	2.2 kohm, ¼ watt, 5% resistor		(276-2068) transistor
R8	1 megohm, ¼ watt, 5% resistor	D1, D2, D3,	1N4004 silicon diode
R9, R10	10 kohm, ¼ watt, 5% resistor	D4, D6, D7	
R11	18.2 kohm, 18 watt, 1% metal film	D8, D19,	1N4004 silicon diode
R12	100 ohm, ¼ watt, 5% resistor	D10, D11	
R13	10 ohm, ¼ watt, 5% resistor	D5	1N4148 silicon diode
R14	68 ohm, ¼ watt, 5% resistor	M1	1 mA dc meter
R15	30.1 kohm, ⅛ watt, 1% metal film	S1	DPST toggle switch
R16	40.9 kohm, ⅛ watt, 1% metal film	S2	12 position, 1 pole rotary switch
R17	49.9 kohm, ⅛ watt, 1% metal film	S3	SPST pushbutton, normally closed
R18, R19	100 kohm, ⅛ watt, 1% metal film	T1	117 volt/12.6 volt, 1.2 center-
R20	150 kohm, ⅛ watt, 1% metal film		tapped transformer
R21, R22	249 kohm, ⅛ watt, 1% metal film	T2	117 volt/9 volt, 300 mA "wall
C1	1000 μF, 16 volt electrolytic capacitor		wart" transformer
C2, C6	0.1 μF, 50 volt polyester capacitor	Miscellaneous	PC board, IC sockets, chassis
C3	0.001 μF, 50 volt polyester capacitor		box, wire, connectors, etc.
C4	100 μF, 16 volt electrolytic capacitor		

Check your work carefully, re-check diode, capacitor, and transistor polarities before applying power to the circuit. When done, check your work carefully, correct any wiring errors, and look at the PC board for cold solder joints or stray wire leads.

Calibration

The final step is to calibrate the unit. Plug transformer T2 into a nearby AC outlet, then set Sl to the "discharge" position and likewise set S2 to the 100 volt position. Set your DMM to its 200 volt dc range and connect it to the binding posts (BP1 and BP2). Flip Sl to the "check" position and the DMM will read somewhere between 85 and 120 volts. If so, the project works and you can go to the calibration.

If you are having problems, disconnect the power and discharge capacitor C5 with a jumper wire, then check over your wiring for errors. Remember—when troubleshooting, always discharge C5 after turning the power off to prevent a dangerous shock.

Calibration is easy to perform. First set Sl to "discharge," and then set S2 to the 100 volt range. Then set your DMM to its 200 volt dc range and connect it across the binding posts. Flip Sl to the "check" position and adjust R2 until the DMM reads 100 volts.

To be on the safe side, you should check the output voltage for each position of S2; it should be within 2% of the panel value. If not, the 1% resistor associated with that position should be checked.

Using the Tester

This unit can provide a dangerous electrical shock, since it provides high voltage. Some typical applications for the project follow. Testing capacitors is easy. Remove the suspect unit from the circuit first, as any external leakage can cause a good capacitor to test leaky. Set Sl to the "discharge" position. Then set S2 to the working voltage of the capacitor. If the voltage is greater than the project can provide, use the 100 volt position. Then use insulated clips to connect the capacitor to the project.

Flip Sl to the "check" position. The meter needle will "kick" upscale and drop to zero for a good capacitor. If desired, press S3 for more sensitivity; that will give you 0–1 mA readings instead of the nominal 0–10 mA readings.

Note that the meter needle may kick upscale many times before settling down. That indicates that the capacitor has excessive leakage at the test voltage and requires "forming." This is a common situation with capacitors that are used at a voltage far lower than the working voltage, like, say, a 50 volt part in a 12 volt circuit, and is not a fault of the project. However, if the meter doesn't stop kicking after several minutes, the capacitor isn't forming and should be replaced.

When the testing is finished, return Sl to the "discharge" position. The meter will read negative, indicating discharge. When the needle returns to zero, remove the capacitor. How much leakage should a good capacitor have? There is no simple answer, since it depends upon the capacitor type and the circuit requirements. However, here are a few rules of thumb you can use. Paper, mica, polyester, and tantalum capacitors should display no leakage. Electrolytic capacitors up to 50 μF should show less than 25 μA of leakage current. Electrolytic capacitors from 51 μF to 500 μF should show less than 50 μA leakage. Electrolytic capacitors from 501 μF to 1000 μF should show less than 100 μA of leakage. Electrolytic capacitors from 1001 μF to 20,000 μF should show less than 500 μA of leakage.

The capacitance leakage tester can also be used to "form" new capacitors. Electrolytic capacitors chemically deteriorate when they sit unused. The capacitor's electrolytic film, essential to capacitor operation, deteriorates, causing very high leakage currents. If you apply a voltage to the capacitor, the film can reform. But if that voltage is too high, that is, close to the device's rating, the film may not be able to form fast enough. So the capacitor will consume power, get hot, and possibly explode. To solve that problem, set Sl to the "discharge" position and S2 to the 3 volt position. Then use insulated clips to connect the capacitor. Flip S1 to the "check" position and note the meter reading. When the meter reads minimum current, change S2 to the 6 volt position. Continue increasing the voltage until the working voltage is reached. Discard any capacitor that takes over five minutes to form or still has high leakage when checked at its working voltage.

You can also use the tester to check the safety or your household appliances. Set Sl to "discharge" and S2 to 100 volts. Assuming that the device to be tested uses a three-wire power cord, use an insulated clip lead to connect the "hot" side of the power cord to the positive binding post. After that, connect the ground lead of the power cord to the negative binding post. The return or common side of the power cord is unconnected. If the appliance does not use a three-wire power cord, connect the negative binding post to any exposed case screws or other metal surfaces on the unit to be tested. To test, flip S1 to the "check" position. Press S3 for more meter sensitivity. The meter must read under 50 μA. For higher readings, repairs are indicated.

Cables may be easily checked for leakage problems. Simply perform the checks the same way you did for appliance leakage. Remember to always connect the positive binding post to the center conductor and the negative binding post to the shield; that prevents a shock hazard. If you need more sensitivity when performing leakage testing, try using your DMM if it has a 200 μA dc range. Simply set it to the 200 μA range and connect it in series with the positive binding post and the cable under test. That technique works well for other applications, except testing high-value capacitors. With those, when you switch to "discharge," the current from the capacitor passes through the DMM, overloading it.

Another use for the project is quickly checking those special high-voltage diodes used in TV sets and microwave ovens. Usually, a DMM cannot check those components because it cannot supply enough voltage to turn on the device. To test the diode, set S1 to "discharge" and S2 to 100 volts. Then remove the diode from the circuit and connect it to the project's binding posts. Flip S1 to "check" and note the reading. Then return S1 to "discharge" and reverse the diode. Flip S1 to "check" again and note the reading. With most silicon diodes, we expect a fall-scale reading when the diode is connected one way and zero when it is reversed. With selenium diodes, the difference should be at least 100 to 1.

Electronic Fuse

How do you troubleshoot power-related problems without blowing fuse after fuse? Just use the electronic fuse! The electronic fuse is a sensitive fast-acting adjustable circuit breaker that will quickly become one of your most useful bench-top accessories. If you have been stumped by a faulty electronic circuit and consumed a number of costly or hard-to-locate fuses, you will appreciate this inexpensive circuit breaker. All you have to do is connect the electronic fuse, shown in Figure 16-11, to the device being repaired and then adjust the current threshold control to the value you need, anywhere from 1/10 to 10 amperes.

Additional applications for the electronic fuse include charging circuits for marine/mobile/ aircraft systems, as well as new circuit designs. The electronic circuit breaker could be used after the design of a new circuit to help choose the correct value fuse. The electronic circuit breaker is connected in place of the original fuse of the device under repair or test. If the breaker "trips," a red LED will light and power is cut off. When you are ready to continue, simply press the reset button.

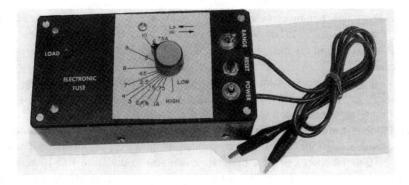

Figure 16-11   Electronic fuse.

Circuit Description

The electronic fuse circuit shown in Figure 16-12, utilizes an op-amp, which amplifies and rectifies the ac input and applies it to U2-a, and an LM339 comparator, which is used to adjust the threshold current via potentiometer R4. A clamp is formed by D3, it holds the input of U2-b to a constant level. A filtered dc output is amplified by U2-b and fed to Q1, a 2N3904 transistor. The transistor changes the output of U2-b

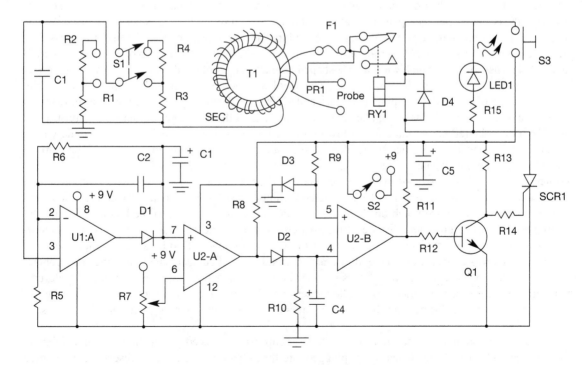

Figure 16-12   Electronic fuse circuit schematic.

to the proper level and polarity in order to trigger the SCR. When the input current exceeds the threshold set by R4, the SCR will turn on. The relay will now open and LED1 will indicate that the circuit has been "tripped." The LED will remain on and the power to the device under test will remain off until the reset button (S3) is pressed.

The two fuse substitution leads are connected in series with the normally closed relay contacts of RY1, a 12 ampere fuse (FI), and the two-turn primary of TI, a toroid transformer. The secondary of TI is wound underneath the primary on the half-inch toroid. The secondary coil is 100 turns of 30 gauge magnet wire with a total resistance of 8 to 10 ohms. The secondary is connected to a high–low "range" switch (S1). The switch connects to a resistor network to provide stability and ease of operation. The low range permits values from 1/10 to 6 amperes, and the high range includes values from 1 to 10 amperes, with overlapping between ranges. Capacitors C1 and C2 form a high-frequency filter to help reduce spikes and line noise.

Current consumption for the electronic fuse is about 10–15 mA at idle and about 100 mA when the relay is pulled in. Both integrated circuits are single-supply types, so any 12 volt battery or power supply can be used.

Construction

Everything except the relay and fuse are mounted on a PC board. If you construct a PC board, you have to drill the holes in the case cover very accurately in order to accept the switches, LED, and potentiometer directly from the circuit board. One way to deal with this problem is to drill a tiny pilot hole on the circuit board in the center of each component location that must come through the top cover. Then place the unpopulated circuit board directly onto the top cover and transfer the holes. The toroid transformer was constructed from a 0.5 inch powdered-iron toroid. A wire spool was made from a scrap of PC-board material, about 1¼ inches long by ¼ inch wide with V-shaped notches cut at both ends; 30 gauge magnet wire was wound on the spool between the two notches. The spool was then pushed in and around the core of the toroid (like a sewing needle), forming a 100 turn coil (T1's secondary) all the way around the entire toroid core (you un-spool the wire as you make the turns). The ends of the 30 gauge magnet wire were stripped and carefully soldered to 24 gauge wires. Five-minute epoxy was then brushed over the secondary coil. After the glue dried, the two splices were glued to the edge of the toroid with another spot of epoxy to reduce the stress on the 30 gauge wires.

The primary coil was wound over the secondary using two turns of 16 gauge wire with insulation heavy enough for about 12 amperes. Heavy line cord can be used for the primary if you like. The toroid was placed over the square notch on the end of the PC board and attached to the board with a plastic strip placed over the toroid and fastened with two screws. One of the 16 gauge wires was connected in series with the 12 ampere fuse; the other end of the fuse was connected to an alligator clip. The other 16 gauge wire was connected to one end of RY1's normally closed contact. The remaining relay contact was connected to another alligator clip. Note that the relay using the prototype is a double-pole unit with the contacts wired in parallel to handle higher current. The circuit board was mounted on the top cover of the plastic enclosure, and the relay was mounted inside the box along with the toroid, as seen in Figure 16-13.

In a later version of the electronic fuse, the alligator clips were replaced with a chassis-mounted female power receptacle. The device under test is plugged into the outlet on the electronic fuse and a 12 ampere

Figure 16-13   Mounted circuit board and relay.

fuse is placed in the fuse holder of the device being tested. The electronic fuse, set at the fuse value of the device being tested, will then fully protect the faulty circuit until you have located the problem. Then, simply replace the original value fuse in the circuit you just repaired. The parts list for the electronic fuse follows.

R1	107.2 kohm, ¼ watt, 1% resistor	U2	LM339 quad comparator
R2	442 kohm, ¼ watt, 1% resistor	D1, D2, D3	1N914 silicon diode
R3	387 kohm, ¼ watt, 1% resistor	D4	1N4004 silicon diode
R4	165 kohm, ¼ watt, 1% resistor	LED1	red LED
R5, R6	300 kohm, ¼ watt resistor	SCR1	NTE-5404 SCR
R7	50 kohm potentiometer	Q1	2N3904 NPN transistor
R8	1.5 kohm, ¼ watt, 5% resistor	T1	toroid on 0.5 inch powered-iron core (see text)
R9	12 kohm, ¼ watt, 5% resistor		
R10	18 kohm, ¼ watt, 5% resistor	S1	DPDT toggle switch
R11	13 kohm, ¼ watt, 5% resistor	S2	SPST toggle switch
R12	4.7 kohm, ¼ watt, 5% resistor	S3	pushbutton switch (normally closed)
R13	2 kohm, ¼ watt, 5% resistor	F1	12 ampere fuse (fast blow)
R14, R15	1 kohm, ¼ watt, 5% resistor	RY1	DPDT 12 volt coil, 12 ampere contacts (use two sets of contacts in parallel)
C1	200 pF, 50 volt ceramic capacitor		
C2	100 pF, 50 volt ceramic capacitor		
C3, C4	1 μF, 50 volt electrolytic capacitor	Miscellaneous	PC board, IC sockets, wire, hardware, 30 gauge magnet wire, alligator clips, plastic shuttle, etc.
C5	100 μF, 50 volt electrolytic capacitor		
U1	LM358 dual op-amp		

Operation

Operation of the electronic fuse is quite simple. The alligator clips connect to the fuse holder of the device under test, essentially substituting the electronic fuse for the fuse that was in the original circuit. First, choose the high- or low-sensitivity position of Sl; the low range covers 1/10 to 6 amperes and the high range covers 1 to 10 amperes with overlap between the two ranges. Next, adjust R7 for the current setting that best represents the desired fuse value. Turn on the power switch S2 and reset the electronic fuse by pressing S3. Now turn on the device being tested; if LED1 lights, the fuse is blown and you must reset the circuit by pressing S3. Continue to troubleshoot until the repair is completed.

Calibration of the electronic fuse was performed by using a 1200 watt heating element coil, but an electric fry pan or toaster could be used instead. The thermostat in a fry pan must be turned up to maximum or disabled. The heater is connected to the output of a variac and the input of the variac is connected in series with an ammeter and the electronic fuse (see Figure 16-14). The variac output is slowly stepped up in small increments. A calibration sheet is placed under R4's adjustment knob. Calibration must be done for both the high and low ranges. Begin by selecting the low range and turn R4 clockwise to about midway. Next, turn on the variac and adjust for about 1 ampere, then rotate R4 to the trip point. Place a pencil mark on the calibration sheet, back down the variac, and reset S3. Bring up the variac to the point you just marked for one ampere and watch the meter to ensure that you are drawing one ampere as the breaker trips. Now proceed with the next value. Adjust R4 past midway, set the variac for two amperes, and rotate R4 down to the trip point. Repeat the procedure for each fuse value in the low and high ranges.

Pulse Generator

If you enjoy doing analog or digital circuit work or digital troubleshooting, you will probably find a need for a pulse generator. Commercial pulse generator units are available but usually cost several hundred dol-

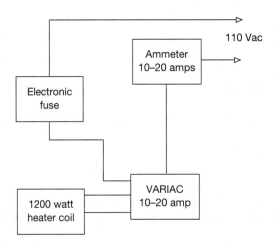

Figure 16-14   *Calibration setup for the electronic fuse.*

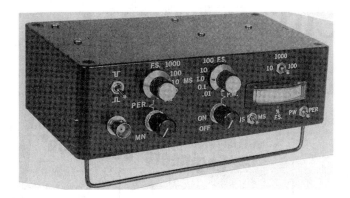

Figure 16-15   *Pulse generator.*

lars—too expensive for many hobbyists. The pulse generator shown in Figure 16-15 can be built for a reasonable cost using standard components. The pulse generator is divided into two parts: the actual pulse generator itself and an analog timer-period display You have option of constructing either section by itself as a stand-alone project or combining both parts into one unit. The pulse generator is powered by a single 9 volt battery.

The pulse generator supplies 5 volt pulses that can range from 1 microsecond to 100 milliseconds. The time period is independently adjustable between 10 microseconds and 1 second. The pulse section and the time-period section each have their own five-position range switch and fine-adjustment controls. The 5 volt amplitude of the pulses is only supplied under no-load conditions.

A specially designed output stage safely sources or sinks up 100 milliamperes to or from the load. The output was designed without any form of short-circuit protection in order to achieve very high speeds. Typical rise and fall times for the output are about 25 nanoseconds and either a positive- or negative-going pulse can be used. Direct interfacing to standard TTL or CMOS logic using 5 volt supplies is quite easy.

The time-period display section uses some unusual techniques to provide an analog display of either the period or pulse width. A 0–100 micro ampere analog panel meter provides the display and simply indicates 0% to 100% of the full-scale range selected. This section has its own individual range controls to cover 1.0 microsecond through 1.0 second with a typical accuracy of a couple of percent of full scale.

About the Circuit

The pulse generator's circuit is shown in Figure 16-16. It uses two ICs, four transistors, and a handful of passive components, which are commonly available. The heart of the circuit is U1, a TLC556 CMOS dual timer. That version of the timer is used because it is much faster than a standard bipolar 556 timer. Two independent timers in one package are needed. One timer section is set up as an astable multivibrator that supplies a basic square wave. That square wave sets the period function of the generator.

The overall period of the waveform is selected by S2 in five steps: 0.1, 1, 10, 100, and 1000 milliseconds. Fine adjustment is done with R4. That lets the period be continuously adjusted from 10% to 100%

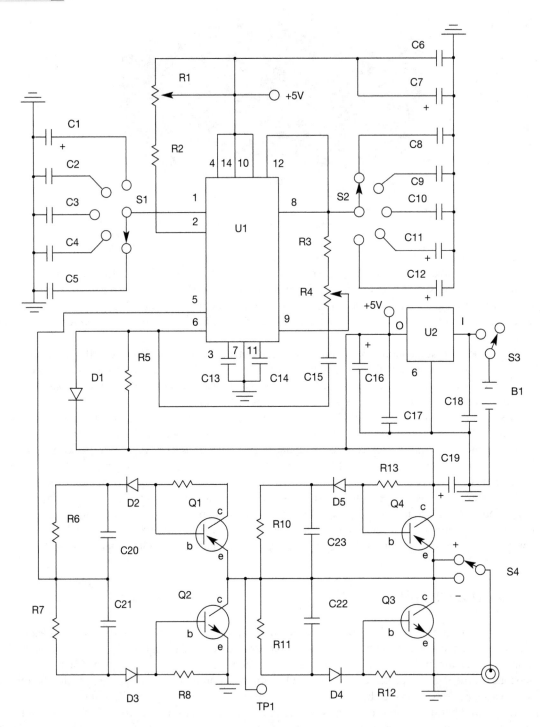

Figure 16-16   Pulse generator circuit.

of the period selection. The output of that section (U1, pin 9) is applied through C15 to the trigger input of the other section of U1.

The second timer section of U1 is configured as a monostable multivibrator. The signal from pin 9 through C15 to pin 6 triggers the pulse output at pin 5. The pulse at that output has a repetition rate set by the astable oscillator and a pulse width set by S1 and R1. The use of S1 is similar to S2 in that it sets the overall pulse width of the waveform in five steps: 0.01, 0.1, 1, 10, and 100 milliseconds. Similar to R4, R1 allows continuous adjustment of the pulse width from 10–100% of the selected range. The final waveform is then applied to the output driver section.

An inverter circuit made from Q1 and Q2 has a very fast rise time (less than 10 nanoseconds). A second identical inverter circuit is made from Q3 and Q4. Those two inverters let the final output be either a positive-going or a negative-going pulse. The polarity of the output pulse is selected with S4.

Display unit

The input for the time-period display circuit, shown in Figure 16-17, comes from TP1, the generator's inverting output from the collectors of Q1 and Q2. If you are building the display as a stand-alone instrument. the input needs a negative-going pulse of the proper width to be measured. The maximum positive voltage should be 5 volts with the minimum no higher than 100 millivolts.

The semiconductors were specially selected for this unusual, analog/digital hybrid-circuit design. CMOS ICs are used throughout to keep current drain to a minimum. This type of integrated circuit is sensitive to electrostatic discharge (ESD), as is U1. Analog panel-meter display MI can be any type of 0 to 100 μA meter that you would like to use.

The input pulse coming from TP1 Goes directly to U3, a CD4013 CMOS dual flip-flop. The flip-flop is configured in its "toggle" mode. Either the output of U3 or the original pulse width can be selected by S5. The selected pulse is applied to Q5, a 2N5117 dual matched-pair PNP transistor configured as a high-speed, gated current mirror. That current-source "integrator" turns on when the pulse signal is low-charging C24 or C25, whichever capacitor discharged is selected by switch S7-a, which determines whether the display will be in microseconds or milliseconds.

When the pulse goes high, the current source turns off and the voltage present on C24 or C25 represents the pulse width's integration period. At that point, one of U4s' one-shots is triggered. Its three microsecond output turns on one of the analog switches in fC5. A "sample" of the voltage on C24 to C25 is sent to C27, a "hold" capacitor through buffer amplifier U6-a. The voltage held by C27 drives MI through U6-b. As soon as the sample is taken, the other half of U4 is triggered, producing a reset pulse to two of the analog switches. The switches, wired in parallel, discharge C24 or C25 rapidly, making the circuit ready for the next pulse. The reset pulse is three microseconds long when S7-b is in the microsecond position; it is one millisecond long when S7-b is in the 1 millisecond position. That gives enough time for C24 or C25 to fully discharge. Meter M1 is a part of the feed- feedback loop for U6-b, which converts the hold voltage from C27 to a current that drives the meter. With 100 μA display, the meter reads directly as a 0–100% display of the full scale range selected by S6 and S7. The meter is protected from overrange currents by R19, and R20 lets the meter be calibrated for precise displays.

The timing diagram in Figure 16-18 shows how the display circuit works in with the controls set to generate and measure a pulse train with a 15 microsecond period and a 5 microsecond pulse width. Those settings readily demonstrate the integration, sampling, and reset functions of the display circuit. With a

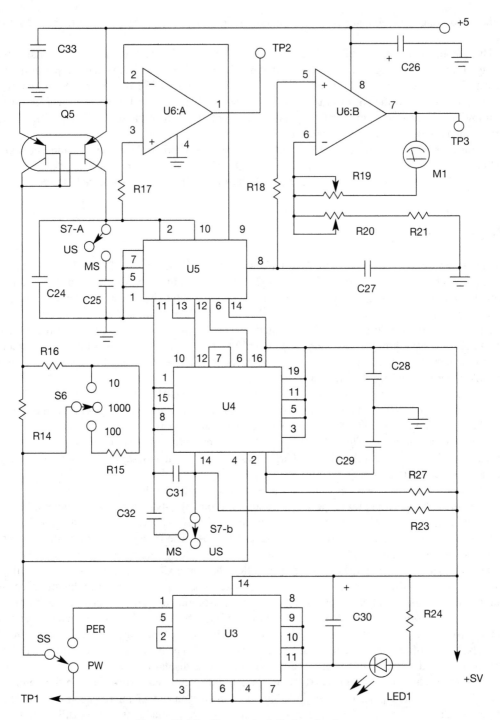

Figure 16-17 Time-period display circuit.

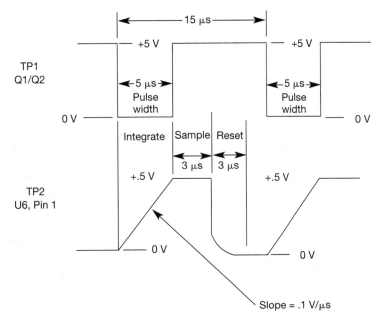

Figure 16-18   *Pulse width diagram.*

dual-trace oscilloscope, the two waveforms can be seen by connecting the scope's probes to TP1 (the inverting-pulse waveform) and TP2 (the sampling waveform).

Construction

The prototype of the pulse generator and the display board were both built on printed-circuit boards. Alternatively, the entire circuit can be built on perf-board using standard construction techniques. It is important. however to keep the lead lengths short. Another important consideration is to keep the passive components as close to their respective ICs as possible in order to prevent cross-coupling of signals. Range capacitors C1–C5 and C8–C12 can be mounted directly on S1 and S2; in that case, an unused lug on the switches can be used as a ground connection. It might be difficult to find a capacitor with the value specified for C8. As an alternative, test a group of 1500 pF units and select the one that has a value closest to the 1400 pF value that is needed. For the best possible accuracy, the actual value of C25 should be exactly 1000 times the value of C24. Also. the values of R14, R15, and R16 should be exact decade multiples of each other. One way to do that is to use parts that are held to tight tolerances. If those are either too expensive or difficult to find, once again select parts from a group of component using an accurate meter to measure the actual values of the parts.

Note that there are many controls mounted on the front panel and many connections will need to be made between the PC board and the front-panel components, so good initial positioning will let the leads remain as short as possible. A 2¼ inch by 4¼ inch by 7Á inch plastic enclosure will hold everything with room to spare. The parts list for the unit follows.

R1, R4	50 kohm, ¼ watt resistor	C18	47 μF, 25 volt electrolytic capacitor
R2	3.3 kohm, ¼ watt resistor		
R3	3.9 kohm, ¼ watt resistor	C19	10 μF, 25 volt electrolytic capacitor
R5, R22, R23	10 kohm, ¼ watt resistor		
R6, R7, R10, R11	1.8 kohm, ¼ watt resistor	C24	0.001 μF, 25 volt polyester capacitor
R8, R9, R12, R13	680 ohm, ¼ watt resistor		
R14	4.7 megohm, ¼ watt resistor	C25	1 μF, 35 volt polyester (see text) capacitor
R15	470 kohm, ¼ watt resistor		
R16	47 kohm, ¼ watt resistor	C27	0.01 μF, 35 volt polyester capacitor
R17, R18	1 kohm, ¼ watt resistor		
R19	50 kohm potentiometer	C29, C31	100 pF, 35 volt ceramic disk capacitor
R20	5 kohm potentiometer		
R21	7.5 kohm, ¼ watt resistor	C32	0.01 μF, 35 volt polyester capacitor
R24	1.2 kohm, ¼ watt resistor		
C1	2.2 μF, 16 volt tantalum capacitor	U1	TLC556 CMOS dual timer IC
		U2	78L05 regulator
C2	0.22 μF, 25 volt polyester capacitor	U3	CD4013 CMOS dual flip-flop
		U4	CD4538 CMOS dual one-shot IC
C3	0.022 μF, 25 volt polyester capacitor		
C4	0.0022 μF, 25 volt polyester capacitor	U5	CD4066 CMOS quad switch
		U6	TLC272 CMOS op-amp IC
C5	180 pF, 35 volt mica capacitor	Q1, Q4	2N3906 PNP transistor
C6, C17, C28, C30, C33	0.1 μF, 35 volt ceramic capacitor	Q2, Q3	2N3904 NPN transistor
		Q5	2N5117 dual matched-pair transistor
C7, C16, C26	1 μF, 35 volt tantalum capacitor		
		D1-D5	1N4148 silicon diode
C8	1400 pF mica /polyester capacitor	LED1	red LED
		S1, S2	single pole, five-position rotary switch
C9	0.015 μF, 35 volt polyester capacitor		
C10	0.15 μF, 35 volt polyester capacitor	S3	SPST switch (toggle)
		S4, S5	SPDT switch (toggle)
C11	1.5 μF, 25 volt polyester capacitor	S6	SPDT, center off (toggle)
		S7	DPDT switch (toggle)
C12	15 μF, 25 volt tantalum capacitor	M1	0–100 μA panel meter
		J1	BNC female connector (chassis mount)
C13, C14	0.01 μF, 35 volt ceramic disk capacitor	B1	9 volt battery
		Miscellaneous	PC board, wire, battery clips, IC sockets, hardware, etc.
C15, C20, C23	220 pF, 35 volt ceramic disk capacitor		

Once the case has the controls mounted in it and the board has been double-checked for wiring accuracy and proper component polarities, install the board on spacers in the case and connect all of the panel components to the board. Install a fresh battery and the pulse generator is ready for testing and calibration.

Calibration

Before turning on the power, set the front panel controls as follows: S1 to 1 millisecond, R1 to its "off" position, S2 to 1 millisecond, and R4 to full clockwise. For the display portion of the circuit, S5 should be set to "period," S6 to "1000," and S7 to "microseconds."

Connect an oscilloscope probe to TP1 and set the scope controls to display a 5 volt square wave with a 1 millisecond period. Turn on power to the unit and advance to its midrotation position. The oscilloscope should be showing a rectangular waveform with 1 millisecond time period. The square wave's amplitude should be 5 volts. If no waveform is present, check pin 9 of U1. If the square-wave is present at that location, double the value of C15 and check TP1 again. If necessary, C15 may be increased to about 600 pF before the overall performance of the circuit starts to fall off. If, on the other hand, there is no waveform present at pin 9, check for any wiring errors or replace U1. With the oscilloscope probe on TP1, adjust R4 for an exact 1 milllisecond time period. Now set S6 to 100 microseconds. That will put M1 in an overload condition. Rotate R19 back and forth until M1 reads somewhere on scale. Advance R19 so that M1's pointer needle pegs gently against its full-scale stop. That setting will prevent M1 from seeing high-current overloads by forcing U6-b's output to operate near its maximum positive value. Set S6 back to 1000 and adjust R20 so that M1 displays exactly 100 (100 μA). As R19 and R20 might interact, recheck both of those last adjustments and readjust as necessary for best results. Leaving S6 set at 1000, set S1 to 0.1 millisecond. Adjust R4 for a waveform time-period display on the oscilloscope of exactly 100 microseconds. The reading on M1 should be 10% or 10 μA. If not, there is an offset voltage error in the meter. It is easily corrected by changing the mechanical setting of M1 for a 10% reading. If you make that adjustment, it will then be necessary to start the display calibration over again.

Now, we will set the portable pulse generator to create the waveforms shown in Figure 16-18 to see how well the unit is working. Set the front-panel controls as follows: S1 to "0.01 milliseconds," R1 to midrotation, S2 to "0.1 milliseconds," R4 to midrotation, S5 to "pulse width," S6 to "10," and S7 to "microseconds." Adjust R1 for a display of 50% on M1, which indicates a pulse width of 5 microseconds. Set S5 to "period" and S6 to 100. Adjust R4 to display 15% on M1 for a time period of 15 microseconds. In order to obtain correct measurements, always set the generator controls for a longer time period than the desired pulse width before attempting any measurements!

With an oscilloscope attached to TP1, the waveform should be the same as in Figure 16-18. You should see a nice negative-going pulse 5 microseconds wide that repeats every 15 microseconds. Set S5 to "pulse width" and attach the oscilloscope probe to TP2 (pin 1 of U6). You will now be able to see the entire integrate/sample/reset waveform sequence.

Examine the waveform at TP3, located at pin 7 of U6. You should see a perfectly straight, horizontal line that represents the output voltage There might be some small glitches occurring at the points where the circuit switches on or off. If the line has a ramp-like appearance, U5 might be too "leaky." Most accuracy problems can be solved by replacing U5 with a good unit. Using a standard, good-quality coaxial cable such as RG-58 or RC-174 with a BNC connector at each end, connect J1 directly to the oscilloscope. The cable should be less than 3 feet in length for best results. The 15 microsecond waveform should be seen. Switching S4 back and forth will change the polarity of the pulse. With a high-quality oscilloscope, you can also see the very fast (about 25 nanoseconds) rise and fall times the pulse. Rotate R1 and R4 back and forth to see that there is a linear change in time period and pulse width.

Additions

If desired, you could also add short-circuit protection to the output driver stage of the pulse generator. One method is to install resistors of about 22 ohms in series with the emitter of Q1–Q5. A simpler method is to install a second output jack that is connected to J1 with a 100 ohm resistor between the center terminals of both jacks. The second jack can be used as a protected output, with J1 remaining as a direct output. Of course, peak output-pulse voltage will drop accordingly when those methods are used.

A standard coaxial cable will provide an excellent output signal from the pulse generator, even without impedance matching. From a practical point of view, most applications only need a pair of standard clip leads that have been connected to either a BNC connector or a dual-binding-post adapter.

Keeping in mind the power supply requirements and the load that will be driven by the pulse generator's battery and select the cable impedance that will be needed to drive the output load. Select an appropriate jack for the additional output jack, either a BNC female jack or an "F" jack for 75 ohm lines. Then connect a proper source resistor (50 ohms or 75 ohms) between the center conductors of J1 and the new output jack. Connect the proper cable between the new output jack and the load, and terminate the cable at the load with the proper terminating resistor. The jack on the generator can be a standard RCA phono jack with clips on the other end of the twin-lead wire. The generator's 9 volt battery is enough to drive that type of load, but keep in mind that when using terminated lines, the peak voltage to the load will be only 50% of the output voltage at J1.

Since the pulse generator runs on a 5 volt supply it can drive standard 5 volt TTL or CMOS logic directly. It is possible to also drive the newer 3 volt logic systems. In that case, simply use one of the balanced-impedance, terminated-cable methods mentioned earlier. That will supply a 2½ volt (maximum) pulse to the load.

As you can see, the pulse generator is a versatile and accurate instrument. It is simple to use and easy to construct, making it a great addition to your test bench.

Inductance/Capacitance (L/C) Meter

The Inductance/capacitance meter or L/C meter will find a prominent place on your electronics workbench from day one. Whether you are testing "junk box" or new components, you will find that the L/C meter shown in Figure 16-19 will help you enormously. The L/C meter uses a PIC microprocessor to perform special-purpose calculations to compute inductance and capacitance. The L/C meter features a 16 character intelligent LCD display with four-digit resolution and five operating modes. You can readily test inductor values from 0.001 μH to 150 mH and capacitor values from .010 pF to 1.5 μF.

The Circuit

The L/C meter is comprised of major blocks or sections: the oscillator, the preprogrammed microprocessor or PIC, and the LCD display. The heart of the L/C meter circuit is the oscillator. The oscillator is centered around U1 the LM311 voltage comparator, shown on the left of the circuit diagram in Figure 16-20. When power is applied to the circuit, the voltage at pin 2 is 2.5 volts, causing the output to be at a level of

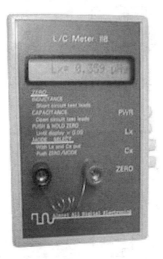

Figure 16-19 *Inductance/capacitance meter. (Courtesy Almost All Digital Electronics.)*

5 volts. This charges capacitor C4 through resistor R4 until the voltage at pin 3 equals 2.5 volts. As it reaches 2.5 volts, the output switches to a low level, inducing a transient into the tank circuit composed of L1 and C1. The transient causes the tuned circuit to ring at its resonant frequency. The ringing causes a square wave at the resonant frequency to appear at the output of the voltage comparator. The square wave is coupled back to the tuned circuit through R3 and C3, sustaining oscillation. For the nominal values of L1 (68 mH) and C1 (680 pF), an increase in L of 1 nHy (0.001 mHy) or an increase in C of 0.01 pF produces a frequency change of slightly more than 5 Hz. A 0.2 second measuring period can resolve 5 Hz and, therefore, 0.001 mHy or 0.01 pF. The oscillator circuit is very reliable, starts up immediately, and can tolerate a large variation in the inductance and capacitance used in the tank circuit.

The preprogrammed PIC microprocessor at U2 receives the output of the oscillator, which is applied to the RTCC (real-time clock counter) at pin 3. This increments an 8 bit counter inside the microcomputer. The microcomputer accumulates the count for a period of 0.4 seconds. The frequency is then the accumulated count divided by the period. Discrete signals from the Lx, Cx and zero switches are input to the microcomputer so it knows what the operator wishes it to do. If you choose not to build the kit version of the L/C meter and you are able to program your own blank microprocessor, then you can refer to the listing in Table 16-1.

To zero Ls, the operator must short-circuit the test leads, press Lx, and then press the zero button. Similarly for capacitors, the operator open-circuits the test leads, presses Cx, and then presses zero. The stored values of Ls and Cs are saved until the operating mode is changed. When measuring components, it is not necessary to rezero between components. When the operating mode is changed from measure to match, these values are reset to zero:

$$Cx = (F1^2/F2^2 - 1)C1$$

You will notice from the above equation that inserting an unknown always causes F2 to be less than F1. If an inductor is inserted when the Cx switch is depressed, the result will be an increase in frequency, F2

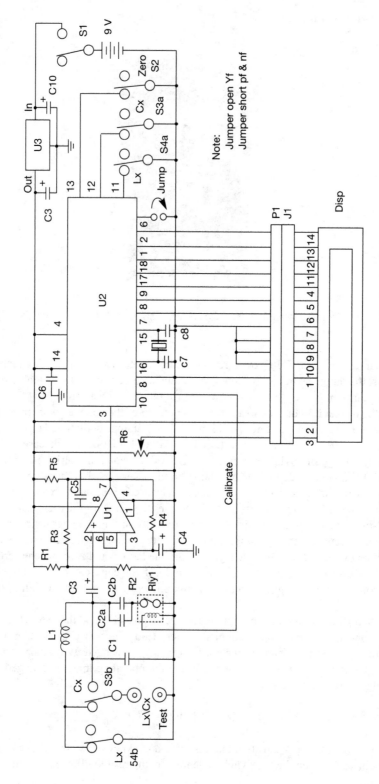

Figure 16-20 Inductance/capacitance meter circuit schematic.

Table 16-1 *L/C meter program listing*

Initialize the CPU and I/O Ports
Initialize the LCD Display
While Lx or Cx are on
      Display "Switch Error"

WEND
(The computer cannot calibrate itself if Lx or Cx are on. The unit waits for the operator to clear the switches.)
Display "Wait" (wait 10 seconds for the oscillator to stabilize).
Calibrate:
      Display "Calibrating"
      Measure F1
      Switch in the calibrating capacitor
      Measure F2
      Switch out the calibration capacitor
      Compute $C1 = F2^2 / F1^2 - F2^2) \, C2$
      Compute $L1 = 1 / (4 \, p^2 \, F1^2 \, C1)$
DO (Loop continuously)
      IF Lx and Cx are off
          IF Zero
              Goto Calibrate (recalibrate the unit)
          ELSE
              Display "Ready" (ready to measure Lx, Cx or be Zeroed)
              Measure and Store F1
          END IF
      ELSEIF Lx on and Cx off
          Measure F2
          IF Zero On
              Measure and Store F1
              Display "0.000"
          ELSE (Zero Off)
              Compute $Lx = (F1^2 / F2^2 -1) \, L1$
              Display "Lx ="
              Display Value in Engineering units
          END IF
      ELSEIF Cx on and Lx off
          Measure F2
          IF Zero On
              Measure and Store F1
              Display "0.000"
          ELSE (Zero Off)
              Compute $Cx = (F1^2 / F2^2 - 1) \, C1$
              Display "Cx ="
              Display Value in Engineering Units
          END IF
      ELSE (Lx and Cx both On)
          Display "Switch Error"
      END IF
Loop

greater than F1, rather than a decrease. This is because the inductor has been placed in parallel with L1 and inductors in parallel always are less than the value of the smallest of the two values. If the unit detects an increase in frequency, it will display "not a capacitor." This does not work for very large values of Lx. The decrease in the effective value of L1 is trivial, whereas the shunt capacitance of the large inductor is significant and the frequency will decrease causing an erroneous reading. The effect of putting a capacitor in when the Lx switch is pressed is similar except that the oscillator tends to stop, causing F2 to be zero. The unit detects this and displays "not an inductor." This is not true for very large values of Cx, in which case the unit may display an erroneous reading. Tables 16-2 and 16-3 illustrate the display options and modes of operation of the L/C meter.

L/C Meter can zero out any value in its range. If a value is inserted and zeroed, the unit will display the difference between it and subsequent components, similar to the MATCHnMODE and MATCHuMODEs. The difference in the MATCHxMODEs is that the range is frozen to the resolution of the initial component. This limits the minimum difference in values to 1 part in 10,000 or 0.01%. The reason for this may not be obvious. The maximum resolution of the unit is four digits at the value of the components being measured. Consider two components, one with an exact value of 5000 pF and the other with an exact value of 5010.25 pF. The difference would be 10.25 pF; however, the unit cannot resolve less than 1 pF at this range and it would be misleading to display the fractional portion of the difference. Note that a positive reading in the matching modes means Lx is greater than Lz, or Cx is greater than Cz, and vice versa.

L/C meter is intended to measure inductors and capacitors "out of the circuit." Inductors must have a reasonable Q for their value and negligible distributed capacitance for their value. I have tested it using commercially available RF chokes ranging from 0.1 microhenry to 1000 micohenry, hash chokes up to 100 micohenry wound on ferrite rods, on Pi-wound RF chokes up to 7.5 millihenry, on toroid-wound inductors up to 150 millihenry (such as the HI-Q series obtainable from Mouser Electronics), and on several slug-tuned inductors from a Coilcraft Slot-10 designer's kit (similar to the TOKO line of tunable inductors).

Construction

The L/C meter was assembled on the glass epoxy circuit board shown in Figure 16-21. Construction is simple. Begin by inserting the integrated circuit sockets for the U1 and U2. Sockets will save you lots of

Table 16-2   L/C meter display options

Inductance	In	Inductance	C	Capacitance	C	Capacitance
nano mode	mi	micro mode	na	nano mode	mi	micro mode
000–999 nH	0.0	0.000–0.999 μH	0.0	0.00–0.99 pF	0.0	0.00–0.99 pF
1.00–9.999 μH	1.0	1.000–9.999 μH	1.0	1.00–9.99 pF	1.0	1.00–9.99 pF
10.00–99.99 μH	10	10.00–99.99 μH	10	10.00–99.99 pF	10	10.00–99.99 pF
100.0–999.9 μH	10	100.0–999.9 μH	10	100.0–999.9 pF	10	100.0–999.9 pF
1.000–1.999 mH	1.0	1.000–1.999 mH	1.0	1.000–9.999 nF	1	1000–9999 pF
10.0–99.99 mH	10	10.00–99.99 mH	10	10.00–99.99 nF	0.0	0.01000–0.09999 mF
100.0–999.9 mH	10	100.0–999.9 mH	10	100.0–999.9 nF	0.1	0.1000–0.9999 mF
				1.0	1.000–9.999 μF	1.000–9.999 μF

Table 16-3 *L/C meter modes*

When the Lx and Cx switches are off, pressing the ZERO button sequences the L/C meter through five different operating modes.

1. READY MEASURE n measures Lx or Cx and displays the result in "nano mode"; i.e.: Lx = 99 nHy, Cx = 12.34 nF.
2. READY MEASURE u measures Lx or Cx and displays the result in "micro mode"; i.e.: Lx = 0.099 uHy, Cx = 0.01234.
3. READY MATCHnMODE first measures a reference component Lz or Cz and displays the value in "nano mode." When the zero button is pressed, this value is stored in RAM and the difference between it and subsequent components is displayed in "nano mode"; i.e.: Lx – Lz = 99 nHy, Cx – Cz = 12.34 nF.
4. READY MATCHuMODE first measures a reference component Lz or Cz and displays the value in "micro mode." When the zero button is pressed, this value is stored in RAM and the difference between it and subsequent components is displayed in "micro mode"; i.e.: Lx – Lz = 0.099 uHy, Cx – Cz = 0.01234 μF.
5. READY MATCH%MODE first measures a reference component Lz or Cz and displays the value in "nano mode." When the zero button is pressed, this value is stored in RAM and the ratio of the difference between it and subsequent components is displayed in percent; i.e.: (Lx – Lz)/Lz × 100 =12.34%, (Cx – Cz)/Cz × 100 = 12.34%.

headaches if the circuit fails at some later date and you need to troubleshoot it. Next, insert the resisters followed by the capacitors. When installing the capacitors, be sure to observe the proper polarity to avoid damaging the meter circuit when first applying power. Next, install the crystal followed by the relay; be careful when installing the relay, as you will have to orient it properly to ensure that the coil and the contact pins are in the correct holes. Next, install inductor L1. Best results are obtained if the terminal connected to the outer winding layer is grounded (toward the bottom of the PCB). It will work and be just as accurate anyway, but will have less stray inductance to zero out; this is probably due to reduced effect of stray capacitance to adjacent components. Note: there is only ⅜ inch space under the display, so leave enough lead length to tip taller parts at an angle so that the vertical dimension does not exceed ⅜ inch.

Precision alignment of all four switches is required for easy installation. Align the pins parallel to each other by eye (particularly the two rows of pins when viewed from the rear of the switch). Start by inserting

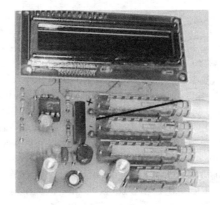

Figure 16-21 *L/C meter circuit board. (Courtesy Almost All Digital Electronics.)*

one or two rows of pins with the switch at an angle to the PCB. Turn the assembly upside down, with the switch resting on top of your bench, and press down on the PCB. Do not press down on the switches as there is a risk of pushing the pins all the way out of the body of the switch. Begin by starting the first two pins in their pads on the PCB. Next, turn the board over on your bench, resting on the top of the switch. Finally, press down with both thumbs, rocking gently back and forth. Use equal pressure to avoid slipping sideways and bending the switch pins. Now, install the voltage regulator U3 in its proper place. Finally, you can install the connector jacks starting with J1, the female square post header, at the top of the PCB. Solder just one pin, then make sure the connector is perpendicular to the PCB and solder the rest. This should complete PCB assembly. The parts list follows.

R1, R2, R3	100 kohm, ¼ watt resistor	U1	LM311N voltage comparator
R4	47 kohm, ¼ watt resistor	U2	PIC16C622 microcomputer
R5	1000 ohm, ¼ watt resistor	U3	LM78L05 voltage regulator
R6	10 kohm potentiometer	RLY1	SPST-N.O. reed relay (has diode)
C1	680 pF, 500 volt ceramic disc (marked 681)	DISP	LM-16151 or equivalent
C2a	1000 pF, 2%, 35 volt capacitor	J1	14 pin square-post socket (built onto the display module)
C2b	5, 10, 15, 20, 24, 27, 33, or 39 pF NPO (to make 1020 pF total)	P1	14 pin square-post plug (install on PCB)
C5, C6	0.1 mF ceramic (monolithic, marked 104)	S1	SPST alternate-action pushbutton (power)
C3	10 µ, 10 volt tantalum	S2	DPDT momentary pushbutton switch (zero)
C4, C9, C10	10 mF, 10 volt electrolytic capacitor (black radial, observe polarity)	S3	DPDT (Lx) alternate-action pushbutton switch
C7, C8	22 pF ceramic (marked 22J)	S4	DPDT (Cx) alternate-action pushbutton switch
X1	8.0 MHz crystal		
L1	68 mH (blue)	Miscellaneous	PC board, test jacks, five-way binding posts, wire

Calibration

If you purchased the L/C meter in kit form, it will be supplied with a small plastic enclosure. You will need to pass the leads from the battery clip through one of the slots in the case and solder them to the appropriate pads of the PCB. Plug in the display, turn the contrast control fully clockwise, and turn on the unit. The unit will display "wait" for 10 seconds, followed by "calibrating" for two seconds, followed by "ready measure x." If so, you are up and running. Adjust the contrast control so the background is just barely visible. Install the PCB in the bottom of the case using three #4 sheet metal screws. Install the top cover of the case and install the binding posts. Test leads should not exceed 4 inches in length with a banana plug at one end and alligator clip at the other. It may be necessary to move the edge of a hole or slot in the case. This is easily done using sandpaper, a file, or a hobbyists knife. Before fitting the test jacks or screws in the back of the case, fit the cover and squeeze the case together while testing the switches for binding to the edge of the slot. Move the edge as required.

Operation of the L/C meter is quite straightforward. To measure inductance, place the unknown across the test leads and depress Lx. To measure capacitance, place the unknown across the test leads and press Cx. The oscillator tends to drift a few hertz during the first few minutes of operation. When measuring very small values, the unit should be allowed to warm up for about five minutes. With a resolution of 5 Hz, thermal drift will always occur, as evidenced by a slowly drifting reading. The first readings after pressing Lx or Cx are the most accurate. Note that the typical stray inductance is 0.04 to 0.06 mH and the typical stray capacitance is 5 to 7 pF. When measuring inductors less than 5 mH or capacitances less than 50 pF, it is advisable to zero the unit first. For larger values, the strays are insignificant to the result. It is difficult to retain a reading of 0.000 pF because of the extreme sensitivity of the unit. Your body capacitance influences the reading. Try zeroing the capacitance and then move your hands around the test leads without touching them. You will find that you can adjust the reading a few hundredths of a pF.

Accuracy and resolution of the L/C meter are specified at 1% of reading. The L/C meter has four-digit resolution, which for small values of L and C are 1 nH and 0.01 pF. You cannot accurately measure values this small as the resolution greatly exceeds the accuracy. You can measure values as small as 0.01 mH and 0.1 pF with about 15% accuracy. You generally won't find components this small. For example, a piece of wire less than one inch long is 0.01 mH. The resolution is, however, relative and can be used for sorting a batch of similar components, as it truly does indicate which are slightly larger or smaller than others. Also, for small values of inductance, the leads will contribute quite a bit to the value. Measuring from the ends of the leads instead of next to the body of the component can add up to .025 mH.

In the event that your L/C meter does not spring to life correctly, here are some hints on where to look.

1. Blank display, contrast control not adjusted correctly. Start with it fully clockwise.
2. Blank display. Check 5V power to CPU and display.
3. Displays 8 black squares. CPU not communicating with display. Check solder around CPU and display. CPU crystal not oscillating. Check with oscilloscope if possible.
4. Displays "wait," then "calibrating," and sticks in "calibrating." (A) Oscillator (LM311) is not oscillating. Check soldering around LM311. LM311 properly installed, parts properly installed? C3 in backward? (B) Zero button stuck in or not soldered. Check continuity to ground from pin 13 of the CPU.
5. Seems to work but readings appear way off from components' marked values. Calibration capacitor not correctly installed or relay in backwards. (relay should be installed with its part number opposite the switches (facing the LM311) and little circle toward the top of PCB).

The L/C meter is available in kit form for $99.00 from Almost All Digital Electronics, 1412 Elm St. S.E., Auburn, WA 98092 http://www.aade.com/lcmeter.htm.

Appendix

Electronics Workbench Resources

All Electronics Corp
P.O. Box 567
Van Nuys, CA 91408
888-826-5432
Fax: 818-781-2653
ww.allcorp.com

Allied Electronics
7410 Prebble Drive
Fort Worth, TX 76118
800-433-5700
www.alliedelec.com

Allstar Magnetics
6205 NE 63rd St.
Vancouver, WA 98661
360-693-0213
www.allstarmagnetics.com

American Design Components
6 Pearl Court
Allendale, NJ 07041
800-803-5857
www.adc-ast.com

American Power Conversion
132 Fairgrounds Road
West Kingston, RI 02892
800-788-2208
www.apcc.com

Amidon Inductive Components
240 Briggs Ave
Costa Mesa, CA 92626
800-898-1883
www.amidon-inductive.com

Anchor Electronics
2040 Walsh Ave
Santa Clara, CA 95050
408-727-4424
www.demoboard.com/anchorstore.htm

Antique Electronic Supply
6221 South Maple Ave
Tempe, AZ 27468
480-820-5411
www.tubesandmore.com

Atlantic Surplus Sales
3730 Nautilus Ave
Brooklyn, NY 11224
718-372-0349

B.G. Micro
555 North 5th Street, Ste 125
Garland, TX 75040
800-276-2206
www.bgmicro.com

Circuit Specialists
220 S. Country Club Drive, Bldg #2
Mesa, AZ 85210
480-464-2485

Coilcraft
1102 Silver Lake Rd
Cary, IL 60013
847-639-6400
www.coilcraft.com

Contact East. Inc
335 Willow Street
North Andover, MA 01845
978-682-9844
www.contacteast.com

Dan's Small Parts
Box 3634
Missoula, MT 59806
406-258-2782
www.fix.net/dans.html

DC Electronics
P.O. Box 3203
2200 N. Scottsdale Rd.
Scottsdale, AZ 85271
800-467-7736
www.dckits.com/index.htm

Digi-Key Corp
701 Brooks Ave South
Thief River Falls, MN 56701
800-344-4539
www.digikey.com

Edlie Electronics, Inc
2700 Hempstead Turnpike
Levittown, NY 11756
800-647-4722
www.edlieelecronics.com

Electric Rainbow, Inc
6227 Coffman Road
Indianapolis, IN 46268
www.rainbowkits.com

Fair Radio Sales Company
2395 St. Johns Road
P.O. Box 1105
Lima, OH 45802
419-227-1313
www.fairradio.com

Gateway Electronics
8123 Page Blvd
St. Louis, MO 63130
800-669-5810
www.gatewayelex.com

Hammond Mfg. Company, Inc
256 Sonwil Drive
Cheektowaga, NY 14225
www.hammondmfg.com

Herbach & Rademan
353 Crider Ave
Moorestown, NJ 08057
www.herbach.com

Hosfelt Electronics
2700 Sunset Blvd
Steubenville, OH 43952
www.hosfelt.com

Howard W. Sams, Inc
5436 W. 78th Street
Indianapolis, IN 46268
800-428-7267
www.samswebsite.con/index.html

Industrial Safety Company
1390 Neubrecht Rd
Lima, OH 45801
800-809-4805
www.indlsafety.com

International Components
175 Marcus Blvd
Hauppauge, NY 11788
www.icc107.com

Jameco Electronics
1355 Shoreway Rd
Belmont, CA 94002
800-831-4242
www.jameco.com

James Millen Electronics
P.O. Box 4215BV
Andover, MA 01810
978-975-2711
www.jamesmillenco.com

JDR Microdevices
1850 South 10th Street
San Jose, CA 95122
www.jdr.com

Kepro Circuit Systems, Inc
3640 Scarlet Oak Blvd
St. Louis, MO 63122
800-325-3878
www.kepro.com

MAI/Prime Parts
5736 N Michigan Rd
Indianapolis, IN 46208
317-257-6811
www.websitea.com/mai/index.html

Marlin Jones & Associates
P.O. Box 12685
Lake Park, FL 33403
800-652-6733
www.mpja.com

Maxim Integrated Products
120 San Gabriel Drive
Sunnyvale, CA 94086
408-737-7194
www.maxim-ic.com

Metal and Cable Corp
9337 Ravenna Rd, Unit C
P.O. Box 117
Twinsburg, OH 44087
330-963-7246
www.metal-cable.com

Mini Circuits Labs
P.O. Box 350166
Brooklyn, NY 11235
800-654-7949
www.minicircuits.com

Motorola Semiconductor
5005 East McDowell Rd
Phoenix, AZ 85008
512-891-2030
www.mot.com

Mouser Electronics
1000 N. Main Street
Mansfield, TX 76063
800-346-6873
www.mouser.com

National Semiconductor Corp
P.O. Box 58090
Santa Clara, CA 95052
800-272-9959
www.national.com

Newark Electronics
4801 N. Ravenswood Ave
Chicago, IL 60640
800-463-9275
www.newark.com

Nuts & Volts Magazine
430 Princeland Court
Corona, CA 92879
800-783-4624
www.nutsvolts.com

Ocean State Electronics
P.O. Box 1458
6 Industrial Ave
Westerly, RI 02891
401-596-3080
www.oselectronics.com

Palomar Engineers
P.O. Box 462222
Escondido, CA 92046
760-747-3343
www.palomar-engineers.com

Pasternak Electronics
P.O. Box 16759
Irvine, CA 92623
949-261-7451
www.pasternak.com

Radio Shack Corporation
100 Throckmorton St, Suite 1800
FT. Worth, TX 76102
817-415-3700
www.radioshack.com

Ramsey Electronics, Inc
793 Canning Parkway
Victor, NY 14564
www.ramseyelectronics.com

Small Parts, Inc
13980 NW 58th Ct.
P.O. Box 4650
Miami Lakes, FL 33014
www.smallparts.com

Solder-it Company
P.O. Box 360
Chargrin Falls, OH 44022

Southern Electronics Supply
1909 Tulane Ave
New Orleans, LA 70112
www.southernele.com

Surplus Sales of Nebraska
1502 Jones Street
Omaha, NE 68102
www.surplussales.com

Toroid Corporation of Maryland
202 Northwood Drive
Salisbury, MD 21801
410-860-0302
www.toroid.com

The Wireman Inc
261 Pittman Road
Landrum, SC 29356
864-895-4195
www.thewireman.com

Zero Surge Inc
889 State Rte 12
Frenchtown, NJ 08825
www.zerosurge.com

Used Test Equipment Sources

www.electrorent.com
www.rentelco.com
www.trsonescource.com
www.techremarketing.com
www.metrictest.com
www.naptech.com

www.testequity.com
www.tucker.com
www.testmart.com
www.techrecovery.com
http://test.labx.com

Tools for Electronics

HMC Electronics
http://www.hmcelectronics.com/cgi-bin/dsp/ht/

Jensen Tools
http://www.jensentools.com/default.asp

Efston Science
http://www.escience.ca/circuitT/

Webtronics
http://www.web-tronics.com/hand-tools-for-elec-tronics.html

Techra Tools
http://www.tecratools.com/

ElecTools
http://www.electools.com/

Techni-Tools, Inc
1547 N. Trooper Rd. Worcester, PA 19490-1117

Harbor Frieght Tools
3491 Mission Oaks Blvd
Camarillo, CA 93011
http://www.harborfreight.com

Contact East
http://www.contacteast.com

Cyberguys
http://www.cyberguys.com

Electronics Workbench Sources On-line

http://www.kasachorych.opole.pl/electronics-workbench—electronics-workbench/
http://www.listaintl.com/files/tekbench.htm
http://www.flexiblefurnature.com
http://www.phoenixdistributing.com
http://work-benches.com

Books—Schematic Reading

How to Read Schematic Diagrams, by D. Mark
How to Read Schematics, by Donald E Herrington
How to Read and Interpret Schematics Diagrams, by J. Richard Johnson
Beginner's Guide to Reading Schematics, by Robert J. Traister

Equipment Manuals

http://www.tech-systems.net (scopes)
http://www.radioera.com (radio)
http://www.w7f9.com/manuals (radio)
http://www.sario.com (radio)
http://www.fullnet.com/u/tomg.manuals.htm (general)

Resistor Color Codes

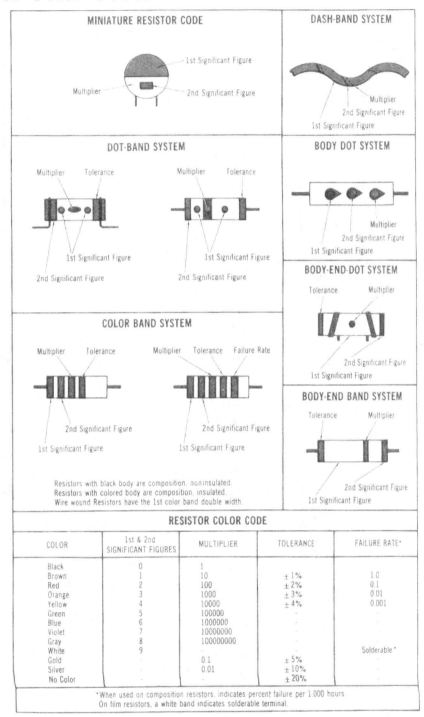

MINIATURE RESISTOR CODE

1st Significant Figure

Multiplier

2nd Significant Figure

DASH-BAND SYSTEM

Multiplier
2nd Significant Figure
1st Significant Figure

DOT-BAND SYSTEM

Multiplier Tolerance

1st Significant Figure

2nd Significant Figure

Multiplier Tolerance

1st Significant Figure

2nd Significant Figure

BODY DOT SYSTEM

Multiplier
2nd Significant Figure
1st Significant Figure

BODY-END-DOT SYSTEM

Tolerance Multiplier

2nd Significant Figure
1st Significant Figure

COLOR BAND SYSTEM

Multiplier Tolerance

2nd Significant Figure

1st Significant Figure

Multiplier Tolerance Failure Rate

2nd Significant Figure

1st Significant Figure

BODY-END BAND SYSTEM

Tolerance Multiplier

2nd Significant Figure
1st Significant Figure

Resistors with black body are composition, noninsulated.
Resistors with colored body are composition, insulated.
Wire wound Resistors have the 1st color band double width.

RESISTOR COLOR CODE

COLOR	1st & 2nd SIGNIFICANT FIGURES	MULTIPLIER	TOLERANCE	FAILURE RATE*
Black	0	1		
Brown	1	10	± 1%	1.0
Red	2	100	± 2%	0.1
Orange	3	1000	± 3%	0.01
Yellow	4	10000	± 4%	0.001
Green	5	100000	.	
Blue	6	1000000	.	
Violet	7	10000000	.	
Gray	8	100000000	.	
White	9			Solderable *
Gold	.	0.1	± 5%	
Silver	.	0.01	± 10%	
No Color	.	.	± 20%	

*When used on composition resistors, indicates percent failure per 1,000 hours
On film resistors, a white band indicates solderable terminal.

Ohm's Law Nomograph

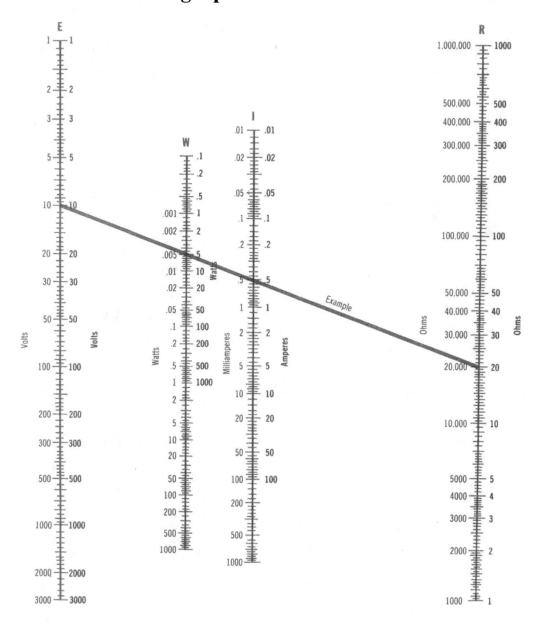

Parallel Resistance Nomograph

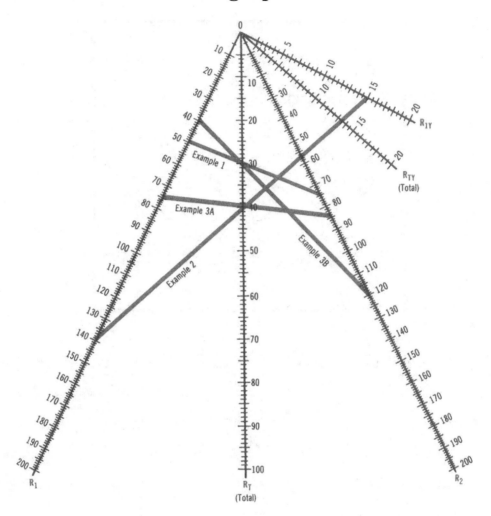

Resistor Formulas

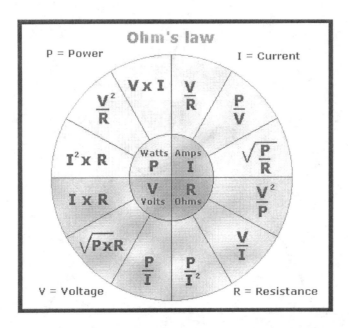

Basic Rules of Symbolic Logic

Symbol	Logic	Switch	Meaning	Circuit
1	True	Closed	The statement is true. The circuit is closed.	
0	False	Open	The statement is false. The circuit is open.	
·	Series	A and B	A is in series with B.	
+	Parallel	A or B	A is in parallel with B.	
Ā or A′	Not A		Opposite of A (if A = 0, Ā = 1; if A = 1, Ā = 0).	

A true statement, and hence a closed circuit, is generally said to have a truth value of 1. Conversely, a false statement, and hence an open circuit, is generally said to have a truth value of 0. Applying the AND and OR (· and +) relations to the truth values (0 and 1) yields the multiplication and addition tables of binary arithmetic.

The various symbols are given in Table 6-5. Table 6-6 summarizes the various logical statements, explains their meanings, and shows the equivalent switch circuits for the statements.

Capacitor Color Codes

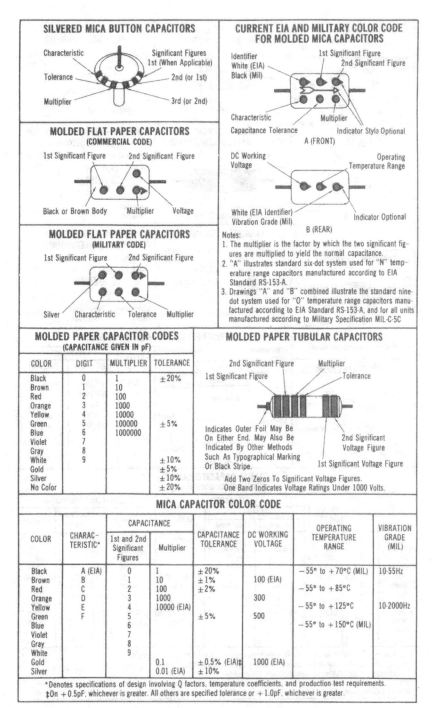

SILVERED MICA BUTTON CAPACITORS

- Characteristic
- Significant Figures 1st (When Applicable)
- Tolerance
- 2nd (or 1st)
- Multiplier
- 3rd (or 2nd)

MOLDED FLAT PAPER CAPACITORS
(COMMERCIAL CODE)

- 1st Significant Figure
- 2nd Significant Figure
- Black or Brown Body
- Multiplier
- Voltage

MOLDED FLAT PAPER CAPACITORS
(MILITARY CODE)

- 1st Significant Figure
- 2nd Significant Figure
- Silver
- Characteristic
- Tolerance
- Multiplier

CURRENT EIA AND MILITARY COLOR CODE FOR MOLDED MICA CAPACITORS

- Identifier White (EIA) Black (Mil)
- 1st Significant Figure
- 2nd Significant Figure
- Characteristic
- Multiplier
- Capacitance Tolerance
- Indicator Style Optional

A (FRONT)

- DC Working Voltage
- Operating Temperature Range
- White (EIA Identifier) Vibration Grade (Mil)
- Indicator Optional

B (REAR)

Notes:
1. The multiplier is the factor by which the two significant figures are multiplied to yield the normal capacitance.
2. "A" illustrates standard six-dot system used for "N" temperature range capacitors manufactured according to EIA Standard RS-153-A.
3. Drawings "A" and "B" combined illustrate the standard nine-dot system used for "O" temperature range capacitors manufactured according to EIA Standard RS-153-A, and for all units manufactured according to Military Specification MIL-C-5C

MOLDED PAPER CAPACITOR CODES
(CAPACITANCE GIVEN IN pF)

COLOR	DIGIT	MULTIPLIER	TOLERANCE
Black	0	1	± 20%
Brown	1	10	
Red	2	100	
Orange	3	1000	
Yellow	4	10000	
Green	5	100000	± 5%
Blue	6	1000000	
Violet	7		
Gray	8		
White	9		± 10%
Gold			± 5%
Silver			± 10%
No Color			± 20%

MOLDED PAPER TUBULAR CAPACITORS

- 1st Significant Figure
- 2nd Significant Figure
- Multiplier
- Tolerance
- Indicates Outer Foil May Be On Either End. May Also Be Indicated By Other Methods Such As Typographical Marking Or Black Stripe.
- 2nd Significant Voltage Figure
- 1st Significant Voltage Figure
- Add Two Zeros To Significant Voltage Figures.
- One Band Indicates Voltage Ratings Under 1000 Volts.

MICA CAPACITOR COLOR CODE

COLOR	CHARAC- TERISTIC*	CAPACITANCE		CAPACITANCE TOLERANCE	DC WORKING VOLTAGE	OPERATING TEMPERATURE RANGE	VIBRATION GRADE (MIL)
		1st and 2nd Significant Figures	Multiplier				
Black	A (EIA)	0	1	± 20%		− 55° to + 70°C (MIL)	10-55Hz
Brown	B	1	10	± 1%	100 (EIA)		
Red	C	2	100	± 2%		− 55° to + 85°C	
Orange	D	3	1000		300		
Yellow	E	4	10000 (EIA)			− 55° to + 125°C	10-2000Hz
Green	F	5		± 5%	500		
Blue		6				− 55° to + 150°C (MIL)	
Violet		7					
Gray		8					
White		9					
Gold			0.1	± 0.5% (EIA)‡	1000 (EIA)		
Silver			0.01 (EIA)	± 10%			

*Denotes specifications of design involving Q factors, temperature coefficients, and production test requirements.
‡0n + 0.5pF, whichever is greater. All others are specified tolerance or + 1.0pF, whichever is greater.

Capacitor Color Codes (*continued*)

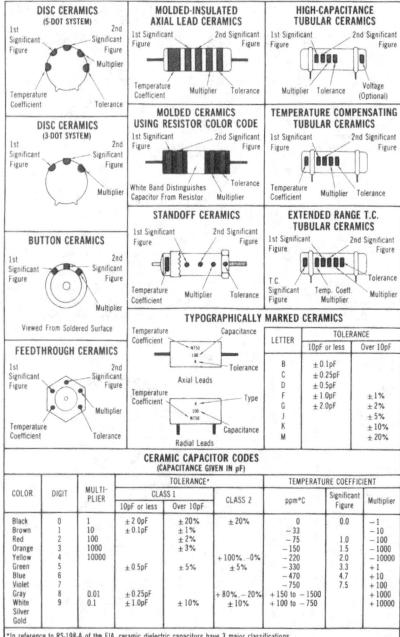

CERAMIC CAPACITOR CODES
(CAPACITANCE GIVEN IN pF)

| COLOR | DIGIT | MULTI-PLIER | TOLERANCE* | | | TEMPERATURE COEFFICIENT | | |
| | | | CLASS 1 | | CLASS 2 | ppm°C | Significant Figure | Multiplier |
			10pF or less	Over 10pF				
Black	0	1	± 2.0pF	± 20%	± 20%	0	0.0	− 1
Brown	1	10	± 0.1pF	± 1%		− 33		− 10
Red	2	100		± 2%		− 75	1.0	− 100
Orange	3	1000		± 3%		− 150	1.5	− 1000
Yellow	4	10000			+ 100%, −0%	− 220	2.0	− 10000
Green	5		± 0.5pF	± 5%	± 5%	− 330	3.3	+ 1
Blue	6					− 470	4.7	+ 10
Violet	7					− 750	7.5	+ 100
Gray	8	0.01	± 0.25pF		+ 80%, − 20%	+ 150 to − 1500		+ 1000
White	9	0.1	± 1.0pF	± 10%	± 10%	+ 100 to − 750		+ 10000
Silver								
Gold								

*In reference to RS-198-A of the EIA, ceramic dielectric capacitors have 3 major classifications
Class 1, temperature compensating ceramics requiring high Q and capacitance stability.
Class 2, where Q and stability of capacitance are not required.
Class 3, low voltage ceramics, where dielectric losses, high insulation resistance, and capacitance stability are not of major importance. Tolerance on Class 3 Ceramic Capacitors is indicated by code, either ± 20% (Code M) or + 80%, − 20% (Code Z).

Electronic Schematic Symbols

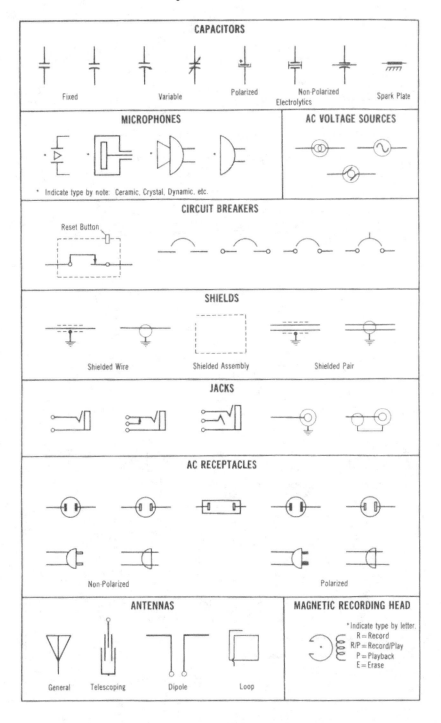

CAPACITORS

Fixed   Variable   Polarized   Non-Polarized
Electrolytics   Spark Plate

MICROPHONES

* Indicate type by note: Ceramic, Crystal, Dynamic, etc.

AC VOLTAGE SOURCES

CIRCUIT BREAKERS

Reset Button

SHIELDS

Shielded Wire   Shielded Assembly   Shielded Pair

JACKS

AC RECEPTACLES

Non-Polarized   Polarized

ANTENNAS

General   Telescoping   Dipole   Loop

MAGNETIC RECORDING HEAD

*Indicate type by letter.
R = Record
R/P = Record/Play
P = Playback
E = Erase

Electronic Schematic Symbols *(continued)*

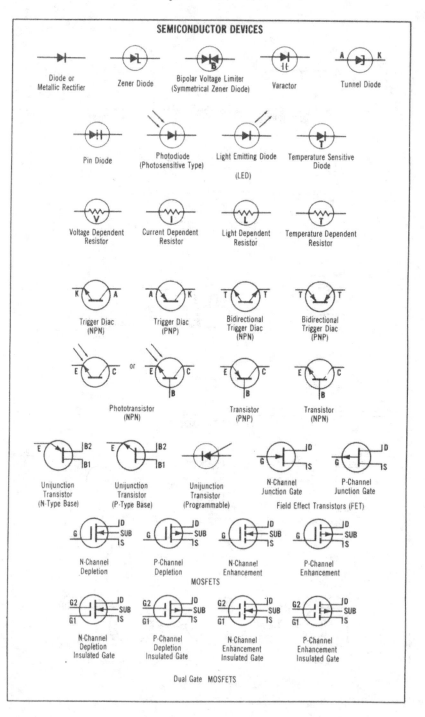

SEMICONDUCTOR DEVICES

Diode or
Metallic Rectifier

Zener Diode

Bipolar Voltage Limiter
(Symmetrical Zener Diode)

Varactor

Tunnel Diode

Pin Diode

Photodiode
(Photosensitive Type)

Light Emitting Diode
(LED)

Temperature Sensitive
Diode

Voltage Dependent
Resistor

Current Dependent
Resistor

Light Dependent
Resistor

Temperature Dependent
Resistor

Trigger Diac
(NPN)

Trigger Diac
(PNP)

Bidirectional
Trigger Diac
(NPN)

Bidirectional
Trigger Diac
(PNP)

Phototransistor
(NPN)

Transistor
(PNP)

Transistor
(NPN)

Unijunction
Transistor
(N-Type Base)

Unijunction
Transistor
(P-Type Base)

Unijunction
Transistor
(Programmable)

N-Channel
Junction Gate

P-Channel
Junction Gate

Field Effect Transistors (FET)

N-Channel
Depletion

P-Channel
Depletion

N-Channel
Enhancement

P-Channel
Enhancement

MOSFETS

N-Channel
Depletion
Insulated Gate

P-Channel
Depletion
Insulated Gate

N-Channel
Enhancement
Insulated Gate

P-Channel
Enhancement
Insulated Gate

Dual Gate MOSFETS

Electronic Schematic Symbols *(continued)*

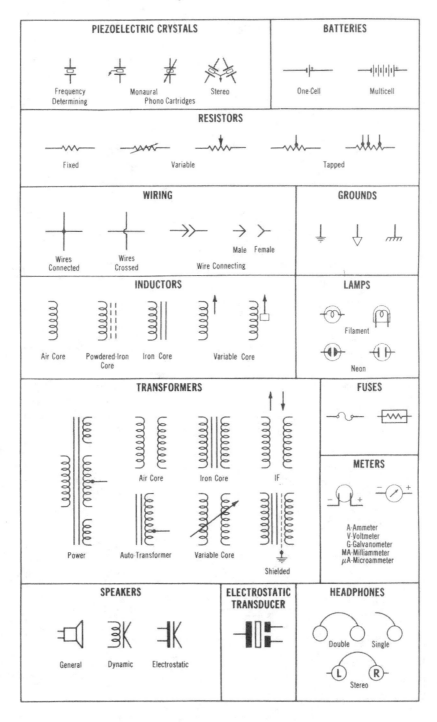

PIEZOELECTRIC CRYSTALS

Frequency Determining   Monaural Phono Cartridges   Stereo

BATTERIES

One-Cell   Multicell

RESISTORS

Fixed   Variable   Tapped

WIRING

Wires Connected   Wires Crossed   Male   Female   Wire Connecting

GROUNDS

INDUCTORS

Air Core   Powdered-Iron Core   Iron Core   Variable Core

LAMPS

Filament   Neon

TRANSFORMERS

Air Core   Iron Core   IF   Power   Auto-Transformer   Variable Core   Shielded

FUSES

METERS

A-Ammeter
V-Voltmeter
G-Galvanometer
MA-Milliammeter
μA-Microammeter

SPEAKERS

General   Dynamic   Electrostatic

ELECTROSTATIC TRANSDUCER

HEADPHONES

Double   Single   L   R   Stereo

Electronic Schematic Symbols *(continued)*

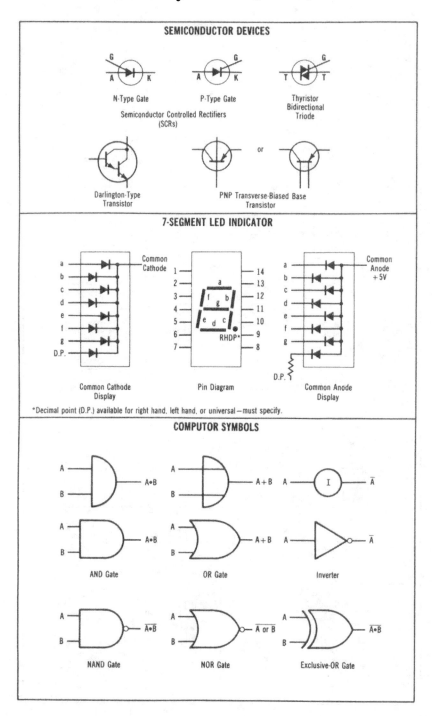

SEMICONDUCTOR DEVICES

N-Type Gate

P-Type Gate

Thyristor
Bidirectional
Triode

Semiconductor Controlled Rectifiers
(SCRs)

Darlington-Type
Transistor

PNP Transverse-Biased Base
Transistor

or

7-SEGMENT LED INDICATOR

Common
Cathode

Common
Anode
+5V

Common Cathode
Display

Pin Diagram

Common Anode
Display

D.P.

RHDP*

*Decimal point (D.P.) available for right hand, left hand, or universal—must specify.

COMPUTOR SYMBOLS

A
B
$A \cdot B$

A
B
$A + B$

A
$\overline{A}$

I

AND Gate
A
B
$A \cdot B$

OR Gate
A
B
$A + B$

Inverter
A
$\overline{A}$

NAND Gate
A
B
$\overline{A \cdot B}$

NOR Gate
A
B
$\overline{A \text{ or } B}$

Exclusive-OR Gate
A
B
$\overline{A \cdot B}$

Electronic Schematic Symbols *(continued)*

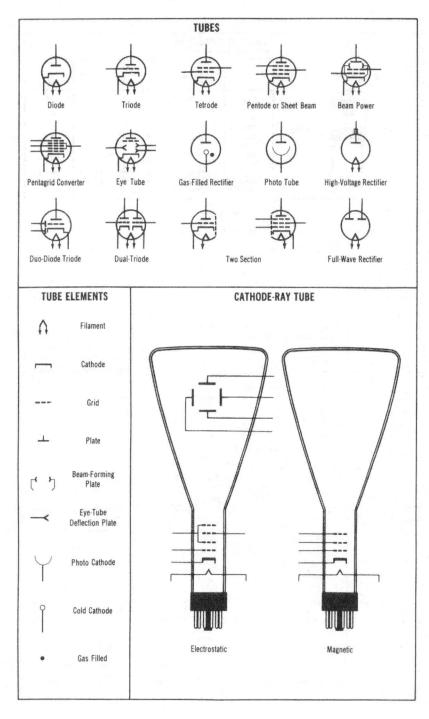

TUBES

Diode	Triode	Tetrode	Pentode or Sheet Beam	Beam Power
Pentagrid Converter	Eye Tube	Gas-Filled Rectifier	Photo Tube	High-Voltage Rectifier
Duo-Diode Triode	Dual-Triode	Two Section		Full-Wave Rectifier

TUBE ELEMENTS

Filament

Cathode

Grid

Plate

Beam-Forming Plate

Eye-Tube Deflection Plate

Photo Cathode

Cold Cathode

Gas Filled

CATHODE-RAY TUBE

Electrostatic

Magnetic

Computer Connector Pinouts

(A)

Parallel Port (DB 25 pin)
Female

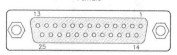

Pin	Signal	Pin	Signal
1	Strobe	10	Acknowledge
2	Data 0	11	Busy
3	Data 1	12	Paper Empty
4	Data 2	13	Select
5	Data 3	14	Auto Feed
6	Data 4	15	Error
7	Data 5	16	Initialize
8	Data 6	17	Select In
9	Data 7	18–25	GND

(B)

Parallel Port (Centronics 36 pin)
Female

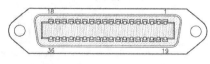

Pin	Signal	Pin	Signal
1	Strobe	13	Select
2	Data 0	14	Auto Feed
3	Data 1	15	N/C (not connected)
4	Data 2	16	Signal GND
5	Data 3	17	Frame GND
6	Data 4	18	+5 V Out
7	Data 5	19–30	GND
8	Data 6	31	Reset
9	Data 7	32	Error
10	Acknowledge	33	External GND
11	Busy	34	N/C
12	Paper Empty	35	N/C
		36	Select In

(C)

Serial Port (DB 9 pin)
Male

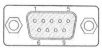

Pin	Signal
1	DCD (Data Carrier Detect)
2	RxD (Receive Data)
3	TxD (Transmit Data)
4	DTR (Data Terminal Ready)
5	GND (Signal Ground)
6	DSR (Data Set Ready)
7	RTS (Request To Send)
8	CTS (Clear To Send)
9	RI (Ring Indicator)

(D)

Serial Port (DB 25 pin)
Male

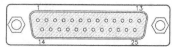

Pin	Signal	Pin	Signal
1	N/C (not connected)	20	DTR (Data Terminal Ready)
2	TxD (Transmit Data)	21	N/C
3	RxD (Receive Data)	22	RI (Ring Indicator)
4	RTS (Request To Send)	23	N/C
5	CTS (Clear To Send)	24	N/C
6	DSR (Data Set Ready)	25	N/C
7	GND (Signal Ground)		
8	DCD (Data Carrier Detect)		
9–19	N/C		

(E)

Ethernet Connector (RJ45–8 pin)
Female

Pin	Signal
1	Output Transmit Data (+)
2	Output Transmit Data (−)
3	Input Receive Data (+)
4	N/C (not connected)
5	N/C
6	Input Receive Data (−)
7	N/C
8	N/C

(F)

Ethernet Connector (RJ45–10 pin)
Female

Pin	Signal
1	DCD (Data Carrier Detect)
2	DTR (Data Terminal Ready)
3	CTS (Clear To Send)
4	GND (Signal Ground)
5	RxD (Receive Data)
6	TxD (Transmit Data)
7	GND (Frame Ground)
8	RTS (Request To Send)
9	DSR (Data Set Ready)
10	RI (Ring Indicator)

(G)

Mouse Port (DB 9 pin)
Male

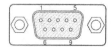

Pin	Signal
1	N/C (not connected)
2	Data
3	Clock
4	N/C
5	GND (Signal Ground)
6	N/C
7	RTS (12–9 V)
8	N/C
9	N/C

(H)

Mouse Port (mini DIN 9 pin)
Female

Pin	Signal
1	+5 V
2	X–A
3	X–B
4	Y–A
5	Y–B
6	Button 1
7	Button 2
8	Button 3
9	GND

(I)

Game/Joystick Port (DB 15 pin)
Female

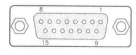

Pin	Signal	Pin	Signal
1	+5 V	10	Button (B–1)
2	Button (A–1)	11	Position (B–X)
3	Position (A–X)	12	GND
4	GND	13	Position (B–Y)
5	GND	14	Button (B–2)
6	Position (A–Y)	15	+5 V
7	Button (A–2)		
8	+5 V		
9	+5 V		

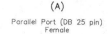

Copper Wire Specifications

Bare and Enamel-Coated Wire

Wire Size (AWG)	Diam (Mils)	Area (CM[1])	Enamel Wire Coating Turns / Linear inch[2]			Feet per Pound Bare	Ohms per 1000 ft 25∞ C	Current Carrying Capacity Continuous Duty[3]			Nearest British SWG No.
			Single	Heavy	Triple			at 700 CM per Amp[4]	Open air	Conduit or bundles	
1	289.3	83694.49				3.948	0.1239	119.564			1
2	257.6	66357.76				4.978	0.1563	94.797			2
3	229.4	52624.36				6.277	0.1971	75.178			4
4	204.3	41738.49				7.918	0.2485	59.626			5
5	181.9	33087.61				9.98	0.3134	47.268			6
6	162.0	26244.00				12.59	0.3952	37.491			7
7	144.3	20822.49				15.87	0.4981	29.746			8
8	128.5	16512.25				20.01	0.6281	23.589			9
9	114.4	13087.36				25.24	0.7925	18.696			11
10	101.9	10383.61				31.82	0.9987	14.834			12
11	90.7	8226.49				40.16	1.2610	11.752			13
12	80.8	6528.64				50.61	1.5880	9.327			13
13	72.0	5184.00				63.73	2.0010	7.406			15
14	64.1	4108.81	15.2	14.8	14.5	80.39	2.5240	5.870	32	17	15
15	57.1	3260.41	17.0	16.6	16.2	101.32	3.1810	4.658			16
16	50.8	2580.64	19.1	18.6	18.1	128	4.0180	3.687	22	13	17
17	45.3	2052.09	21.4	20.7	20.2	161	5.0540	2.932			18
18	40.3	1624.09	23.9	23.2	22.5	203.5	6.3860	2.320	16	10	19
19	35.9	1288.81	26.8	25.9	25.1	256.4	8.0460	1.841			20
20	32.0	1024.00	29.9	28.9	27.9	322.7	10.1280	1.463	11	7.5	21
21	28.5	812.25	33.6	32.4	31.3	406.7	12.7700	1.160			22
22	25.3	640.09	37.6	36.2	34.7	516.3	16.2000	0.914		5	22
23	22.6	510.76	42.0	40.3	38.6	646.8	20.3000	0.730			24
24	20.1	404.01	46.9	45.0	42.9	817.7	25.6700	0.577			24
25	17.9	320.41	52.6	50.3	47.8	1031	32.3700	0.458			26
26	15.9	252.81	58.8	56.2	53.2	1307	41.0200	0.361			27
27	14.2	201.64	65.8	62.5	59.2	1639	51.4400	0.288			28
28	12.6	158.76	73.5	69.4	65.8	2081	65.3100	0.227			29
29	11.3	127.69	82.0	76.9	72.5	2587	81.2100	0.182			31
30	10.0	100.00	91.7	86.2	80.6	3306	103.7100	0.143			33
31	8.9	79.21	103.1	95.2		4170	130.9000	0.113			34
32	8.0	64.00	113.6	105.3		5163	162.0000	0.091			35
33	7.1	50.41	128.2	117.6		6553	205.7000	0.072			36
34	6.3	39.69	142.9	133.3		8326	261.3000	0.057			37
35	5.6	31.36	161.3	149.3		10537	330.7000	0.045			38
36	5.0	25.00	178.6	166.7		13212	414.8000	0.036			39
37	4.5	20.25	200.0	181.8		16319	512.1000	0.029			40
38	4.0	16.00	222.2	204.1		20644	648.2000	0.023			
39	3.5	12.25	256.4	232.6		26969	846.6000	0.018			
40	3.1	9.61	285.7	263.2		34364	1079.2000	0.014			
41	2.8	7.84	322.6	294.1		42123	1323.0000	0.011			
42	2.5	6.25	357.1	333.3		52854	1659.0000	0.009			
43	2.2	4.84	400.0	370.4		68259	2143.0000	0.007			
44	2.0	4.00	454.5	400.0		82645	2593.0000	0.006			
45	1.8	3.10	526.3	465.1		106600	3348.0000	0.004			
46	1.6	2.46	588.2	512.8		134000	4207.0000	0.004			

Index

A/D conversion process, 48
A/D converter, 46, 87
A/D oscilloscope PC Card, 80
ac measurements, 51
ac meter, 43
ac signal, 14
ac-to-dc conversion, 53
Alignment, 282
Allergies, 8
Ammeter, 30, 35
  multirange, 38
  simple, 37
Ampere, 17
Amplifiers, 285
  dc coupled, 286
  gain, 285
Amplitude distortion, 279
Analog-to-digital (A/D) conversion, 46
Analog-to-digital converter, 85
AND gate, 21
Antistatic wristband, 9
Arbitrary waveform generator, 97, 142
Arcing, 282
Audio amplifier circuit, 23
Audio generator, 94
Audio-signal generator, 141

Batteries, 14, 156, 186, 257
  "smart" lead–acid battery charger, 192
  alkaline, 186
  capacity, 189
  charging, 190, 191
  chemical and other hazards, 188
  deep-cycle, 187
  discharge planning, 190
  discharging, 190
  internal resistance, 189
  lead–acid, 186
  lithium, 186
  lithium ion cells, 188
  lithium–thionyl–chloride, 186
  nickel–cadmium, 187
  nickel–metal hydride, 187
  primary, 186
  rechargeable, 186
  safety, 196, 204
  secondary, 186
  sensing, 198
  solar electric battery charger, 196
  under Load, 258
Battery charger, 156
Battery tester, 155
Bridge-type rectifier, 43

Building your own test equipment, 360, 365
  capacitance leakage meter, 376
  continuity tester, 366
  diode/transistor curve tracer, 371
  electronic fuse, 382
  inductance/capacitance meter, 394
  logic probe, 367
  pulse generator, 386
  transistor tester, 372
  wire-tracer circuit, 368
  zener diode tester, 369
Buying equipment, components, and tools, 353, 354, 360

Cable tester, 146
Calibration and standards, 290
Capacitance, 14
Capacitance leakage meter, 376
Capacitance meter, 136
Capacitor, 14, 214
  adjustable, 219
  air-core, 219
  ceramic, 217
  charging, 216
  codes, 217
  color Codes, 413
  double-layer, 219
  electrolytic, 218
  epoxy, 219
  in parallel, 220
  in series, 219
  leakage current, 243
  leaky, 242
  measurements, 239
  measuring an electrolytic capacitor, 242
  metalized polyester film, 218
  multilayer, 219
  polyester film, 218
  polypropylene, 218
  polystyrene, 218
  relative amount of capacitance, 240
  shorted, 242
  silver–mica, 218
  supercapacitors, 219
  tantalum, 219
  trimmer, 219
  tuning, 219
  types, 217
Cathode-ray tube (CRT), 61
Charger, 201
Chassis fabrication, 348
  cutting and bending sheet metal, 349

  drilling techniques, 350
  finishing aluminum, 349
  painting, 352
  rectangular holes, 351
  working, 349
Circuit ground, 14
Clamp-on ammeter, 145
Contamination, 282
Continuity, 245
Continuity Tester, 366
Control circuitry, 286
Cooling circuit, 181
Copper wire specifications, 421
"Crowbar" circuits, 170
Current, 17
  consumption, 17
  drain, 259
Current input conditioner, 53

D'Arsonval type movement, 30
Dedicated space, 3
Desoldering station, 158, 319
Digital DVM display, 50
Digital signal processing (DSP), 88
Digital storage oscilloscope, 89
  applications, 89
Diode, 14, 229, 248
Diode and transistor tester, build your own, 263
Distortion meter, 281
Double-sided tape, 301
Dual-slope converter, 49
Duty cycle, 78
DVM, basic, 45

Earth, 14
Electric power, 7
  ground system, 8
Electrical tape, 301
Electronic circuits, 326
Electronic fuse, 12, 382
Electronic hardware, 12
  binding posts, 12
  connectors, 12
  jacks, 12
  nuts, 12
  plastic storage boxes for, 12
  plugs, 12
  rubber foot chassis boxes, 12
  screws, 12
  spade lugs, 12
  terminal strips, 12
Electronic schematic symbols, 13, 415
Electronics construction techniques, 327

point-to-point, 328
Electronics workbench resources, 404
Electrostatic discharge, 230
    protection, 326
Emergency power system, 200
Equipment manuals, 408
ESR capacitance meter, 137

Farad, 14
FET, 253
Field-strength meter, 150
First-aid kit, 309
Frequency counter, 11, 112, 140
    1 Ghz frequency counter project, 124
    ccessories, 122
    accuracy, 117, 120
    active preamp probes, 123
    aging, 120
    architecture, 113
    calibration, 121
    coupling loop, 123
    differences between resolution and
        accuracy, 117
    low-frequency measurements, 119
    measurement techniques, 119
    noise triggering, 118
    offset, 122
    parts per million, 122
    preselector, 124
    probe or clip leads, 123
    reference, 122
    resolution, 117
    sensitivity, 117
    spurious frequency count, 118
    stability, 120
    standard, 122
    terms, 120
    time Base, 115
    trigger accuracy, 119
    tuned antenna, 124
    whip antenna, 123
Frequency distortion, 280
Full-wave rectifier, 163
Function generator, 11, 94, 142
    amplitude, 100
    analog, 95
    applications, 99
    arbitrary waveform generator, 97
    build your own, 103
    controls, 101
    digital, 95
    direct digital synthesis, 96
    duty cycle, 100
    frequency characteristics, 95

period, 100
    purchase considerations, 102
    sequence generator, 98
    specifications, 95
    terms, 99
Fuses, 12, 14, 304
    electronic, 12
    fast blow, 12
    slow blow, 12

Gain, 285
Glue, 303
Graticule, 64
Ground rod connection, 8
Ground-plane construction, 329
Guitar amplifier, 23

Half-digit display, 51
Half-wave doubler, 166
Half-wave rectifier, 162
Handheld microscope/magnifier, 7
Heat-shrink tubing, 300
Henry, 14

IC voltage regulation, 170
Impedance, 244
Inductance, 14, 244
    measurements, 244
    units of, 223
Inductance/capacitance (L/C) meter,
        137, 394
Inductor, 14, 220
    parallel, 225
    series, 225
    types, 222
Insulation tester, 146
Integrated circuit, 230
Isolation transformer, 154

JFET, 254

Kilohms, 13

Lighted magnifier lamp, 7
Lighting, 5, 7
Linear regulator, 168
Logic analyzer, 153
Logic probe, 153, 367
Logic pulser, 152

Measuring semiconductor devices, 248
Megohm, 13
Meter shunts, 38
Microfarad, 14

Microhenry, 14
Micromicrofarad, 14
Microprocessor programmer, 154
Millihenry, 14
Mini jumper wires, 12
Miniature light or lamp, 14
MOSFET, 255
Multimeter, 30, 134, 234
    accuracy, 55
    analog, 30, 31
    audible Continuity, 57
    autopolarity, 56
    autoranging, 56
    characteristics, 32
    comparison of analog to digital, 45
    conductance measurements, 57
    digital, 30, 45
    DVM characteristics, 45
    internal meter resistance, 33
    measuring meter sensitivity, 33
    meter accuracy, 33
    meter damping, 34
    meter movement sensitivity, 32
    protection, 56
    range, 55
    reading storage, 57
    resolution, 55
    response time, 56
    semiconductor testing, 57
    speech output, 57
Multimeters, 10
Multirange ac voltmeter, 44
Multitester, 30

Noise, 280

Ohm, 13
Ohm's Law, 17
    nomograph, 410
Ohmmeter, 30, 39
    basic, 39
    series, 41
    shunt-type, 42
Op-amp, 17, 18, 19, 20, 21
Operational amplifier (*see* Op-amp)
Oscillations, 278
Oscillators, 286
Oscilloscope, 11, 60, 150, 234
    "add" mode, 64
    "dual" mode, 64
    accuracy, 63
    ac–dc coupling switch, 66
    A/D PC card
        advanced features, 78

Oscilloscope *(continued)*
  analog, 61
  analog-to-digital converter, 87
  archive, 89
  back-light, 65
  bandwidth, 63, 88
  battery test, 260
  beam intensity, 65
  buying used, 91
  care of, 90
  channel 2 invert, 68
  controls, 64
  CRT, 61
  digital, 80
  digital signal processing (DSP), 88
  digital storage, 82, 90
  display contrast, 65
  displays, 88
  dual-channel, 61
  dual-trace, 62
  duty cycle measurement, 78
  focus, 65
  graticule, 64
  graticule illumination, 65
  ground switch, 67
  horizontal position, 69
  horizontal section, 68
  horizontal sweep rate, 68
  horizontal sweep rate control, 70
  input circuits, 66
  input impedance, 64
  limitations, 79
  maximum input, 63
  measuring voltage gain, 261
  memory, 89
  multichannel, 64
  PC, 80
  PC card, 80
  phase measurements, 74
  power switch, 64
  probes, 72
  pulse width measurement, 77
  range, 63
  rise time, 79
  seconds (time) per division, 69
  signal measurements, 73
  sine-wave measurement, 74
  specifications, 63
  square-wave measurement, 76
  start-up, 71
  testing with, 259
  time measurements, 74
  trace rotation, 66
  transformer testing, 260

  trianglular wave measurement, 75
  trigger, 87
  trigger level control, 70
  trigger source, 64, 70
  triggered-sweep, 61
  vertical attenuator control, 67
  vertical deflection, 63
  vertical mode, 67
  vertical position, 66
  vertical section, 66
  voltage measurements, 73
  volts per division, 68
Oven-controlled Xtal oscillator, 121
Overcurrent protection, 169

Parallel resistance nomograph, 411
Parts, 307
Parts organization, 296
PC board, 331, 335
  artwork, 339
  assembly Techniques, 345
  chassis boxes, 350
  cleanliness, 345
  commerical, 344
  complex, 342
  design considerations, 342
  design software, 337
  double-sided, 341
  drilling, 344
  etchant, 340
  installing components, 346
  iron-on resist, 340
  off-board wiring, 346
  paint, 339
  photographic process, 339
  plating, 341
  production, 343
  production costs, 344
  ready-made, 332
  resist pens, 338
  resists, 338
  rub-on transfer, 339
  soldering, 346
  stock, 338
  tape, 338
  utility, 332
Perf-board, 329
Photoconductor, 239
Picofarad, 14
Pictorial diagram, 23
Pinouts, 420
Potentiometer, 14
Power consumption, 17
Power equation, 17

Power inverter, 202
Power line spikes, 7
Power losses, 203
Power protection, 8
Power strips, 8
Power supplies, 157, 160
  build your own bench/lab power
    supply, 174
  choosing, 160
  components, 162
  dc-to-dc, 173
  dual-voltage, 171
  filters, 165
  linear, 160
  linear regulator, 168
  purchasing, 161
  regulation, 167
  secondary, 186
  single-voltage, 161
  switching, 160
  voltage multiplier, 165
  voltage quadrupler, 166
  voltage tripler, 166
  voltage doubler, 166
  zener diode regulation, 167
ProtoBoard, 12, 328
Pulse generator, 143, 386
Pulse width, 77

Reading schematics, 13
Resistance, 17
Resistance input conditioner, 53
Resistance testing of transformer
    windings, 247
Resistance/capacitance decade box, 138
Resistor, 13, 206
  axial leads, 213
  carbon composition, 206
  carbon film, 207
  color codes, 236, 409
  configuration, 210
  dual inline package (DIP), 214
  filament, 208
  flat pack, 214
  foil, 208
  formulas, 412
  frequency response, 212
  fuse, 210
  fuse clip mounting, 214
  identification, 210
  measuring a fixed resistor, 238
  measuring a variable resistor, 238
  measurements, 235
  metal film, 208

mounting, 206, 213
noise, 212
parameters, 211
PC mounting, 214
power film, 208
power rating, 213
power wire-wound, 209
precision wire-wound, 208
radial leads, 213
reliability, 213
"shifty", 239
single inline packaging (SIP), 214
sizes, 206, 213
stability, 212
surface mount, 213
temperature coefficient of resistance, 212
temperature rating, 213
thermally intermittent, 239
thermocouple effect, 212
types, 206
variable, 14
voltage coefficient, 212
RF impedance bridge, 149
RF signal generator, 143
Ring shunt, 38
Rise time, 79
Rough layout—manual method, 336

Safety, 308
Schematic, 335
Schematic reading, books on, 408
SCR, 256
Semiconductor, 228
doping silicon, 228
N-Type, 228
P-Type, 228
Shop materials, 307
Shorts, 247
primary, 247
secondary, 247
Signal conditioner, 52
Signal generator, 94
Signal injection, 276
equipment, 276
procedure, 276
Signal injector, 144
Signal tracer, 144
Signal tracing, 274
Sine wave, 74
Solder, 312
60/40 rosin-core, 312
acid-core, 312
flux, 312

lead–indium–silver, 312
silver-bearing, 312
tin-bearing, 312
Solder bridge, 283
Soldering, 311
desoldering, 319
how to, 316
making a good mechanical joint, 316
preparing work, 315
printed circuit boards, 317
unsoldering, 318
ventilation, 320
Soldering gun, 314
Soldering iron, 313
caring for, 320
preparing, 315
using, 315
Soldering station, 158, 313
build your own, 321
Solderless breadboard, 12
Solder-smoke or fume removal, 8
Spectrum analyzer, 151
Spike protection power strips, 8
Square wave, 76
Staircase converter, 48
Static buildup, 9
Static electricity, 9
Surface-mount construction, 333
SWR or VSWR bridge, 150

Temperature-compensated crystal oscillator, 121
Temperature-controlled soldering station, 158
Test equipment, 10
Testing, 234
alignment, 282
amplifier gain, 285
amplifiers, 285
amplitude distortion problems, 279
arcing problems, 282
contamination problems, 282
distortion measurement, 281
dividing and conquering, 277
frequency and distortion problems, 280
gate tests, 288
intuitive approach, 277
logic levels, 288
microprocessor troubles, 287
noise problems, 280
oscillation problems, 278
power supplies, 283
switching power supplies, 284

tristate devices, 288
typical symptoms and caults, 283
voltage levels, 278
within intermediate stages, 278
Thermistor, 239
Tone test set, 143
Tools, 12, 294, 407
buying, 363
care of, 294
crimping, 300
deburring, 299
dental pick, 12, 303
drill bits, 299
electronic workshop tool list, 295
files, 298
hammers, 12, 306
hemostat, 12
hot-glue gun, 303
knives, 304
locking-grip pliers, 298
magnetic screw starters, 12
magnifying glass, 302
measuring devices, 306
needle-nose pliers, 12, 298
new versus used, 363
nibbling, 299
organization, 294
pliers, 12, 298
proper use, 297
right-angle pliers, 12
screw starter, 305
screwdrivers, 12, 297, 306
sharpening, 296
socket punch, 300
solder gun, 12
soldering iron, 12
sources of, 307
specialized, 299
vises, 305
wire cutters, 12, 298
wire marker, 305
Transformer, 14, 226
basics, 246
ideal, 246
step-down, 247
step-up, 247
Transistor curve tracer, 147
Transistor tester, 146, 372
Transistors, 14, 230, 249
bipolar lunction, 249
FET, 253
field-effect, 253
JFET, 254
MOSFET, 255

Transistors *(continued)*
  SCR, 256
  triac, 257
  UJT, 252
  unijunction, 252
Triac, 257
Trianglular wave, 75
Troubleshooting, 268
  approaches, 273
  components, 289
  connectors, 289
  defining the problem, 271
  documentation, 270
  equipment, 274
  fuses, 289
  look for obvious problems, 272
  procedures, 275
  safety, 269
  signal injection, 276
  signal tracing, 274
  simplify the problem, 272
  test equipment, 269
  using all your senses, 270
  wires, 290
Tube tester, 148
TV pattern generator, 146
Twelve volt power-distribution panel,
  202

Used equipment, 354
  cleaning, 359
  equipment manuals, 359
  guidelines for buying, 356
  hidden costs, 357
  oscilloscopes, 357
  sources of, 407

Vacuum tube voltmeter (VTVM), 135
Variable resistors, 14
Variac, 154
Varistor, 239
Velcro, 301
Ventilation, 8
  spot, 8
Visual diode/transistor curve tracer, 371
Voltage, 17
Voltage divider, 17
Voltage input conditioner, 52
Voltage multipliers, 165
Voltage quadrupler, 166
Voltage tripler, 166
Voltage-controlled crystal oscillator,
  121
Voltmeter, 30, 34
  multiplier resistance, 34
  multirange, 34
  multirange ac, 44

  sensitivity, 35
Volt–ohm meter, 30
VOM, 30

Wattage, 17
Watts, 17
Waveform cycle frequency, 98
Waveform generator, 94
Wheatstone bridge, 139
Wire-tracer circuit, 368
Wire-wrap construction, 332
Work mats, 9
  antistatic, 9
Workbenches, 3
  buy or build, 3
  computer table, 5
  electric power for, 7
  metal, 3
  metal desk, 4
  metal work table, 4
  surplus, 4
  two-tier, 3
  wood, 5
  wood-door, 5
  wooden, 3

Zener diode, 167
  tester, 369

ABOUT THE AUTHOR

Thomas Petruzzellis is an electronics engineer currently working with the geophysical field equipment in the geology department at the State University of New York, Binghamton. Tom has over 30 years of experience in electronics and he is a veteran author who has written extensively for industry publications, including *Electronics Now, Modern Electronics, QST, Microcomputer Journal, Circuit Cellar,* and *Nuts & Volts.* The author of three earlier books—*STAMP 2 Communications and Control Projects; Optoelectronics, Fiber Optics, and Laser Cookbook; Alarm, Sensor, and Security Circuit Cookbook—* all from McGraw-Hill, he lives in Vestal, New York.